Fuzzy Classifier Design

Studies in Fuzziness and Soft Computing

Editor-in-chief
Prof. Janusz Kacprzyk
Systems Research Institute
Polish Academy of Sciences
ul. Newelska 6
01-447 Warsaw, Poland
E-mail: kacprzyk@ibspan.waw.pl
http://www.springer.de/cgi-bin/search_book.pl?series=2941

Vol. 3. A. Geyer-Schulz
Fuzzy Rule-Based Expert Systems and Genetic Machine Learning, 2nd ed. 1996
ISBN 3-7908-0964-0

Vol. 4. T. Onisawa and J. Kacprzyk (Eds.)
Reliability and Safety Analyses under Fuzziness, 1995
ISBN 3-7908-0837-7

Vol. 5. P. Bosc and J. Kacprzyk (Eds.)
Fuzziness in Database Management Systems, 1995
ISBN 3-7908-0858-X

Vol. 6. E.S. Lee and Q. Zhu
Fuzzy and Evidence Reasoning, 1995
ISBN 3-7908-0880-6

Vol. 7. B.A. Juliano and W. Bandler
Tracing Chains-of-Thought, 1996
ISBN 3-7908-0922-5

Vol. 8. F. Herrera and J.L. Verdegay (Eds.)
Genetic Algorithms and Soft Computing, 1996
ISBN 3-7908-0956-X

Vol. 9. M. Sato et al.
Fuzzy Clustering Models and Applications, 1997
ISBN 3-7908-1026-6

Vol. 10. L.C. Jain (Ed.)
Soft Computing Techniques in Knowledge-based Intelligent Engineering Systems, 1997
ISBN 3-7908-1035-5

Vol. 11. W. Mielczarski (Ed.)
Fuzzy Logic Techniques in Power Systems, 1998
ISBN 3-7908-1044-4

Vol. 12. B. Bouchon-Meunier (Ed.)
Aggregation and Fusion of Imperfect Information, 1998
ISBN 3-7908-1048-7

Vol. 13. E. Orłowska (Ed.)
Incomplete Information: Rough Set Analysis, 1998
ISBN 3-7908-1049-5

Vol. 14. E. Hisdal
Logical Structures for Representation of Knowledge and Uncertainty, 1998
ISBN 3-7908-1056-8

Vol. 15. G.J. Klir and M.J. Wierman
Uncertainty-Based Information, 2nd ed., 1999
ISBN 3-7908-1242-0

Vol. 16. D. Driankov and R. Palm (Eds.)
Advances in Fuzzy Control, 1998
ISBN 3-7908-1090-8

Vol. 17. L. Reznik, V. Dimitrov and J. Kacprzyk (Eds.)
Fuzzy Systems Design, 1998
ISBN 3-7908-1118-1

Vol. 18. L. Polkowski and A. Skowron (Eds.)
Rough Sets in Knowledge Discovery 1, 1998
ISBN 3-7908-1119-X

Vol. 19. L. Polkowski and A. Skowron (Eds.)
Rough Sets in Knowledge Discovery 2, 1998
ISBN 3-7908-1120-3

Vol. 20. J.N. Mordeson and P.S. Nair
Fuzzy Mathematics, 1998
ISBN 3-7908-1121-1

Vol. 21. L.C. Jain and T. Fukuda (Eds.)
Soft Computing for Intelligent Robotic Systems, 1998
ISBN 3-7908-1147-5

Vol. 22. J. Cardoso and H. Camargo (Eds.)
Fuzziness in Petri Nets, 1999
ISBN 3-7908-1158-0

Vol. 23. P.S. Szczepaniak (Ed.)
Computational Intelligence and Applications, 1999
ISBN 3-7908-1161-0

Vol. 24. E. Orłowska (Ed.)
Logic at Work, 1999
ISBN 3-7908-1164-5

continued on page 315

Ludmila I. Kuncheva

Fuzzy Classifier Design

With 113 Figures
and 81 Tables

Physica-Verlag
A Springer-Verlag Company

Dr. Ludmila I. Kuncheva
School of Informatics
University of Wales, Bangor
Dean Street, Bangor
Gwynedd LL57 1UT
United Kingdom
E-mail: L.I.Kuncheva@bangor.ac.uk

ISSN 1434-9922
ISBN 3-7908-1298-6 Physica-Verlag Heidelberg New York

Cataloging-in-Publication Data applied for
Die Deutsche Bibliothek – CIP-Einheitsaufnahme
Kuncheva, Ludmila I.: Fuzzy classifier design: with 81 tables / Ludmila I. Kuncheva. – Heidelberg; New York: Physica-Verl., 2000
(Studies in fuzziness and soft computing; Vol. 49)
ISBN 3-7908-1298-6

Physica-Verlag is a company in the BertelsmannSpringer publishing group.

Printed in Germany

Hardcover Design: Erich Kirchner, Heidelberg

SPIN 10765173 88/2202-5 4 3 2 1 0 – Printed on acid-free paper

Preface

Fuzzy sets were first proposed by Lotfi Zadeh in his seminal paper [366] in 1965, and ever since have been a center of many discussions, fervently admired and condemned. Both proponents and opponents consider the arguments pointless because none of them would step back from their territory. And still, discussions burst out from a single sparkle like a conference paper or a message on some fuzzy-mail newsgroup. Here is an excerpt from an e-mail message posted in 1993 to fuzzy-mail@vexpert.dbai.twvien.ac.at, by somebody who signed "Dave".

> '... Why then the "logic" in "fuzzy logic"? I don't think anyone has successfully used fuzzy sets for logical inference, nor do I think anyone will. In my admittedly neophyte opinion, "fuzzy logic" is a misnomer, an oxymoron. (I would be delighted to be proven wrong on that.)... I came to the fuzzy literature with an open mind (and open wallet), high hopes and keen interest. I am very much disillusioned with "fuzzy" per se, but I did happen across some extremely interesting things along the way."

Dave, thanks for the nice quote! Enthusiastic on the surface, are not many of us suspicious deep down? In some books and journals the word *fuzzy* is religiously avoided: fuzzy set theory is viewed as a second-hand cheap trick whose aim is nothing else but to devalue good classical theories and open up the way to lazy ignorants and newcomers. This view is sometimes attributed to the conservatism of the "western culture" compared to the enterprising and practical "eastern culture" that took fuzzy sets on board right from the very start [342]. The other face of this coin is the "fuzzy" euphoria going under the motto "fuzzify the unfuzzifiable", sometimes with little or no rationale. Despite the unfinished philosophic debate there are a remarkable number of research papers and engineering applications involving fuzzy sets – some good and some bad, like in any other branch of science.

What are fuzzy classifiers and where do they stand in this stream? Do we need fuzzy classifiers or are we satisfied with pattern recognition "classics"? Are fuzzy classifiers better in some sense? In which cases and by how much are they better? Are fuzzy classifiers just "fuzzified" versions of non-fuzzy classifiers or are they an original contribution?

The difficulty in writing a coherent text on fuzzy classifiers roots in the diversity of the field itself. A paragraph taken from the Introduction by John Hartigan to the book "Clustering and Classification" [20], 1996, reads:

> 'Let us agree that classification does not fit easily into any of the academic disciplines in which its practitioners live. We remain too diffuse and dispersed in our methods and applications to form a coherent academic discipline ourselves.'

The author speaks about clustering and classification, but insert *fuzzy* in front of these two words, and the statement is enhanced by orders of magnitude.

Fuzzy Classifier Design is an attempt to address and answer some of the above questions without giving a comprehensive account of the area. My intention was not to take sides in the pro-con argument but to highlight some issues which would let you, my reader, make your own educated choice about fuzzy classifier design.

Bangor, Wales
February, 2000

Ludmila Ilieva Kuncheva

Contents

1. Introduction

1.1 What are fuzzy classifiers?

Fuzzy pattern recognition is sometimes identified with fuzzy clustering or with fuzzy if-then systems used as classifiers. In this book we adopt a broader view: fuzzy pattern recognition is about *any* pattern classification paradigm that involves fuzzy sets. To a certain extent fuzzy pattern recognition is dual to classical pattern recognition, as delineated in the early seventies by Duda and Hart [87], Fukunaga [100], Tou and Gonzalez [324], and thereby consists of three basic components: clustering, classifier design and feature selection [39]. Fuzzy clustering has been the most successful offspring of fuzzy pattern recognition so far. The fuzzy c-means algorithm devised by Bezdek [34] has admirable popularity in a great number of fields, both engineering and non-engineering. Fuzzy feature selection is virtually absent, or disguised as something else. This book is about the third component *fuzzy classifier design*.

The diversity of applications in the studies retrieved upon the keyword "fuzzy classifier" is amazing. Remote sensing; environmental studies; geoscience; satellite and medical image analysis; speech, signature and face recognition are few examples of highly active areas. Even more curious are the concrete applications such as grading fish products and student writing samples; analysis of seasonal variation of cloud parameters; speeding up fractal image compression; development of metric-based software; classification of odours, road accidents, military targets and milling tool ware; estimating a crowding level in a scene; tactile sensing; glaucoma monitoring; and even quality evaluation of biscuits during baking. It seems that applications of fuzzy pattern recognition are far ahead of the theory on the matter. This book aims at systematizing and hopefully a better understanding of the theoretical side of fuzzy classifiers.

1.1.1 Three "fuzzy" definitions of a fuzzy classifier

What are **fuzzy classifiers**? It is difficult to propose a clear-cut definition. Let $\mathbf{x}$ be a vector in an n-dimensional real space $\Re^n$ (the feature space), and let $\Omega = \{\omega_1, \ldots, \omega_c\}$ be a set of class labels[1]. A (crisp) *classifier* is any mapping

$$D : \Re^n \to \Omega \tag{1.1}$$

In a broad sense, we can define a fuzzy classifier as follows

Definition 1.1.1. *A* **fuzzy** *classifier is any classifier which uses fuzzy sets either during its training or during its operation*

Bezdek et al. [38] define a **possibilistic** classifier as the mapping

$$D_p : \Re^n \to [0,1]^c - \mathbf{0}, \tag{1.2}$$

i.e., instead of assigning a class label from Ω, D_p assigns to $\mathbf{x} \in \Re^n$ a *soft* class label with degrees of membership in each class (by convention, the zero vector is excluded from the set of possible soft labels). We can think of the components of the output vector as degrees of support for the hypothesis that $\mathbf{x}$ belongs to the respective class. Denote by $\mu(\mathbf{x}) = [\mu_1(\mathbf{x}), \ldots, \mu_c(\mathbf{x})]^T$ the classifier output calculated via (1.2). Then, according to [38],

Definition 1.1.2. *A* **fuzzy** *or* **probabilistic** *classifier, is any possibilistic classifier for which*

$$\sum_{i=1}^{c} \mu_i(\mathbf{x}) = 1. \tag{1.3}$$

Thus, the crisp classifier D (1.1) is a special case of the fuzzy classifier D_p. A third definition which is implicitly assumed in most publications on fuzzy classifiers is that

Definition 1.1.3. *A* **fuzzy** *classifier is a fuzzy if-then inference system (a fuzzy rule-base system) which yields a class label (crisp or soft) for* $\mathbf{x}$.

These three definitions are pictured in the Venn diagram in Figure 1.1. Definition 1.1.3 is the most specific one, and since it is based explicitly on fuzzy sets, it lies inside Definition 1.1.1. Most probabilistic classifiers have as output the posterior probabilities for the classes ($P(\omega_i|\mathbf{x})$). These designs will be labelled as fuzzy classifiers by Definition 1.1.2, but not by Definition 1.1.1, because fuzzy sets are not involved in their design or operation. On the other hand, some classifiers that use fuzzy sets, e.g., fuzzy k-nearest neighbor methods, do not necessarily produce class labels that sum up to

[1] This chapter uses some notions and notations with a "flying start", e.g. class, feature, class label, error rate, training and testing sets, etc. These are introduced and explained in detail in the ensuing chapters. The readers who are not familiar with the (fuzzy) pattern recognition jargon can skip the details in this chapter at no loss. It was important for reference purposes to put together the three "fuzzy" definitions of a fuzzy classifier and the description of the data sets used throughout the book.

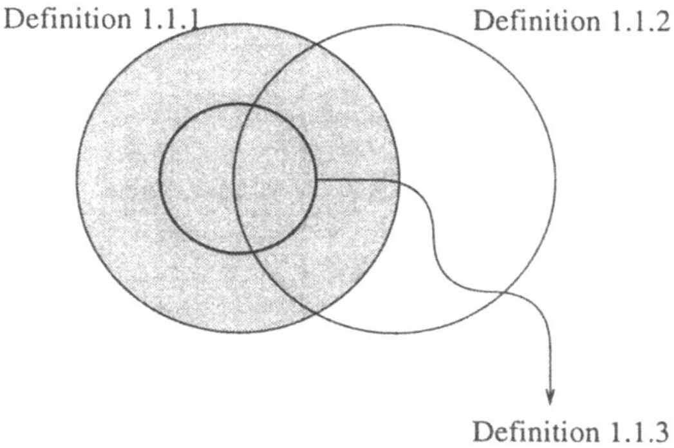

Fig. 1.1. The scope of the three definitions of a *fuzzy classifier*

one, nor are they rule-based. Hence, there is a scope covered by Definition 1.1.1 which is not accounted for by either Definition 1.1.2 or Definition 1.1.3. A fuzzy if-then system may or may not produce labels that sum up to one, therefore Definition 1.1.3 also covers designs outside the scope of Definition 1.1.2. Throughout this book we shall use Definition 1.1.1 (the shaded circle) as **fuzzy classifier** and will consider separately if-then and non-if-then fuzzy classifiers.

1.1.2 Why should we use fuzzy classifiers?

- In some problems, there is **insufficient information** to properly implement classical (e.g., statistical) pattern recognition methods. Such are the problems where we have no data set.
- Sometimes the user needs **not only the class label** of an object but also some additional information (e.g., how typical this object is, how severe the disease is, how desirable the option is).
- Sometimes characteristics of objects or class labels are **conveniently represented** in terms of fuzzy sets. For example, in a medical inquiry we may wish to quantify the "degree of pain" or "the extent of alcohol abuse" with numbers in [0,1].
- Fuzzy set theory gives a mathematical tool for including and processing **expert opinions** about classification decisions, features and objects.
- Fuzzy classifiers based on if-then rules might be **"transparent"** or **"interpretable"**, i.e., the end user (expert) is able to verify the classification paradigm. For example, such verification may be done by an expert judging the plausibility, consistency or completeness of the rule-base in fuzzy if-then classifiers. This verification is appropriate for small-scale systems, i.e., systems which do not use a large number of input features and big rule bases.

1.1.3 What obstructs using fuzzy classifiers?

- There is **no rigorous theory** (e.g., a theory specifying conditions for optimality of a fuzzy classifier) and therefore there is no theoretical methodology to design a fuzzy classifier for every instance.
- Fuzzy classifiers which are entirely based on expert opinion are **difficult** to design because of the so-called "knowledge acquisition bottleneck". This used to be a popular issue in Artificial Intelligence and refers to the difficulty in eliciting verbal reasoning rules using the help of a domain expert.
- Fuzzy if-then classifiers do not offer an easy way to handle complex dependencies between the features. To ensure transparency (interpretability) we use linguistic reasoning, thereby "granulating" the feature space. In many cases this leads to **sacrificing accuracy**. This delicate point is addressed throughout the book.
- Interpretability makes sense only when we use **small number of features** (e.g., up to 3 or 4) and small number of linguistic labels defined on these features (e.g., $\{small, medium, large\}$ or $\{low, high\}$). In problems of a higher dimensionality, interpretation might not be feasible [322].

When and how shall we use fuzzy classifiers? Assume we collaborate with a domain expert on a certain pattern recognition problem. We wish to include in our model the knowledge and the insight of our expert about the problem and its possible solutions. A linguistically-based fuzzy classifier is a natural choice for this case. However, to succeed in this task, beside the respective mathematical toolbox, we also need excellent intuition, a bag with clever heuristics, a lot of patience and good luck. In most cases we only have a labeled data set and no expert. Then the fuzzy paradigm is not enforced by the circumstances and we can use a non-fuzzy classifier instead. Shall we use a fuzzy classifier anyway? Do we need the transparency or shall we use an opaque (and still fuzzy!) classifier? If we have both data and expertise, shall we use both (how?), shall we use the expertise only (how?), or the data only (how?)? In most of the recent fuzzy classifier models the domain expert is no longer a part of the design process, nether in the setup nor in the evaluation. Then what is the point trading off accuracy for transparency which nobody needs? Can we gain accuracy from the fuzzy "expertless" model?

In the course of writing this book I realized that bringing together two distinct areas such as *pattern recognition* and *fuzzy sets* requires an introduction into both areas. A pattern recognition background is vitally needed in fuzzy classifier design. Whenever this is overlooked, we are often witnessing or participating in the reinvention of the bicycle. Well, it is not hazardous but is pointless. On the other hand, the pattern recognition community has not always appreciated highly ad-hoc fuzzy classifiers. Thus, a better understanding is needed. I tried to adhere to the concepts that are jointly used in non-fuzzy and fuzzy classifier design. The field of fuzzy classifiers is pretty amorphous on its own, thereby making my systematization task even more difficult. Some topics will be revisited at different places in the text. To facilitate understanding, the text contains simple examples, illustrations and explanations. Knowledge of elementary probability and set theory would be helpful. The book also contains some original research. Inevitably, some excellent works and ideas will be left unmentioned either because there has been no room; because they have not been tightly integrable within this bit of fuzzy classifier design that I have cut out of the huge field; or simply because I have not been aware of these works. Models that are not in the book are fuzzy tree-wise classifiers, fuzzy ARTMAP classifiers and fuzzy classifiers with a reject option. The aim of this book is to give you a toolbox of fuzzy and non-fuzzy designs, and hopefully a hint about which department you should search for the problem that *you* have to solve.

1.2 The data sets used in this book

We use three types of data throughout the book:

1.2.1 Small synthetic data sets

Small artificial 2-dimensional data sets are used for illustrating basic calculations and ideas. Such is the **15-point** dataset depicted in Figure 1.2 and displayed in Table 1.1. The features are denoted by x_1 and x_2, and the 15 points are $\mathbf{z}_1$ to $\mathbf{z}_{15}$. The class labels are ω_1 for the squares and ω_2 for the snowflakes. We use this set to explain the term "classification region", the idea of k-nearest neighbor method, Voronoi diagrams, fuzzy if-then classifiers, etc.

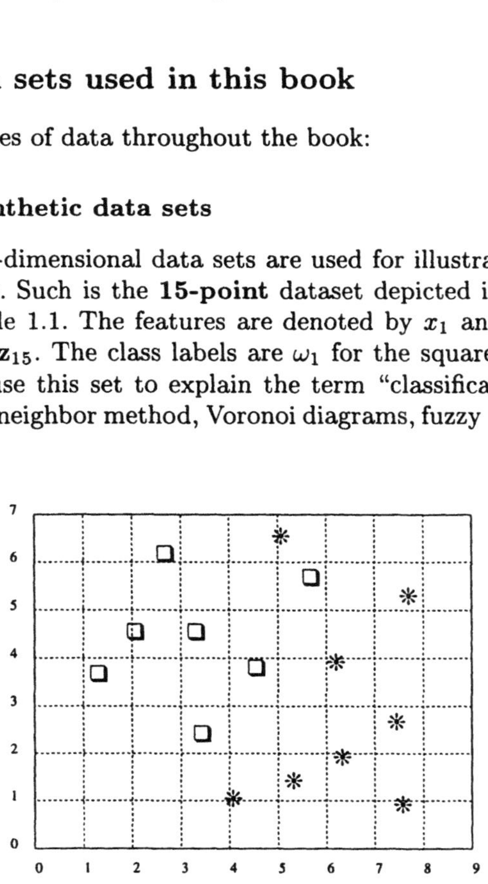

Fig. 1.2. The **15-point** two-class example

Other small synthetic data sets are also used wherever necessary.

Table 1.1. The labeled 15-point set Z

	$\mathbf{z}_1$	$\mathbf{z}_2$	$\mathbf{z}_3$	$\mathbf{z}_4$	$\mathbf{z}_5$	$\mathbf{z}_6$	$\mathbf{z}_7$	$\mathbf{z}_8$	$\mathbf{z}_9$	$\mathbf{z}_{10}$
x_1	1.3	2.1	2.7	3.3	3.4	4.0	4.5	5.0	5.4	5.7
x_2	3.7	4.6	6.2	4.6	2.4	1.1	3.8	6.6	1.4	5.7
class	ω_1	ω_1	ω_1	ω_1	ω_1	ω_2	ω_1	ω_2	ω_2	ω_1

	$\mathbf{z}_{11}$	$\mathbf{z}_{12}$	$\mathbf{z}_{13}$	$\mathbf{z}_{14}$	$\mathbf{z}_{15}$
x_1	6.1	6.3	7.4	7.5	7.6
x_2	3.9	1.9	2.7	0.9	5.3
class	ω_2	ω_2	ω_2	ω_2	ω_2

1.2.2 Two benchmark synthetic data sets

We use two benchmark synthetic data available in the literature or on the Internet. These data are again 2-dimensional for illustration and didactic purposes but have a moderate sample count.

- **Normal-mixtures** data (Figure 1.3).

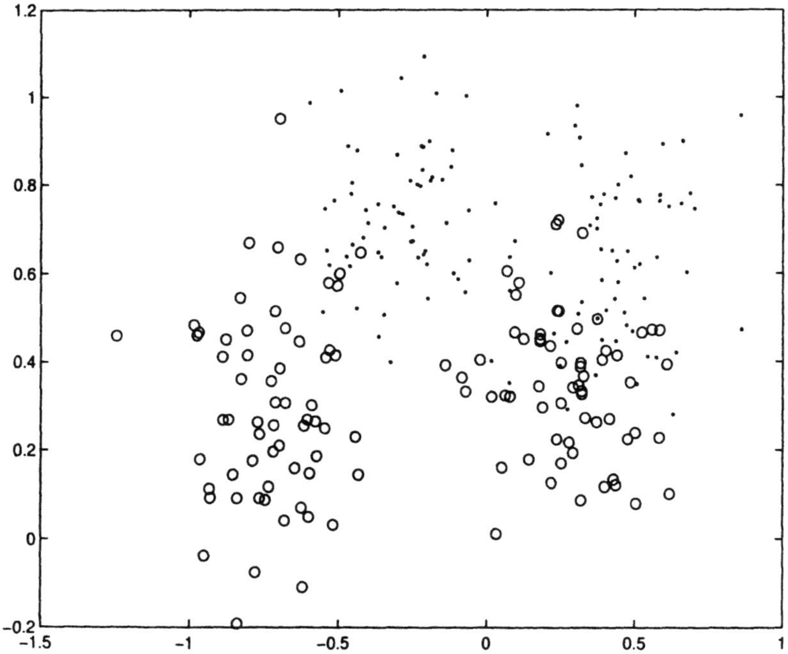

Fig. 1.3. The **Normal-mixtures** data.

This dataset is used in [280] for illustrating classification techniques. The training data consists of 2 classes with 125 2-d points in each class. The points in each class come from a mixture of two normal distributions with the same covariance matrices. The data set is available on http://www.stats.ox.ac.uk/~ripley/PRNN/. A testing set containing 1000 more points drawn from the same distribution is also provided. Ripley [280] points out that the class distributions have been chosen to allow a best possible error rate of about 8 %.

- **Cone-torus** data (Figure 1.4).

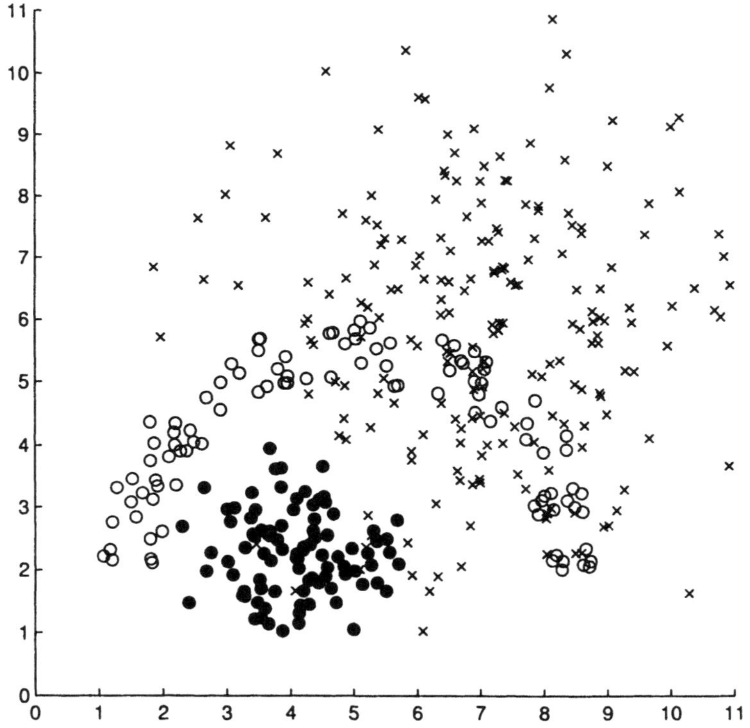

Fig. 1.4. The **Cone-torus** data.

This is a three class dataset with 400 2-d points generated from three differently shaped distributions: a cone, half a torus, and a normal distribution (Example 2.4.1 on page 22 in Chapter 2) with prior probabilities for the three classes 0.25, 0.25, and 0.5. This data set is available on http://www.bangor.ac.uk/~mas00a/Z.txt. A separate data set (for testing) with 400 more points generated from the same distribution is also available as the file Zte.txt.

1.2.3 Two real data sets

Two data sets from the ELENA database are used throughout the book, the Satimage data and the Phoneme data. They are available at anonymous ftp at ftp.dice.ucl.ac.be, directory pub/neural-nets/ELENA/databases.

- The **Satimage** data has been generated from the Landsat Multi-Spectral Scanner image data. It consists of 6435 patterns (pixels) with 36 attributes (4 spectral bands $\times$ 9 pixels in neighborhood). Pixels are classified in 6 classes, and are presented in random order in the database. The classes are: red soil (23.82 %), cotton crop (10.92 %), grey soil (21.10 %), damp grey soil (9.73 %), soil with vegetation stubble (10.99 %), and very damp grey soil (23.43 %). What makes this database attractive is: large sample size; numerical, equally ranged features; no missing values; compact classes of approximately equal size, shape and prior probabilities. Figure 1.5 shows the scatterplot of the 6 Satimage classes on features # 17 and # 18.

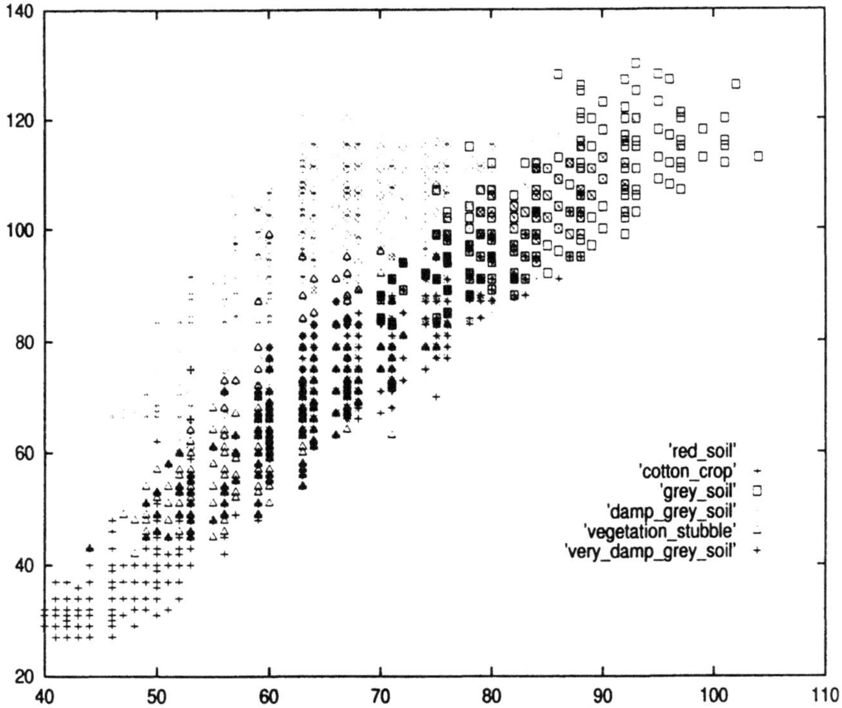

Fig. 1.5. Satimage data on features # 17 and # 18.

- The **Phoneme** data consists of 5404 five-dimensional vectors characterizing two classes of phonemes: nasals (70.65 %) and orals (29.35 %). The scatterplot on features # 3 and # 4 of 800 randomly selected data points is shown in Figure 1.6. A series of classification results with the Phoneme data are presented in [136]. The test classification error varies between 11 and 25 % with the first half of the data set used for training and the second half, for testing.

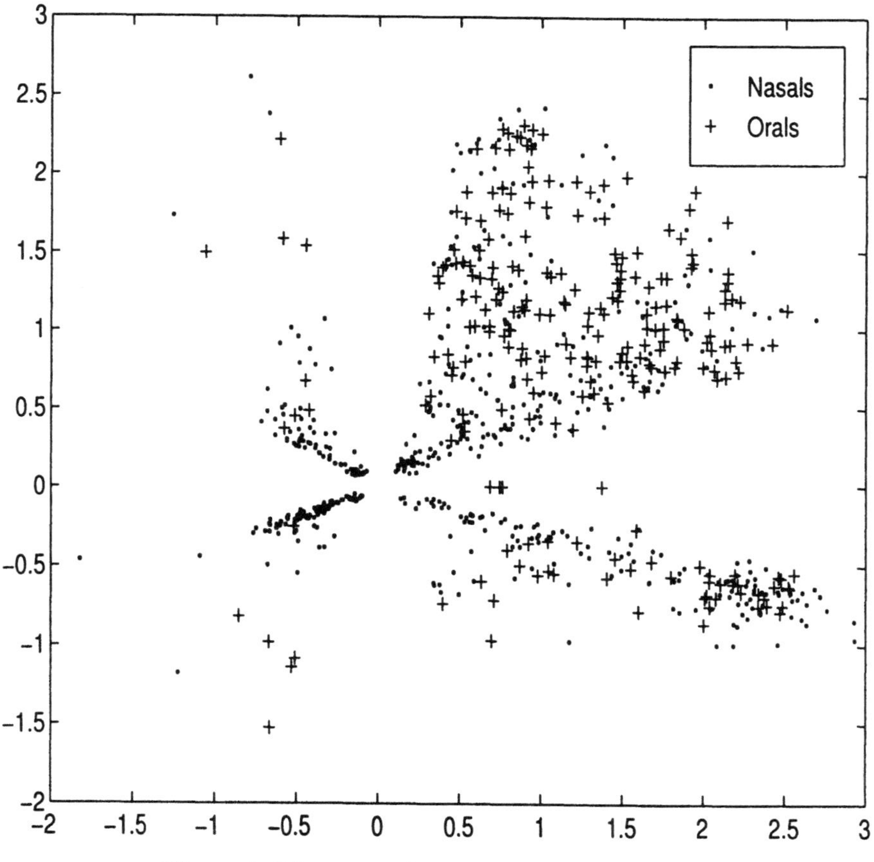

Fig. 1.6. Phoneme data on features # 3 and # 4

N.B. In all experiments in the book, the training and testing parts are formed in the same way. With the two synthetic data sets: Cone-torus and Normal-mixtures the two parts are used for training and for testing, as designated. With the Satimage and Phoneme data, the **first 500** elements of each data sets are used for training, and the remaining part is used for testing. So, the testing sample for Satimage consists of 5935 elements and for Phoneme, of 4904 elements. We restricted the Satimage data set to four dimensions by using only features # 17 to # 20 from the original 36 features.

1.3 Notations and acronyms

Generally, scalars are denoted by lower case italics, such as a, i, etc.; vectors (assumed to be column vectors), by boldface letters, e.g., $\mathbf{x}$, $\mathbf{z}$; vector components are sub-indexed, e.g., $\mathbf{x} = [x_1, \ldots, x_n]^T$. Capital letters are used for matrices and sets, and sometimes for scalars too. Probability density functions are denoted by small $p(\cdot)$, and probabilities, by $P(\cdot)$. "Hat" denotes an estimate, e.g., $\hat{\zeta}$ is an estimate of ζ. Closed intervals are denoted as $[a, b]$, and open intervals as (a, b). Standard symbols for set operations are used, e.g., $\cup, \cap, \in, \subset, \subseteq$. $\forall$ means "for all"; $\exists$, "there exists"; $\emptyset$ is the empty set; $\Longleftrightarrow$ is used as "if and only if", abbreviated also as "iff"; and $\Rightarrow$ for "it follows". Several commonly used notations are given in Table 1.2. (They are explained at the first occurrence in the text but in the ensuing chapters the reader might find this reference helpful.) The end of examples is marked with "▬", and the end of proofs, with "■".

Table 1.3 shows the acronyms most used in the book.

1.4 Organization of the book

The target audience are academic researchers, graduate and postgraduate students in mathematics, engineering, computer science and related disciplines.

Chapter 2 is a brief, reference-like detour through the classics of statistical pattern recognition. The basic notions are introduced and explained along with the underlying Bayes classification model. Special attention is given to the experimental comparison of classifiers.

Chapter 3 details several approaches to statistical classifier design. Parametric and nonparametric classifiers are derived from the Bayesian classifier model. Finding prototypes for the k-nearest neighbor and nearest prototype classifier is a special accent of this chapter. Three popular neural network models are introduced: multi-layer perceptron (MLP), radial-basis function network (RBF) and learning vector quantization (LVQ) networks.

Chapter 4 introduces fuzzy set theory to the extent that is needed for understanding fuzzy classifier designs thereafter. The emphasis is on basic operations on fuzzy sets, especially fuzzy aggregation methods. Practical issues such as determining the membership functions are also discussed. This chapter is not related to the previous two, so the reader who is familiar with statistical pattern recognition may start with Chapter 4.

Chapter 5 explains how fuzzy if-then systems work. The Mamdani-Assilian (MA) and Takagi-Sugeno-Kang (TSK) models are explained and

Table 1.2. Some common notations

$\mathbf{X} = \{X_1, \ldots, X_n\}$	the set of features
$\Re^n$	the feature space spanned by the features from $\mathbf{X}$
$\mathbf{x} = [x_1, \ldots, x_n]^T \in \Re^n$	a feature vector
$\Omega = \{\omega_1, \ldots, \omega_c\}$	the set of class labels
c	number of classes
$g_i(\mathbf{x}), i = 1, \ldots, c$	discriminant functions
$\boldsymbol{\mu}(\mathbf{x}) = [\mu_1(\mathbf{x}), \ldots, \mu_c(\mathbf{x}]^T$	(fuzzy) classifier output
$\mathbf{Z} = \{\mathbf{z}_1, \ldots, \mathbf{z}_N\}$	the data set (unlabeled or labeled in the c classes)
$\mathbf{z}_j = [z_{1j}, \ldots, z_{nj}]^T \in \Re^n$	an element of the data set $\mathbf{Z}$
$l(\mathbf{z}_j) \in \Omega$	the crisp class label of $\mathbf{z}_j$
$\mathbf{l}(\mathbf{z}_j) \in [0,1]^c - \mathbf{0}$	the soft class label of $\mathbf{z}_j$
$l_i(\mathbf{z}_j) \in [0,1]$	the degree of membership of $\mathbf{z}_j$ in class ω_i
N	the number of elements of $\mathbf{Z}$ (cardinality of $\mathbf{Z}$)
N_i	the number of elements of $\mathbf{Z}$ from class ω_i
$Ind(\mathbf{z}_j, \omega_i)$	a binary indicator function with value 1 if $\mathbf{z}_j$ is from ω_i
$p(\mathbf{x})$	probability density function (p.d.f) of $\mathbf{x}$
$p(\mathbf{x}\|\omega_i)$	class-conditional p.d.f of $\mathbf{x}$ given ω_i
$P(\omega_i)$	prior probability for class ω_i
$P(\omega_i\|\mathbf{x})$	posterior probability for class ω_i given $\mathbf{x}$
$U = \{u_1, \ldots, u_m\}$	universal set
$\mu_A(u_i)$	the degree of membership of $u_i \in U$ in the fuzzy set A
$\mathcal{P}(U)$	the class of all subsets of U (the power set of U)

translated into pattern classifiers. The last section of Chapter 5 investigates some theoretical properties of fuzzy if-then models. The (well-proven already) universal approximation by fuzzy TSK systems is revisited with respect to the pattern classification task. A caveat is indicated: fuzzy if-then classifiers could be simple look-up tables in disguise.

Various options for training of fuzzy if-then classifiers are explored in Chapter 6. While some of them are only sketched, (e.g., using neuro-fuzzy models), others are explained in more details (e.g., min-max or hyperbox designs) with numerical examples and experiments.

Chapter 7 presents non if-then fuzzy classifiers. Many such models appeared in the 80s but were then overrun by the more successful if-then stream. Some early models are summarized in a succinct manner at the beginning of the chapter. The two most successful non-if-then designs are outlined next: fuzzy k-nearest neighbor and fuzzy prototype classifiers. Ten fuzzy k-nn variants are tested with the four data sets (the two synthetic sets, Satimage and Phoneme). The Generalized Nearest Prototype Classifier (GNPC) is introduced as a common framework for a number of fuzzy and non-fuzzy classifier models.

The combination of multiple classifiers is discussed in Chapter 8. Various fuzzy and non-fuzzy schemes for classifier fusion and classifier selection are described, 28 of which are also illustrated experimentally. Majority vote over dependent classifiers is analyzed on a synthetic example. The designs chosen for comparison (as well as some of the designs in the previous chapters) are given with enough algorithmic details to be reproducible from the text.

1.5 Acknowledgements

I would like to thank Prof. Janusz Kacprzyk, the Editor of the series and my friend, for inviting me to write this book and trusting me to see it through. I am grateful to my colleagues from the School of Mathematics, University of Wales, Bangor for the wonderfully creative and friendly academic atmosphere. Sincere thanks to my special friends Chris Whitaker and Tim Porter for having the patience to read and correct the draft, and for staying friends with me even after that. I wish to thank my husband Roumen and my daughters, Diana and Kamelia, for putting up with my constant absence from home and with my far too frequent excuse "Leave me alone! I've got a book to write!" for sneaking away from housework.

Table 1.3. Some common acronyms

p.d.f (p.d.f's)	probability density function(s)
k-nn	k-nearest neighbor(s)
HCM	hard c-means (clustering)
NN (NN's)	neural network(s)
LDC	linear discriminant classifier
QDC	quadratic discriminant classifier
MLP	multi-layer perceptron
RBF	radial basis function (NN)
OLS	orthogonal least squares (training of RBF networks)
LVQ	learning vector quantization
SISO	singe-input single-output (system)
MISO	multiple-input single-output (system)
MIMO	multiple-input multiple-output (system)
MA	Mamdani-Assilian (fuzzy if-then model)
TSK	Takagi-Sugeno-Kang (fuzzy if-then model)
COG	center-of-gravity (defuzzification)
MOM	mean-of-maxima (defuzzification)
MSE	minimum squared error
GA	genetic algorithms
GNPC	generalized nearest prototype classifier
BKS	behavior knowledge space (classifier fusion)
DTs	decision templates (classifier fusion)
C	crisp (scheme for classifier fusion)
CC	class-conscious (scheme for classifier fusion)
CI	class-independent (scheme for classifier fusion)

2. Statistical pattern recognition

2.1 Class, feature, feature space

Pattern recognition problems emerge constantly in our everyday life. The ring of the telephone triggers an on-line pattern recognition problem: who might this be? Hearing the voice on the line we are almost always able to tell who this is, no matter that the line might be noisy, or the person at the other end might have a cold. Humans easily identify faces, speakers, smells – tasks that are still a challenge for a computer. When we are able to instruct the computer *how* to label the objects into the prespecified groups, the problem becomes routine. Pattern recognition is about those problems that are still not algorithmically clear-cut. Examples of pattern recognition problems are: classification of crops and soil types from remote-sensing images; detection of clustered microcalcifications on mammograms; optical character recognition (OCR); classification of airmass for predicting a thunderstorm flood; discrimination between stars and galaxies in sky images; etc.

Class is a core notion in pattern recognition. Let Ω be a set of class labels $\Omega = \{\omega_1, \omega_2, \ldots, \omega_c\}$. The term **class** symbolizes a group of objects with a common characteristic or common meaning. In speaker verification, for example, each class is associated with one speaker: all voice patterns uttered by speaker ω_i bear class label ω_i.

Classes can be mutually exclusive or not mutually exclusive, exhaustive (full group) or non-exhaustive, crisp or soft, etc. The letters of the Bulgarian alphabet (Cyrillic) form a set of 30 exhaustive and mutually exclusive classes. Thus, a printed sample of a Cyrillic letter belongs to one and only one of these classes. A set of speakers consists of mutually exclusive classes, but most often, not exhaustive. The two classes of left-handed and right-handed people are exhaustive but they are not mutually exclusive: there can be dextrous people that belong to both classes. If we attach a degree to this class membership showing the left hand and right hand skill, the classes become "soft".

Features (variables) are used to describe the objects numerically. The feature values for a given object are arranged as an n-dimensional vector $\mathbf{x} = [x_1, \ldots, x_n]^T \in \Re^n$. The real space $\Re^n$ is called **feature space**, each axis corresponding to a physical feature. In many real problems we have both qualitative and quantitative features. The diagram in Figure 2.1 illustrates the types of features. Discrete variables with a large number of possible values are treated as quantitative. Qualitative (categorical) features are these with small number of possible values, either with or without gradations. A branch of pattern recognition called Syntactic Pattern Recognition is based on qualitative features [99].

Real-number representation ($\mathbf{x} \in \Re^n$) requires a methodology to convert qualitative features into quantitative. Typically, such methodologies are highly subjective and heuristic. For example, sitting an exam is a methodology to quantify students learning progress. There are also "unmeasurable" features that we, as humans, can assess intuitively but hardly explain, like: sense of humor, intelligence, beauty. In the remaining part of this book we place quantitative, qualitative and "unmeasurable" features into the same pool provided the latter two have some numerical expression. It is another question how suitable this representation is.

The information to design a classifier is usually in the form of a labeled **data set** $\mathbf{Z} = \{\mathbf{z}_1, \ldots, \mathbf{z}_N\}$, $\mathbf{z}_j \in \Re^n$. The class label of $\mathbf{z}_j$ is denoted by $l(\mathbf{z}_j) \in \Omega$, $j = 1, \ldots, N$.

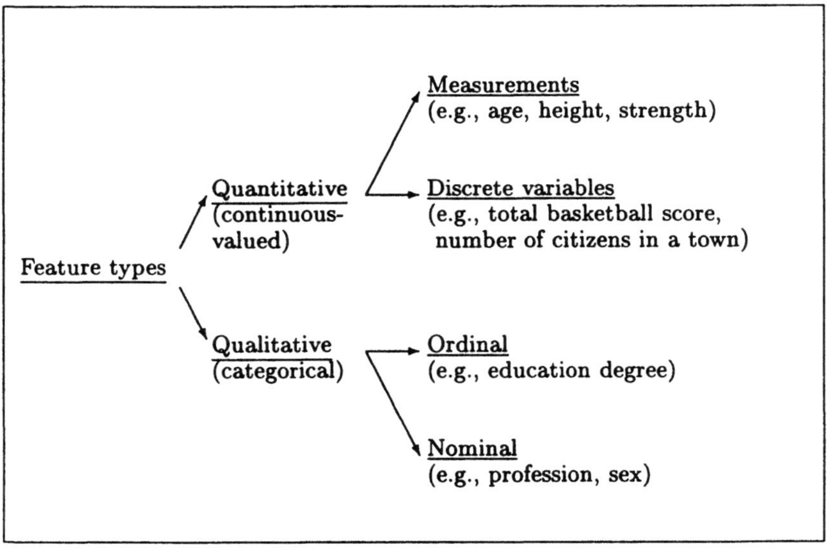

Fig. 2.1. Types of features

2.2 Classifier, discriminant functions, classification regions

A **classifier** is any function:

$$D : \Re^n \to \Omega. \tag{2.1}$$

In the "canonical model of a classifier" [87] shown in Figure 2.2 we consider a set of c **discriminant functions** $G = \{g_1(\mathbf{x}), \ldots, g_c(\mathbf{x})\}$,

$$g_i : \Re^n \to \Re, \ \ i = 1, \ldots, c, \tag{2.2}$$

each one yielding a score for the respective class. Typically (and most naturally), $\mathbf{x}$ is labeled in the class with the highest score, called the **maximum membership rule**, i.e.,

$$D(\mathbf{x}) = \omega_{i^*} \in \Omega \quad \Longleftrightarrow \quad g_{i^*}(\mathbf{x}) = \max_{i=1,\ldots,c} \{g_i(\mathbf{x})\}. \tag{2.3}$$

The maximum membership rule should be equipped with a tie-break extension. Most often ties are broken randomly, i.e., $\mathbf{x}$ is assigned randomly to one of the classes with the highest $g_i(\mathbf{x})$'s.

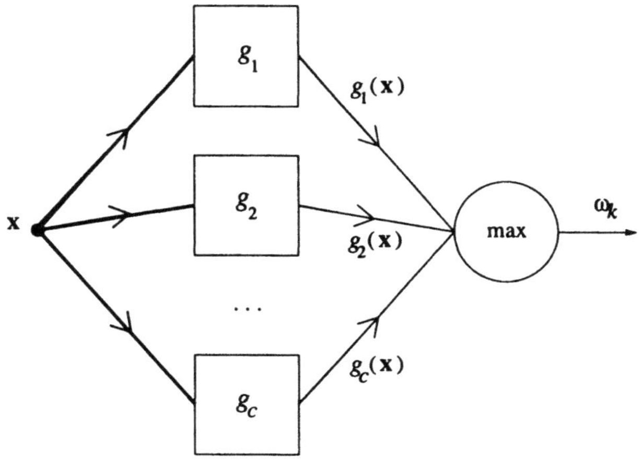

Fig. 2.2. Canonical model of a classifier. The thick vectors denote the n-dimensional input $\mathbf{x}$, the output of the boxes are the discriminant functions $g_i(\mathbf{x})$, and the output of the circle is the class label $\omega_k \in \Omega$ assigned according to the maximum membership rule.

The discriminant functions partition the feature space $\Re^n$ into c (not necessarily compact) **decision regions** or **classification regions** denoted by $\mathcal{R}_1, \ldots, \mathcal{R}_c$

$$\mathcal{R}_i = \left\{ \mathbf{x} \,\middle|\, \mathbf{x} \in \Re^n, g_i(\mathbf{x}) = \max_{k=1,\ldots,c} g_k(\mathbf{x}) \right\}, \quad i = 1, \ldots, c. \tag{2.4}$$

That is, the decision region for class ω_i is the set of points for which the ith discriminant function has the highest score. According to the maximum membership rule (2.3), all points in decision region $\mathcal{R}_i$ are assigned in class ω_i. The decision regions are specified by the classifier D, or, equivalently, by the discriminant functions G. The boundaries of the decision regions are called **classification boundaries**, and contain the points for which the highest discriminant function votes tie. A point on the boundary can be assigned to any of the bordering classes. If a decision region $\mathcal{R}_i$ contains data points from the labeled set $\mathbf{Z}$ with true class label ω_j, $j \neq i$, the classes ω_i and ω_j are called **overlapping**. Note that overlapping classes for a particular partition of the feature space (defined by a certain classifier D) can be non-overlapping if the feature space was partitioned in another way. If in $\mathbf{Z}$ there are no identical points with different class labels, we can *always* partition the feature space into classification regions so that the classes are non-overlapping. Generally, the smaller the overlapping, the better the classifier.

Example 2.2.1. A 15-point two-class problem is depicted in Figure 2.3. The feature space $\Re^2$ is divided into two classification regions: $\mathcal{R}_1$ is shaded (class ω_1: squares) and $\mathcal{R}_2$ is not shaded (class ω_2: snowflakes). For two classes we can use only one discriminant function instead of two:

$$g(\mathbf{x}) = g_1(\mathbf{x}) - g_2(\mathbf{x}), \tag{2.5}$$

and assign class ω_1 if $g(\mathbf{x})$ is positive and class ω_2, if it is negative. For this example, we have drawn the classification boundary produced by the linear discriminant function

$$g(\mathbf{x}) = -7x_1 + 4x_2 + 21 = 0. \tag{2.6}$$

■

Notice that *any* line in $\Re^2$ is a linear discriminant function for any 2-class problem in $\Re^2$. Generally, *any* set of functions $g_1(\mathbf{x}), \ldots, g_c(\mathbf{x})$, (linear or non-linear) is a set of *discriminant functions*. It is another matter how successfully these discriminant functions separate the classes.

Let $G^* = \{g_1^*(\mathbf{x}), \ldots, g_c^*(\mathbf{x})\}$ be a set of *optimal* (in some sense) discriminant functions. We can obtain infinitely many sets of *optimal* discriminant functions from G^* by applying any transformation $f(g_i^*(\mathbf{x}))$ that preserves the order of the function values for every $\mathbf{x} \in \Re^n$. For example, $f(\zeta)$ can be a $\log(\zeta)$, $\sqrt{\zeta}$ for positive definite $g^*(\mathbf{x})$, a^ζ, for $a > 1$, etc. Using the maximum membership rule (2.3), $\mathbf{x}$ will be labeled to the same class by any of the discriminant function sets.

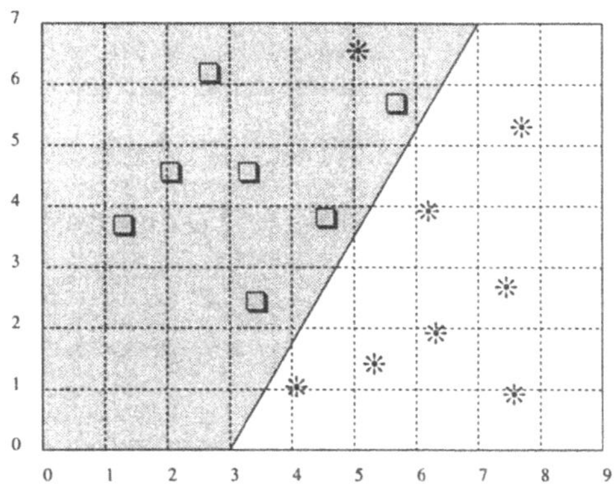

Fig. 2.3. A two-class example with a linear discriminant function

If the classes in **Z** can be separated completely from each other by a hyperplane (a point in $\Re$, a line in $\Re^2$, a plane in $\Re^3$), they are called **linearly separable**. The two classes in Example 2.2.1 are not linearly separable because of the snowflake at (5,6.6) which is on the "wrong" side of the discriminant function (in the shaded region). The classes are therefore overlapping with respect to the partition depicted in Figure 2.3.

2.3 Clustering

Clustering is an important topic in pattern recognition [38, 87, 110]. **Crisp clustering** means to partition the data set **Z** into nonintersecting subsets, thereby identifying groups in the data. In this process we disregard the data labels and treat **Z** as unlabeled. Therefore clustering is also called *unsupervised learning*. Ideally, the groups should contain "similar" objects, and the objects from different groups should be "dissimilar". The most intuitive concept of similarity is associated with distance. Thus, objects which are close to each other in $\Re^n$ should be put in the same group. There are many different ways to define the distance from an object to a group of objects. Two such ways define the following two popular clustering methods.

2.3.1 Hard *c*-means

Define the distance between some $\mathbf{z} \in \Re^n$ and a group of objects in $\Re^n$ as the distance between **z** and the mean of the group. This idea has resulted in the **hard *c*-means** clustering algorithm (HCM)[1] shown in Figure 2.4.

[1] The word "hard" is used to distinguish this model from the fuzzy c-means

HARD c-MEANS

1. **Pick the number of clusters c. Initialize the cluster centers (e.g., by randomly selecting c points from $\mathbf{Z}$ as the centers).**
2. **Label all points in $\mathbf{Z}$ with respect to their distance to the cluster centers: each point is assigned to the cluster with the closest center.**
3. **Calculate the new centers as the means of the points from $\mathbf{Z}$ assigned to the respective clusters.**
4. **Loop through steps 2 and 3 until no change in the centers occurs.**

Fig. 2.4. The hard c-means clustering procedure

Example 2.3.1. Figure 2.5 shows the clustering of the 15-point data in two and three clusters. Each partition is obtained by one run of the HCM algorithm. The points in the same cluster are enclosed together. Class labels are also shown, although not used in the clustering. ■

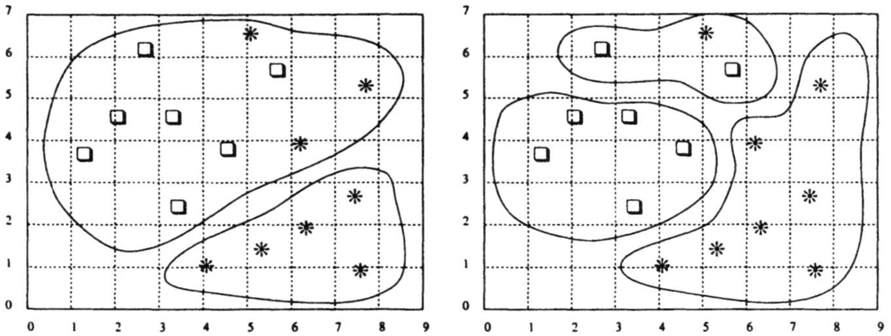

2 clusters 3 clusters

Fig. 2.5. Clustering of the 15-point data sets in 2 and 3 clusters using HCM.

Since HCM starts with random initialization, different c clusters can be found at each run.

2.3.2 Single linkage clustering

This algorithm stems from another concept of a distance between a point $\mathbf{z}$ and a cluster defined as the distance between $\mathbf{z}$ and its *nearest neighbor* from the cluster. The distance between two clusters is the distance between the pair of closest elements, one from each cluster. The single linkage procedure belongs to a family of hierarchical clustering procedures [87]. We start with all points of $\mathbf{Z}$ as individual clusters and proceed to find the whole set $\mathbf{Z}$ as one cluster. For a fixed number of clusters $c < N$, we stop the clustering when exactly c clusters are found. The single linkage clustering algorithm is shown in Figure 2.6.

SINGLE LINKAGE CLUSTERING

1. Pick the number of clusters c. Initialize the procedure by defining an individual cluster for each point in $\mathbf{Z}$.
2. Identify the two closest clusters and join them as a new cluster, dismissing the initial two.
3. Repeat step 2 until c clusters are found.

Fig. 2.6. The single linkage clustering procedure

Example 2.3.2. Figure 2.7 shows the clustering of the 15-point data in two and three clusters by the single linkage algorithm. The points in the same cluster are enclosed together. Class labels are also shown, although not used in the clustering. ■

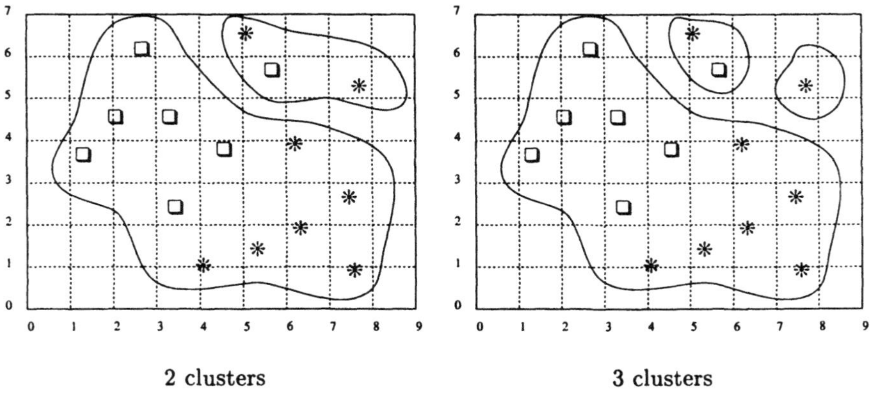

Fig. 2.7. Clustering of the 15-point data sets in 2 and 3 clusters using single linkage.

Unlike HCM the single linkage method is not likely to find "cloud-shaped" clusters. Single linkage is apt if we expect the clusters to be shaped like strings.

There are a number of debatable issues in clustering with no ready solutions [110]. Example questions are: How many clusters we shall look for? What is a good clustering criterion function? What type of distance shall we use? Do we need to transform the data? There is a huge amount of literature on clustering. For the purpose of classifier design, we shall need only the generic idea.

2.4 Prior probabilities, class-conditional probability density functions, posterior probabilities

Let $\mathbf{x}$ be a random variable in $\Re^n$ coming from one of c mutually exclusive classes. A classification problem is characterized completely by

1. $P(\omega_i), \omega_i \in \Omega$, the prior probabilities for the classes ($\sum_{i=1}^{c} P(\omega_i) = 1$), and
2. $p(\mathbf{x}|\omega_i)$, the class-conditional probability density functions (p.d.f's)

The integral of the p.d.f $p(\mathbf{x}|\omega_i)$ over a certain region $\mathcal{S} \subset \Re^n$ gives the probability of $\mathbf{x}$ falling in this region of the feature space, given that the true class is ω_i, i.e.,

$$P(\mathbf{x} \in \mathcal{S}|\omega_i) = \int_{\mathcal{S}} p(\mathbf{x}|\omega_i)\, d\mathbf{x}. \tag{2.7}$$

A p.d.f has integral one over the whole feature space $\Re^n$.

Example 2.4.1. Let $\Omega = \{\omega_1, \omega_2, \omega_3\}$, $\mathbf{x} \in \Re^2$. Shown in Figure 2.8 are the three (peculiarly shaped) p.d.f's

- a cone

$$p(\mathbf{x}|\omega_1) = \begin{cases} \frac{3}{8\pi}\left[2 - \sqrt{(x_1-4)^2 + (x_2-2)^2}\right], & \text{if } (x_1-4)^2 + (x_2-2)^2 < 4 \\ 0, & \text{otherwise;} \end{cases} \tag{2.8}$$

- half a torus

$$p(\mathbf{x}|\omega_2) = \begin{cases} \frac{2}{7\pi}, & \text{if } \; 9 \le (x_1-5)^2 + (x_2-2)^2 < 16 \;\text{ and }\; x_2 > 2 \\ 0, & \text{otherwise;} \end{cases} \tag{2.9}$$

- a Gaussian

$$p(\mathbf{x}|\omega_3) = \frac{1}{8\pi} \exp\left\{ -\frac{1}{8}\left[(x_1-7)^2 + (x_2-6)^2\right] \right\}. \tag{2.10}$$

We assume that $\mathbf{Z}$ contains N labeled, independent, and identically distributed samples (i.i.d) drawn from the distributions of the model. Figure 1.4 in Chapter 1 plots the **Cone-torus** set $\mathbf{Z}$ sampled from these three distributions with $P(\omega_1) = 0.25$, $P(\omega_2) = 0.25$, $P(\omega_3)=0.5$, and N=400. ■

Using $p(\mathbf{x}|\omega_i)$ and $P(\omega_i)$ we can calculate the posterior probabilities by the Bayes formula:

$$P(\omega_i|\mathbf{x}) = \frac{P(\omega_i)\, p(\mathbf{x}|\omega_i)}{p(\mathbf{x})}, \; i = 1, \ldots, c, \tag{2.11}$$

where $p(\mathbf{x})$ is the unconditional probability density function

$$p(\mathbf{x}) = \sum_{k=1}^{c} P(\omega_k)\, p(\mathbf{x}|\omega_k). \tag{2.12}$$

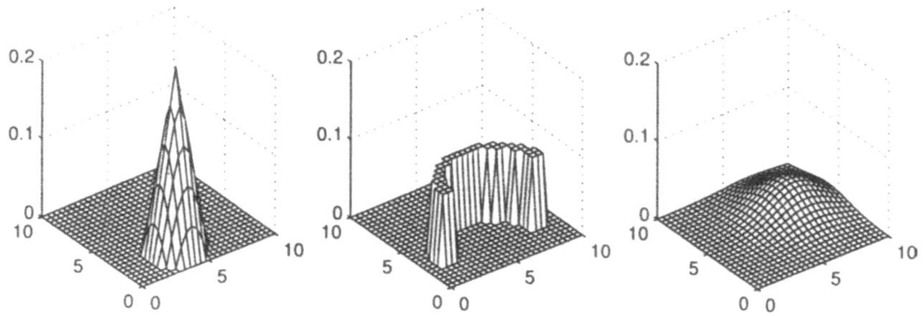

Fig. 2.8. Three p.d.f's.: a cone, half a torus and a Gaussian

2.5 Minimum error and minimum risk classification. Loss matrix.

The total probability of error of a classifier D over the whole feature space $\Re^n$ is

$$P_D(error) = \int_{\Re^n} P_D(error|\mathbf{x})\ p(\mathbf{x})\ d\mathbf{x}. \tag{2.13}$$

where $P_D(error|\mathbf{x})$ is the probability of error of classifier D for a given $\mathbf{x}$. If we know the posterior probabilities $P(\omega_i|\mathbf{x}), i = 1, \ldots, c$, we can minimize the error by choosing the class label for $\mathbf{x}$ corresponding to the highest posterior probability. Then the probability of error for this $\mathbf{x}$ will be

$$P(error|\mathbf{x}) = 1 - \max_{\omega_k \in \Omega} P(\omega_k|\mathbf{x}). \tag{2.14}$$

Therefore, any choice of another class label will produce a higher probability of error for $\mathbf{x}$ and in total. Notice that in the case of tied posterior probabilities, *any* of the tied class labels leads to the same (smallest) error. Thus, the optimal (Bayes) classifier D_B is:

$$D_B(\mathbf{x}) = \omega_i^* \in \Omega \quad \Longleftrightarrow \quad P(\omega_i^*|\mathbf{x}) = \max_{\omega_i \in \Omega} P(\omega_i|\mathbf{x}). \tag{2.15}$$

D_B is also called the **minimum-error classifier**. Practically we never have enough information to design D_B. It is important to distinguish this classifier from what some authors call "The Bayes classifier". Equation (2.15) implies the knowledge of the true p.d.f's and prior probabilities, and therefore, the true posterior probabilities. When we use approximations of these, the classifier should be called "pseudo-Bayes". In the cases where we assume independence of the features, thereby significantly simplifying the problem, better names for this classifier are "naive Bayes" or even "idiotic Bayes" [280].

In some problems the set Ω is augmented by another class (we denote it ω_{c+1}) for vectors which D "refuses to label". This is important in the cases where misclassifications are expensive. A refuse-to-decide option can be further split into two: $D(\mathbf{x})$ meaning either "one of these c classes but I don't know which one" or "none of these classes" (when Ω is not a full group). For example, consider the real-time classification of radar images of flying objects. Assume that assigning class label $\omega_s \in \Omega$ implies that the object should be immediately destroyed (e.g., ω_s is "enemy missile"). In this problem it may pay off to postpone assigning label ω_s if the decision is not "absolutely sure", and put the object in the class "don't know". This classification strategy involves an individual weight (or cost) for each decision, including the "don't know" one. We call Λ the **loss matrix** with entries λ_{ij}, $i = 1, \ldots, c(+1)$, $j = 1, \ldots, c$, denoting the loss incurred by assigning $\mathbf{x}$ to class ω_i if the true class is ω_j.

The conditional risk in making decision $D(\mathbf{x}) = \omega_i$ for a given $\mathbf{x} \in \Re^n$ is

$$R(\omega_i|\mathbf{x}) = \sum_{k=1}^{c} \lambda_{ik}\, P(\omega_k|\mathbf{x}), \quad i = 1, \ldots, c(+1). \tag{2.16}$$

The total risk using classifier D is

$$R_D = \int_{\Re^n} R\,(D(\mathbf{x})|\mathbf{x})\; p(\mathbf{x})\; d\mathbf{x}. \tag{2.17}$$

To minimize the total risk we have to minimize the conditional risk for every $\mathbf{x}$. The **minimum risk** classifier chooses a class label for $\mathbf{x}$ corresponding to the smallest conditional risk. A zero-one (0-1) loss matrix is defined as $\lambda_{ij} = 0$ for $i = j$ and $\lambda_{ij} = 1$ for $i \neq j$, i.e., all errors are equally costly. Then the conditional risk becomes

$$R(\omega_i|\mathbf{x}) = \sum_{k=1, k\neq i}^{c} P(\omega_k|\mathbf{x}) = 1 - P(\omega_i|\mathbf{x}). \tag{2.18}$$

The minimum risk classifier with a 0-1 loss matrix assigns $\mathbf{x}$ to ω_{i^*} iff

$$\begin{aligned} R(\omega_{i^*}|\mathbf{x}) &= \min_{\omega_i \in \Omega} R(\omega_i|\mathbf{x}) \\ &= 1 - \max_{\omega_i \in \Omega} P(\omega_i|\mathbf{x}), \end{aligned} \tag{2.19}$$

i.e., iff

$$P(\omega_{i^*}|\mathbf{x}) = \max_{\omega_i \in \Omega} P(\omega_i|\mathbf{x}), \tag{2.20}$$

which is the same decision as with the minimum error classifier.

2.6 Performance estimation

The **performance** of a classifier D has several issues. The major issue is the classification accuracy of D. Secondary components of the term performance are: the rejection rate and the corresponding accuracy (e.g., represented on a rejection-accuracy plot); the ability to identify outliers; the performance with noisy features and missing data; the performance with mixed variables as inputs [220].

2.6.1 Classification accuracy, error rate, training, testing, and generalization

The most important goal in designing a classifier is to achieve the highest possible **classification accuracy** or, equivalently – its dual characteristic – the lowest possible **error rate**. Ideally, the estimates of these characteristics should be robust, i.e., the classifier should achieve similar performance both on the data it has been trained on, and on unseen data. The **training** performance is measured on that part of $\mathbf{Z}$ which has been used to design the classifier. The **testing** performance should be measured on an independent sample drawn from the same distribution. Large values of N (preferably thousands) are often desirable. The training estimate of accuracy (or the error) is usually optimistically biased. The testing performance of a classifier is also called **generalization**. If the training accuracy is significantly higher than the testing accuracy (respectively the training error is significantly lower than the testing error), the classifier is said to be **overtrained**. To prevent overtraining, the training data set is sometimes divided into **training** and **validation** subsets, and the latter acts as a "pseudo-test". An indication of overfitting is a continuously decreasing training error and correspondingly increasing "pseudo-testing" error. At this point the training should be stopped. It is worth mentioning that "pseudo-testing" reduces the risk of overfitting but cannot eliminate it completely. Moreover, the validation set cuts off a piece of the training set, which could be otherwise used for a more accurate estimation of the classifier parameters.

Estimating the classification error is a difficult problem. Using (2.13) with D_B as D we obtain the smallest possible error (called the **Bayes error**)

$$P_B = 1 - \int_{\Re^n} \max_{j=1,\dots,c} \{P(\omega_j)p(\mathbf{x}|\omega_j)\}\, d\mathbf{x}, \qquad (2.21)$$

or also using the classification regions $\mathcal{R}_1, \dots, \mathcal{R}_c$,

$$P_B = 1 = \sum_{i=1}^{c} \int_{\mathcal{R}_i} P(\omega_i)p(\mathbf{x}|\omega_i)d\mathbf{x}. \qquad (2.22)$$

Four possible cases for N and D are summarized in Table 2.1. In practice we usually have a finite data set $\mathbf{Z}$ and an imperfect classifier D, shown as Case 4. The question about how to find a "good" estimate $\hat{P}_D$ splits into the following two subquestions:

$\clubsuit$. How to calculate the value of $\hat{P}_D$?

Table 2.1. Four cases for estimating classification error

Case	Limit size of **Z**	Classifier	What can we assess (theoretically)?
1	∞	D_B	**Bayes error (P_B):** Having an infinite sample we can estimate $P(\omega_i)$ and $p(\mathbf{x}\|\omega_i)$, $i = 1, \dots, c$, and therefore the Bayes error with arbitrary precision.
2	N	D_B	**Approximation of the Bayes error ($\hat{P}_B$):** The distributions can be estimated up to a certain level of precision.
3	∞	Any D	**Exact value of the error of this classifier (P_D)**
4	N	Any D	**Approximate value of the error of this classifier ($\hat{P}_D$)**

$\diamondsuit$. How to make the best use of **Z**?

$\clubsuit$. Let $\bar{\mathbf{Z}}$ be the set of $\bar{N}$ labeled vectors on which we will calculate $\hat{P}_D$. We assume the following model: The classifier D commits an error with probability P_D on any object $\mathbf{x} \in \Re^n$. Then the number of errors has a binomial distribution with probability P_D. Thus, an estimate of P_D is:

$$\hat{P}_D = \frac{N_{error}}{\bar{N}}, \qquad (2.23)$$

where N_{error} is the number of misclassifications committed by D on the elements of $\bar{\mathbf{Z}}$. This is called **counting estimator** of the error rate because it is based on the count of misclassifications. It can be rewritten as

$$\hat{P}_D = \frac{1}{\bar{N}} \sum_{j=1}^{\bar{N}} I_D(\mathbf{z}_j), \qquad (2.24)$$

where $I_D(\cdot)$ is an indicator function taking value 1 if the argument is misclassified by D, and 0, otherwise. $\hat{P}_D$ is also called the **apparent error rate.** Dual to this characteristic is the apparent classification accuracy which is calculated by

$$1 - \hat{P}_D = \frac{\bar{N} - N_{error}}{\bar{N}} = \frac{N_{correct}}{\bar{N}}. \qquad (2.25)$$

If we have a large enough $\bar{N}$, we can approximate the binomial distribution by a normal distribution and find the 95 % **confidence interval** for the error rate

$$\left[\hat{P}_D - 1.96\sqrt{\frac{\hat{P}_D(1-\hat{P}_D)}{\bar{N}}},\ \hat{P}_D + 1.96\sqrt{\frac{\hat{P}_D(1-\hat{P}_D)}{\bar{N}}}\right].$$

There are so called **smooth modifications** [107] of the counting estimator. The binary indicator function $I_D(\cdot)$ in (1) is replaced by a "smoothing" function taking values in the interval $[0,1] \subset \Re$. The smoothing modification is designed to reduce the variance of the estimate of P_D.

◇ How to make the best use of **Z** [325]?

- **Resubstitution** (R-method). Design the classifier D on **Z** and test it on **Z**. $\hat{P}_D$ is optimistically biased.
- **Hold-out** (H-method). Traditionally, split **Z** into halves, use one half for training D, and the other half for calculating $\hat{P}_D$. $\hat{P}_D$ is pessimistically biased. Splits in other proportions are also used. We can swap the two subsets, get another estimate $\hat{P}_D$ and average the two. A version of this method is the **data shuffle** where we do L random splits of **Z** into training and testing parts and average all L estimates of P_D calculated on the respective testing parts.
- **Cross-validation** (called also "rotation method"). We choose an integer K (preferably a factor of N) and randomly divide **Z** into K subsets of size N/K. Then we use one subset to test the performance of D trained on the union of the remaining $K-1$ subsets. This procedure is repeated K times choosing a different part for testing each time. To get the final value of $\hat{P}_D$ we average the K estimates. The most popular version is called **leave-one-out** (or U-method), where $K = N$.
- **Bootstrap** [152]. This method is designed to correct the optimistic bias of the R-method. This is done by randomly generating L sets of cardinality N from the original set **Z**, with replacement. Then we assess and average the error rate of the classifiers built on these sets.

How are the errors distributed over the classes? The class-conditional error rates can be estimated from **confusion matrices**. The entry a_{ij} in the confusion matrix denotes the number of elements from $\bar{\mathbf{Z}}$ whose true class is ω_i, and which are assigned by D to class ω_j.

Example 2.6.1. The confusion matrix for the linear classifier from Example 2.2.1 (Figure 2.3) is

$D(\mathbf{x})\rightarrow$ True class ↓	ω_1	ω_2
ω_1	7	0
ω_2	1	7

■

2.7 Experimental comparison of classifiers

Although seemingly an easy task, comparing classifiers is perhaps one of the most deceiving and tricky problems in classifier design. To start with, we have to always keep in mind that **there is no such thing as the "best" classifier**. Classifiers applied to different problems and trained by different algorithms perform differently [89, 153]. The asymptotic behavior of some classifiers is known, as we shall see later. However, nice asymptotic properties do not guarantee good performance for finite N. Not surprisingly, some experimental studies find k-nearest neighbor classifiers better than neural classifiers [12, 237, 45], others find the opposite [125, 91, 345]. Typically, such studies are based on extensive experimental evidence using a number of simulated and real data sets. Who is right and who is wrong? The answers is that they all are right – for some problems some classifiers are better suited than others, and therefore, a variety of schemes should be kept in the toolbox.

Can we make a "fair" comparison of a group of classifiers? Some issues are discussed below.

2.7.1 Error rate comparison

We can calculate the error on the training and testing set as discussed earlier. Some authors prefer to show the classification accuracy instead of the error. Prechelt [277] argues that the *error rate* is the better of the two because an accuracy of 98 % is actually twice as good as one of 96 % which is clear if we looked at the error rates: 2 % and 4 %. The eyeball comparison, however, may be misleading. Suppose we have two classifiers trained and tested on the same data sets, giving testing accuracy of 98 % and 96 %, respectively. Can we claim that the first classifier is better? In other words, are the two figures *significantly* different?

Looney [223] proposed a method to compare statistically multiple classification rates. His method is based on the application of repeated measures analysis techniques to dichotomous data.

Let $\{D_1, \ldots, D_L\}$ be a set of L classifiers tested on the same data set $\mathbf{Z} = \{\mathbf{z}_1, \ldots, \mathbf{z}_N\}$. The results are stored in a binary $N \times L$ matrix $\mathbf{M}$ whose ijth entry m_{ij} is 0 if classifier D_j has misclassified vector $\mathbf{z}_i$, and 1, if D_j has produced the correct class label $l(\mathbf{z}_i)$.

Example 2.7.1. Consider the Cone-torus data as $\mathbf{Z}$ ($N = 400$) and the resubstitution results of 3 classifiers: Linear Discriminant Classifier (LDC), Quadratic Discriminant Classifier (QDC) and the Logistic Classifier (LOG)[2]. The number of possible combinations of zeros and ones for this example is $2^L = 8$. Table 2.2 shows the combinations and the number of times they occur in $\mathbf{Z}$. The estimate of the classification accuracy (by the counting estimator (2.25)) is shown in the bottom row of the Table. ■

[2] LDC and QDC are discussed in the next chapter. Here we are interested only in their performance on the Cone-torus data.

Table 2.2. Resubstitution results (matrix **M**) from the three classifiers on the Cone-torus data: "0" means misclassification, "1" means a correct classification

D_1 =LDC	D_2 =QDC	D_3 =LOG	Number
1	1	1	281
1	1	0	7
1	0	1	2
1	0	0	11
0	1	1	10
0	1	0	0
0	0	1	23
0	0	0	66
0.7350	0.8050	0.7500	-

Let P_{D_i} denote the estimate of the classification accuracy of classifier D_i[3]. Using an F-test from statistics we shall test the hypothesis for no difference between the classification accuracies:

$$H_0 : P_{D_1} = P_{D_2} = \ldots = P_{D_L}, \tag{2.26}$$

using the data in Table 2.2.[4]

Denote by $\bar{P}$ the averaged accuracy. We calculate the sum of squares for the classifiers:

$$SSA = N \sum_{i=1}^{L} P_{D_i}^2 - N\, L\, \bar{P}^2 \tag{2.27}$$

(here $L = 3$), and the sum of squares for the objects

$$SSB = L \sum_{j=1}^{N} \left(\frac{1}{L} \sum_{i=1}^{L} m_{ij} \right) - L\, N\, \bar{P}^2. \tag{2.28}$$

The total sum of squares is

$$SST = N\, L\, \bar{P}(1 - \bar{P}), \tag{2.29}$$

and the sum of squares for classification-object interaction is

$$SSAB = SST - SSA - SSB. \tag{2.30}$$

The *calculated* F value is obtained by

$$MSA = \frac{SSA}{(L-1)}; \quad MSAB = \frac{SSAB}{(L-1)(N-1)}; \quad F_{cal} = \frac{MSA}{MSAB}. \tag{2.31}$$

---

[3] Notice that for convenience here we use the notation P_{D_i} for the accuracy (not error!)

[4] Readers not familiar with statistics can take the below sequence of calculations as a cooking recipe for a statistical comparison of L classifiers.

At this point we can choose a level of significance α and check the validity of H_0 by comparing our F_{cal} with the tabulated value of an F-distribution with degrees of freedom $(L-1)$ and $(L-1)(N-1)$. If F_{cal} is greater than the tabulated F-value we reject the null hypothesis H_0 and can further search for pairs of classifiers that differ significantly. Looney suggests we use the same F_{cal} but with *adjusted* degrees of freedom for deciding about H_0 (called the F^+ test). The adjustment uses $\mathbf{S}$, the $L \times L$ covariance matrix of the data in $\mathbf{M}$. Let s_{ij} be the ij-th entry of $\mathbf{S}$, $\bar{s}_{ii}$ be the mean of the main diagonal entries of $\mathbf{S}$, $\bar{s}_{i\cdot}$ be the mean of row (column) i, and $\bar{s}$ be the total mean. The adjusting factor is calculated as

$$\epsilon = \frac{L^2\ (\bar{s}_{ii} - \bar{s})^2}{(L-1)\left(\sum_{i=1}^{L}\sum_{j=1}^{L}\ s_{ij}^2 - 2L\sum_{i=1}^{L}\bar{s}_{i\cdot}^2 + L^2\bar{s}^2\right)}, \tag{2.32}$$

and the adjusted degrees of freedom are $(L-1)\epsilon$ and $(L-1)(N-1)\epsilon$.

If the null hypothesis is rejected by the F^+ test, pairwise comparisons are recommended to find out which classifiers differ significantly. Denote by $\mathbf{p}$ the vector with classifier accuracies $\mathbf{p} = [P_{D_1}, \ldots, P_{D_L}]^T$. For values of $N > 30$, the large-sample multiple comparison method is based on the χ^2 characteristic

$$\chi^2_{cal} = \frac{N\left(\mathbf{b}^T\mathbf{p}\right)^2}{\mathbf{b}^T\mathbf{S}\mathbf{b}}, \tag{2.33}$$

where $\mathbf{b}$ is a vector with contrasts. For example, let $L = 5$. To compare classifiers D_3 and D_5, we calculate (2.33) with $\mathbf{b} = [0, 0, 1, 0, -1]^T$ and match χ^2_{cal} against a χ^2 distribution with $L = 5$ degrees of freedom (using the same α that we used for the F^+ test). We can use other contrasts too, e.g., $\mathbf{b} = [1/2, -1/3, -1/3, 1/2, -1/3]^T$ checking the hypothesis

$$H_0 : \frac{P_{D_1} + P_{D_4}}{2} = \frac{P_{D_2} + P_{D_2} + P_{D_5}}{3}. \tag{2.34}$$

Example 2.7.2. For the three classifiers LDC, QDC, and LOG (Table 2.2), using the above procedure, we obtain:

SSA	=	1.0867	MSA	=	0.5433
SSB	=	181.4533	$MSAB$	=	0.0429
SST	=	216.7867			
$SSAB$	=	34.2467	F_{cal}	=	**12.6605**

The (unadjusted) degrees of freedom are (2, 798) and the p-value for the F-test is **0.0000**. Choosing $\alpha = 0.99$, the obtained p-value suggests rejection of H_0. For the adjusted degrees of freedom:

$$\mathbf{S} = \begin{bmatrix} 0.1953 & 0.1387 & 0.1566 \\ & 0.1574 & 0.1165 \\ & & 0.1880 \end{bmatrix}$$

$$\epsilon = 0.9124$$
$$\text{d.f.1} = 1.8247$$
$$\text{d.f.2} = 728.0557$$

The *p*-value for the F^+ test is again **0.0000**, which suggests rejection of H_0. In this example the adjustment did not lead to changing the statistical decision. The decision might have been different if N was small. Using (2.33) for pairwise comparisons we get

Classifiers	Contrasts	χ^2_{cal}	d.f.	p
LDC, QDC	$[1, -1, 0]$	26.0333	3	**0.000**
LDC, LOG	$[1, 0, -1]$	1.2866	3	0.2378
QDC, LOG	$[0, 1, -1]$	10.7790	3	**0.006**

For the chosen $\alpha = 0.99$, two pairs of classifiers appear to be significantly different: LDC and QDC; and QDC and LOG, QDC having the highest accuracy of the three. ■

Usually, the designer conducts a series of experiments with each classifier (a good practice is at least a 10-fold cross-validation). The averaged accuracy/error is a background for claiming that some classifier outperforms the others for a given pattern recognition problem. It is advisable to provide more information, e.g., the standard deviation and the minimal and maximal accuracy/error obtained for each classifier.

2.7.2 Experiment design

In a cute little correspondence titled "Candide's practical principles of experimental pattern recognition" [244] George Nagy postulates a series of invaluable principles and recommendations. Two of them are literally reproduced below.

> **Comparison of classification accuracies.** *Comparisons against algorithms proposed by others are distasteful and should be avoided. When this is not possible, the following Theorem of Ethical Data Selection may prove useful.*
>
> *Theorem: There exists a set of data for which a candidate algorithm is superior to any given rival algorithm. This set may be constructed*

by omitting from the test set any pattern which is misclassified by the candidate algorithm.

Replication of experiments. *Since pattern recognition is a mature discipline, the replication of experiments on new data by independent research groups, a fetish in the physical and biological sciences, is unnecessary. Concentrate instead on the accumulation of novel, universally applicable algorithms.*

Casey's Caution: Do not ever make your experimental data available to others; someone may find an obvious solution that you missed.

Speaking seriously now, to avoid bias, the experiments for all classifiers should be carried out using the same testing method (e.g., cross-validation) and on the same split of the data into training and testing.

Testing should be done on previously *unseen* data. A common mistake in presenting results from classification experiments is that the *best testing* results are reported [89].

Example 2.7.3. Let $D(k)$ be a classifier with a parameter $k \in \{1, 2, \ldots, N\}$. Assume that $D(k)$ has been applied to some hypothetical 2-class problem for $k = 1, 3, \ldots, 11$ and the leave-one-out accuracy on the training set, and the accuracy on the test set are shown in the table below.

k	1	3	5	7	9	11
Training error [%]	15	12	10	9	7	6
Testing error [%]	19	14	11	13	13	15

Suppose the results from applying a linear discriminant classifier to the same problem yields: 11 % resubstitution error on the training set, and 12 % error on the test set. Looking at the test error it is very tempting to choose $k = 5$ and conclude that $D(k)$ outperforms LDC. A fair comparison, however, requires to take the best *training* result (6 % with $k = 11$) and the *corresponding* test result of 15 % which changes the odds in this example in favor of LDC. ∎

The above table shows also the "overtraining" phenomenon where the training error becomes misleadingly small while testing error starts increasing with k.

2.7.3 Classifier complexity

Simple models are less likely to be overtrained but they are not versatile enough to fit complex classification boundaries (they are said to have high bias [43]). Consider the linear discriminant model: it trains very quickly, has few parameters, but cannot reach high recognition accuracy. More complex models (e.g., neural networks and prototype-based classifiers) have a better flexibility but require more system resources and are prone to overtraining (high variance [43]). Usually the designer needs to seek a balance: the simplest possible model with the best possible performance. What do "simple" and "complex" mean in this context? We consider here several aspects of complexity [12].

- Training time and training complexity.
- Memory requirements (e.g., the number of the parameters of the classifier that are needed for its operation)
- Running complexity.

Lippmann [220] and Lee [213] use the following parameters for comparison of classifiers: (i) error rate; (ii) free parameters of the classifier; (iii) training time; (iv) classification time.

At a first glance, in the light of the fast industrial developments towards more powerful hardware and software, these items seem to be not as relevant as the accuracy of the classifier. Tasks seemingly infeasible 20 years ago are executable in seconds now. On the other hand, more powerful computer facilities open up space for more complex pattern recognition problems such as face, voice, and speaker recognition; signature verification; satellite image analysis. In such problems the computational demand may be the leading component in designing a classifier.

Complexity should be analyzed with respect to the problem at hand. For example, if we design a classifier for supporting a clinical diagnostic decision, seconds are not crucial – many clinical tests take hours or days. Therefore we may afford to use a complex model. In other applications (e.g., industrial real-time systems) the dominating factor is operating time and we can agree on a slightly degraded performance in order to keep the classifier simple and fast [213]. In both examples training time and training complexity are not essential. However, these will count when the classifier is a part of an *adaptive* system. The word "adaptive" is sometimes misused in the neural network literature. By "adaptive" we mean a classifier whose parameters are subject to adjustment during the operation of the system. That is, the classifier is equipped with a tuning routine that changes the parameter values as new data is being processed. Notice that these changes take place in the real application, out of the control of the designer.

In summary: as advocated by many authors [252, 237, 12], simple models should be tried first.

2.7.4 Comments and recommendations

A point raised by Duin [89] is that the performance of the classifier depends on the expertise and the **willingness** of the designer. There is not much to be done for classifiers with fixed structure and training procedure (called "automatic" classifiers in [89]), e.g., LDC. With the neural network classifiers, however, we can make them work or fail due to the infinitely many design options. Keeping in mind that there are no rules for the fair comparison of classifiers, here are a few (non-Candide's) guidelines:

1. Use fixed procedures for designing all classifiers. Pick these procedures in advance, prior to training the classifiers. When publishing, explain them in enough detail so that they are reproducible by other researchers.
2. Compare modified versions of classifiers with the *original schemes of the same type*. For example, a distance-based modification of k-nearest neighbors (k-nn) should be compared with the standard k-nn first, and then with other classifier models, e.g., neural networks. A common mistake is to compare a slight modification of a certain model with a totally different classification paradigm. It is not clear then who deserves the credit: the modification or the original model itself.
3. Make sure that the information about the data is utilized by all classifiers if the models can incorporate such information. For example, a judicious initialization of a prototype based classifier such as the learning vector quantization design (LVQ) can rate that design highly among classifiers of the same group which are initialized purely at random.
4. Report the test result corresponding to the best *training* result. Training results are also interesting on their own account for assessing the magnitude of the difference (the training-testing gap) and the correlation between training and testing accuracy.
5. If possible, give also the complexity of the classifier: training and running times, memory requirements, etc.

2.8 A taxonomy of classifier design methods

Practically, we never know the true prior probabilities and the class-conditional p.d.f's and therefore we can only design a flawed version of D_B. Statistical pattern recognition provides a variety of methods for designing a classifier: refer to [87, 100, 268, 324, 75] and more recent sources which also include neural networks [280, 43, 222].

One solution is to try to estimate $P(\omega_i)$ and $p(\mathbf{x}|\omega_i), i = 1, \dots, c$, from $\mathbf{Z}$ and substitute the estimates $\hat{P}(\omega_i)$ and $\hat{p}(\mathbf{x}|\omega_i), i = 1, \dots, c$, in the Bayes formula (2.11). Approximating $p(\mathbf{x}|\omega_i)$ as a function of $\mathbf{x}$ divides classifier design methods into two big groups: *parametric* and *non-parametric*. Figure 2.9 shows one possible taxonomy of methods for classifier design. Exemplar classifiers are placed in boxes. Some of the models are presented in the next chapters while others are mentioned for completeness. The acronyms are explained earlier, on page 13.

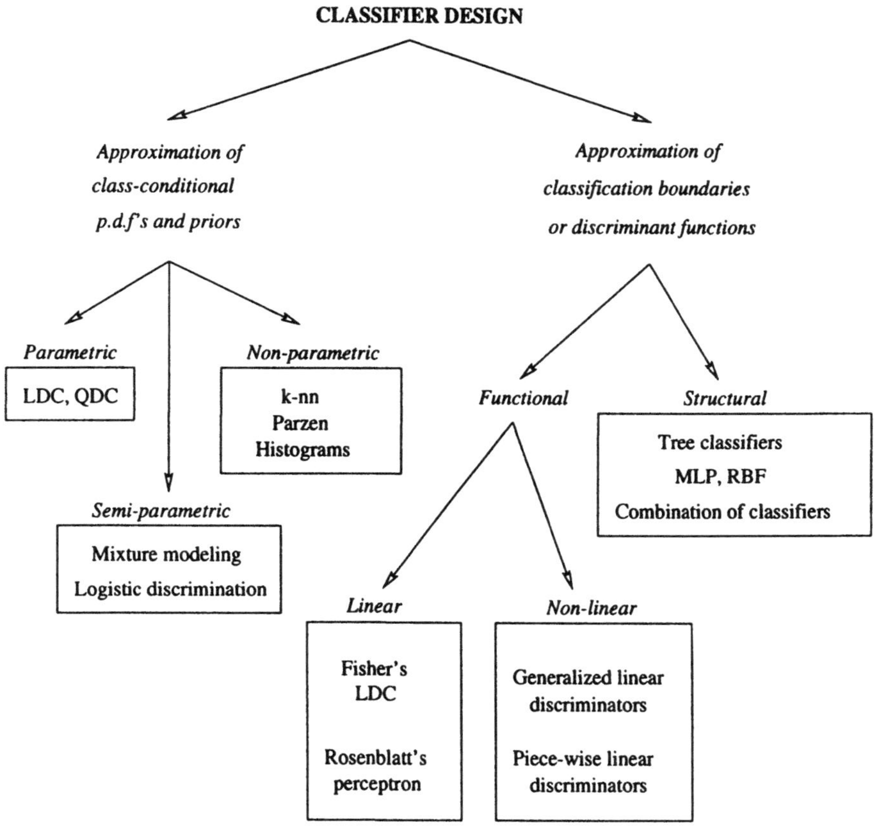

Fig. 2.9. A taxonomy of methods for classifier design

The distinction between the groups is not clear-cut. For example, Radial Basis Function (RBF) networks from the group of structural approximation of the discriminant functions can be moved to the group of functional approximation, or even to the group of semiparametric p.d.f modeling [326]. Similarly, the k-nearest neighbor method, although theoretically linked with the non-parametric p.d.f estimation, produces a direct estimate of the discriminant functions and can be put under the heading of structural designs for approximating the discriminant functions.

There is no consensus on a single taxonomy, or even about the definition of "parametric" and "nonparametric" classifiers. Lippmann [220] lists 5 types of classifiers:

- Probabilistic (LDC, QDC, Parzen);
- Global (MLP);
- Local (RBF NN's);
- Nearest-neighbor type (k-nn, LVQ);

- Rule-forming (binary decision trees, rule-based systems);

Holmström et al. [136] consider another grouping

- Classifiers based on density estimation:
  - parametric (LDC, QDC);
  - nonparametric (k-nn, kernel methods, finite mixtures, RBF).
- Classifiers based on regression:
  - parametric (linear regression, logistic regression, MLP);
  - nonparametric (projection pursuit, additive models).
- Other classifiers (e.g., prototype-based: LVQ, k-nn for small k)

Some authors distinguish between neural and non-neural classifiers, local and global classifiers, etc.

Fuzzy classifier designs are even more difficult to summarize. The reason is that in fuzzy pattern recognition there is no *benchmark* classifier like the Bayes classifier in statistical pattern recognition. Due to this, fuzzy models are developing more in a parallel rather than a subordinate way as we shall see later.

3. Statistical classifiers

3.1 Parametric classifiers

To design a Bayes-optimal classifier for a certain problem we need the prior probabilities ($P(\omega_i)$) and the class-conditional p.d.f's ($p(\mathbf{x}|\omega_i)$) for all classes $\omega_1, \ldots, \omega_c$.

Non-biased estimates of the prior probabilities are found as the proportion of elements of $\mathbf{Z}$ from class ω_i

$$\hat{P}(\omega_i) = \frac{N_i}{N} = \frac{1}{N}\sum_{j=1}^{N} Ind(\mathbf{z}_j, \omega_i), \tag{3.1}$$

where $Ind(\mathbf{z}_j, \omega_i)$ is a binary indicator function[1] such that

$$Ind(\mathbf{z}_j, \omega_i) = \begin{cases} 1, & \text{if } l(\mathbf{z}_j) = \omega_i, \\ 0, & \text{otherwise} \end{cases} \tag{3.2}$$

where $l(\mathbf{z}_j) \in \Omega$ is the class label of $\mathbf{z}_j$. The parametric class of models assume the class-conditional p.d.f's are of a hypothetical form, and only the set of p.d.f parameters are estimated.

3.1.1 Normal distribution. Dependent features.

A typical guess for the p.d.f's is the *normal distribution*. In the one-dimensional case, the normal distribution denoted $p(x|\omega_i) \sim N(m_i, s_i^2)$, (where $x, m_i, s_i \in \Re$), and is defined by

$$p(x|\omega_i) = \frac{1}{\sqrt{2\pi}\, s_i} \exp\left\{-\frac{1}{2}\left(\frac{x - m_i}{s_i}\right)^2\right\}, \tag{3.3}$$

where m_i is the mean and s_i is the standard deviation for class ω_i. For the multidimensional case, $p(\mathbf{x}|\omega_i) \sim N(\mathbf{m}_i, S_i)$ is

[1] Remember this function for later

$$p(\mathbf{x}|\omega_i) = \frac{1}{(2\pi)^{\frac{n}{2}}\sqrt{|S_i|}} \exp\left\{-\frac{1}{2}(\mathbf{x}-\mathbf{m}_i)^T S_i^{-1}(\mathbf{x}-\mathbf{m}_i)\right\}, \quad (3.4)$$

where $\mathbf{m}_i \in \Re^n$ is the mean vector for class ω_i, and S_i is the covariance matrix.

The normal distribution is the most natural assumption reflecting the following situation: there is an "ideal prototype" (a point in $\Re^n$) of class ω_i and all class members are distorted versions of it. Slight distortions are more likely to occur than big distortions, causing more objects to be located in the close vicinity of the ideal prototype than far away from it. The prototype is represented by the mean $\mathbf{m}_i$ and the scatter of the points around it is associated with the covariance matrix S_i.

Figure 3.1 shows four 2-dimensional data sets generated from normal distributions with different covariance matrices as displayed in the upper right corner of each scatterplot. The two top plots are generated with **independent** (non-interacting) features, i.e., the data cloud is "stretched" along the coordinate axes. This is manifested by off-diagonal entries of the covariance matrix being zeros. The lower two plots represent cases where the features are **dependent**. In the case of independent features we can decompose the n-dimensional p.d.f's as a product of n one-dimensional p.d.f's. Let $\mathbf{s}_i = [s_{i1}^2, \ldots, s_{in}^2]^T$ be the vector of the diagonal entries of the covariance matrix S_i. Then

$$p(\mathbf{x}|\omega_i) = \frac{1}{(2\pi)^{\frac{n}{2}}\sqrt{|S_i|}} \exp\left\{-\frac{1}{2}(\mathbf{x}-\mathbf{m}_i)^T S_i^{-1}(\mathbf{x}-\mathbf{m}_i)\right\} =$$

$$= \prod_{k=1}^{n}\left(\frac{1}{\sqrt{(2\pi)}\, s_{ik}} \exp\left\{-\frac{1}{2}\left(\frac{x_k - m_{ik}}{s_{ik}}\right)^2\right\}\right). \quad (3.5)$$

Parametric methods are fast because only the parameters of the c distributions need to be estimated from $\mathbf{Z}$. If we hypothesize normally distributed classes, the total number of parameters is $c(n + n(n+1)/2)$. The estimates of $\mathbf{m}_i$ and S_i are obtained as

$$\mathbf{m}_i = \frac{1}{N_i} \sum_{\mathbf{z}_j, l(\mathbf{z}_j)=\omega_i} \mathbf{z}_j \quad (3.6)$$

where N_i is the total number of elements of $\mathbf{Z}$ from class ω_i, and[2]

$$S_i = \frac{1}{(N_i - 1)} \sum_{\mathbf{z}_j,\, l(\mathbf{z}_j)=\omega_i} (\mathbf{z}_j - \mathbf{m}_i)(\mathbf{z}_j - \mathbf{m}_i)^T. \quad (3.7)$$

[2] In the maximum likelihood (ML) estimate of the covariance matrix, the denominator is N_i.

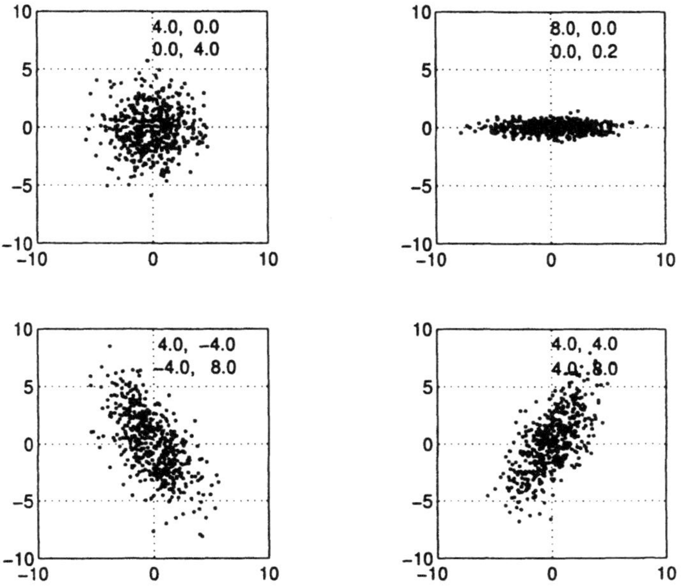

Fig. 3.1. Normally distributed data sets with mean $[0,0]^T$ and different covariance matrices (shown in the upper right corners).

Example 3.1.1. In the Normal-mixtures example (Figure 1.3) each of the two classes is a mixture of two Gaussians. The histograms of the distributions of the points from the two classes on axis x_2 are shown in Figure 3.2 (top two plots). Shown in the bottom plot in Figure 3.2 is the normal approximation of the two class-conditional p.d.f's. ∎

As discussed in t he previous chapter, for the minimum error classifier (Bayes classifier) we can use the posterior probabilities (2.11) as a set of optimal discriminant functions. Equivalently, we can disregard the denominator of (2.11) which has the same value for all c classes (for a given $\mathbf{x}$), and use the following set G

$$g_i(\mathbf{x}) = \log\left(P(\omega_i)\, p(\mathbf{x}|\omega_i)\right), \quad i = 1, \ldots, c, \tag{3.8}$$

which will produce the same optimal classification as if we used $P(\omega_i|\mathbf{x})$ in (2.3). If the classes are normally distributed, (3.8) takes the form

$$g_i(\mathbf{x}) = \log(P(\omega_i)) - \frac{n}{2}\log(2\pi) - \frac{1}{2}\log(|S_i|) - \frac{1}{2}(\mathbf{x}-\mathbf{m}_i)^T S_i^{-1}(\mathbf{x}-\mathbf{m}_i). \tag{3.9}$$

3.1.2 Linear discriminant classifier (LDC)

Let the classes be normally distributed with different means and the same covariance matrix, i.e., $p(\mathbf{x}|\omega_i) \sim N(\mathbf{m}_i, S)$; $i = 1, \ldots, c$. Opening the paren-

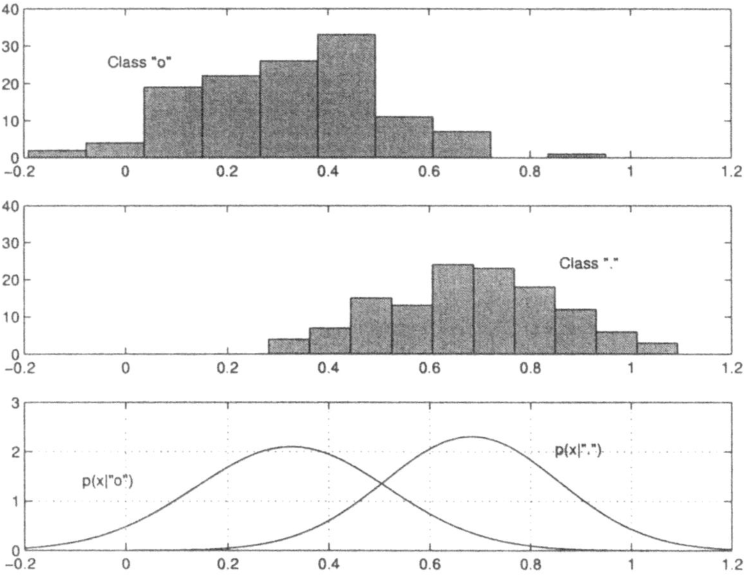

Fig. 3.2. Histograms of the two classes on feature x_2 (top two plots) and the approximation of the class-conditional p.d.f's (bottom plot, $p(\mathbf{x}|\circ) \sim N(0.3258, 0.1901)$, $p(\mathbf{x}|\cdot) \sim N(0.6830, 0.1731)$).

theses and discarding all terms that do not depend on ω_i in (3.9) we obtain a new set of discriminant functions

$$g_i(\mathbf{x}) = \log(P(\omega_i)) - \frac{1}{2}\mathbf{m}_i^T S^{-1}\mathbf{m}_i + \mathbf{m}_i^T S^{-1}\mathbf{x} = w_{i0} + \mathbf{w}_i^T \mathbf{x}, \qquad (3.10)$$

where $w_{i0} \in \Re$ and $\mathbf{w}_i \in Re^n$ are coefficients.

Let $\mathbf{x}^a = [1, x_1, \ldots, x_n]^T$ be the augmented feature vector whose first component is always 1. Using the augmented vector $\mathbf{w}_i^a = [w_{i0}, w_{i1}, \ldots, w_{in}]^T$ (3.10) becomes

$$g_i(\mathbf{x}) = (\mathbf{w}_i^a)^T \mathbf{x}^a. \qquad (3.11)$$

This linear model is used in the perceptron and multi-layer perceptron designs. If the classes do have normal distributions with equal covariance matrices (not necessarily with independent features), (3.10) gives a set of Bayes-optimal discriminant functions, or D_B. More often, neither are the classes normally distributed nor are the true values of $\mathbf{m}_i$ and S_i known. There are numerous procedures to estimate $\mathbf{w}_i^a$, $i = 1, \ldots, c$, from data [87].

Example 3.1.2. The results from applying LDC to the training part of the Normal-mixtures data (Figure 1.3) are shown in Figure 3.3. Ripley [280] points out that the true error rate for this set is about 8 %. The training error was 14.5 %, and the testing error was 10.8 %. ■

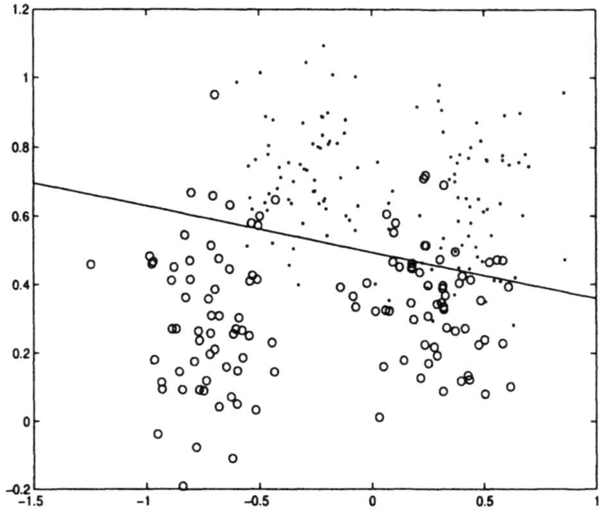

Fig. 3.3. Classification boundary on the Normal-mixtures data obtained by LDC (14.5 % training (resubstitution) error and 10.8 % testing error)

3.1.3 Quadratic discriminant classifier (QDC)

Again let the classes be normally distributed, but now with class-specific covariance matrices, i.e., $p(\mathbf{x}|\omega_i) \sim N(\mathbf{m}_i, S_i)$. The set of optimal discriminant functions is quadratic on $\mathbf{x}$ and is obtained from (3.9) as

$$g_i(\mathbf{x}) = w_{i0} + \mathbf{w}_i^T\ \mathbf{x} + \mathbf{x}^T\ W_i\ \mathbf{x}, \tag{3.12}$$

where

$$w_{i0} = \log(P(\omega_i)) - \frac{1}{2}\mathbf{m}_i^T S_i^{-1}\mathbf{m}_i - \frac{1}{2}\log(|S_i|), \tag{3.13}$$

$$\mathbf{w}_i = S_i^{-1}\mathbf{m}_i, \tag{3.14}$$

and

$$W_i = -\frac{1}{2}S_i^{-1}. \tag{3.15}$$

Example 3.1.3. The classification boundaries for the quadratic classifier are also quadratic functions of $\mathbf{x}$. Figure 3.4 shows 4 data sets, each with two normally distributed classes. The optimal classification boundaries found by QDC, which *is* the Bayes-optimal classifier for this case, are also plotted. The QDC classification boundary for the Normal-mixtures data is depicted in Figure 3.5. The training error is 13.6 %. The QDC regions for the Cone-torus data are shown in Figure 3.6 (19.5 % training error and 16.57 % testing error). ■

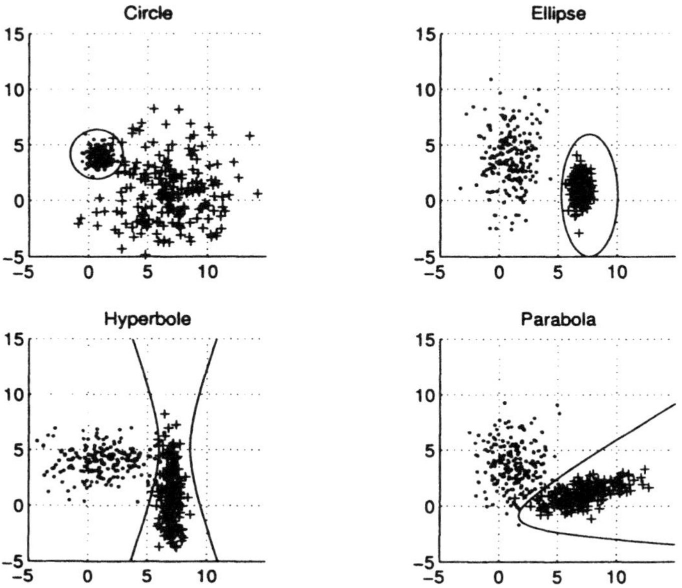

Fig. 3.4. Classification boundaries obtained by QDC.

3.2 Nonparametric classifiers

3.2.1 Preliminaries

In nonparametric designs we do not assume any model of the p.d.f's. The p.d.f $p(\mathbf{x}|\omega_i)$ is estimated in the vicinity of $\mathbf{x}$ in $\Re^n$. The probability p^* that $\mathbf{x}$ is in a certain region $R \subset \Re^n$ is

$$p^* = P(\mathbf{x} \in R) = \int_R p(\mathbf{u})d\mathbf{u}. \tag{3.16}$$

Assume that N samples are drawn from the distribution in question. Using the Binomial distribution, the probability that exactly k samples fall in R is given by

$$p_k = \binom{N}{k} (p^*)^k (1 - p^*)^{N-k}, \tag{3.17}$$

and the value p^* can be estimated as the proportion of the points in R with respect to the total number of samples, i.e.,

$$p^* \approx \frac{k}{N}. \tag{3.18}$$

Let $\mathbf{x}$ be a point inside region R. For a small R, assuming that $p(\mathbf{u})$ is approximately constant in R, p^* can be approximated as

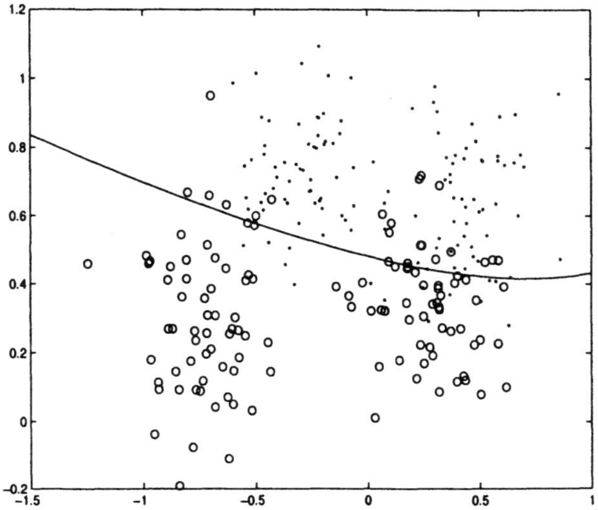

Fig. 3.5. Classification boundary on the Normal-mixtures data obtained by QDC (13.6 % training (resubstitution) error and 10.6 % testing error)

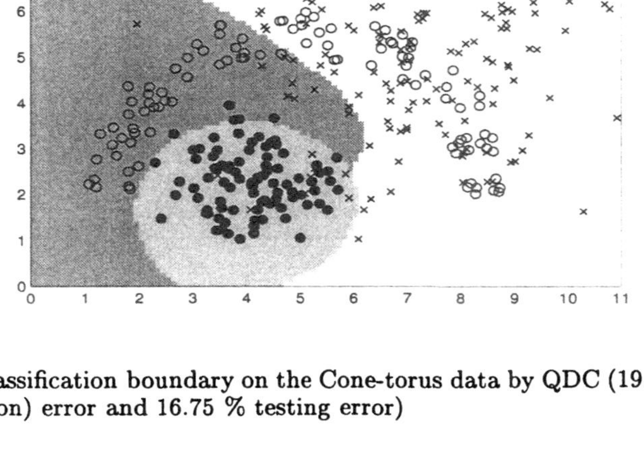

Fig. 3.6. Classification boundary on the Cone-torus data by QDC (19.5 % training (resubstitution) error and 16.75 % testing error)

$$p^* \approx p(\mathbf{x}) \int_R d\mathbf{u} = p(\mathbf{x}) V_R, \tag{3.19}$$

where V_R is the volume of R in $\Re^n$. Joining (3.18) and (3.19), and solving the equation for $p(\mathbf{x})$ we get

$$p(\mathbf{x}) = \frac{k}{NV_R}. \tag{3.20}$$

When N tends to infinity and the region R shrinks to a point ($V_R \to 0$), equation (3.20) produces the exact $p(\mathbf{x})$. The above result is an important point of departure of various nonparametric classification methods.

Nonparametric classification is often associated with the notion of **prototype**. In the light of the above we can think of a prototype as a representative element (usually a point) for the region $R \subset \Re^n$. Thus, an input $\mathbf{x}$ gets the class labels of one or more prototypes to which $\mathbf{x}$ is **similar**. Usually, similarity is defined in a geometrical sense, i.e., based on a certain distance. The smaller the distance, the higher the similarity between $\mathbf{x}$ and the prototype. The concept of prototype is discussed later in this Chapter and also underpins the Generalized Nearest Prototype Classifier model described in Chapter7.

3.2.2 Method of histograms

To derive the method of histograms from (3.20), we fix N and V, and calculate k from $\mathbf{Z}$. The feature space is divided into bins (cells). Consider a bin B containing m points from $\mathbf{Z}$. Let m_i be the number of points in B from class ω_i, $i = 1, \ldots, c$, $m = m_1 + \ldots + m_c$. According to (3.20)

$$p(\mathbf{x}) \approx \frac{m}{NV_B}, \tag{3.21}$$

where V_B is the volume of the bin, and

$$p(\mathbf{x}|\omega_i) \approx \frac{m_i}{N_i V_B}. \tag{3.22}$$

Here N_i is the number of elements from class ω_i in $\mathbf{Z}$. Using (3.1) for estimating $P(\omega_i)$'s, the posterior probabilities are obtained as

$$P(\omega_i|\mathbf{x}) = \frac{p(\mathbf{x}|\omega_i)P(\omega_i)}{p(\mathbf{x})} \approx \frac{\frac{m_i}{N_i V_B}\frac{N_i}{N}}{\frac{m}{NV_B}}, \tag{3.23}$$

hence

$$P(\omega_i|\mathbf{x}) \approx \frac{m_i}{m}. \tag{3.24}$$

The approximation of the Bayes classifier is obtained by using the approximated values of the posterior probabilities for the classes as the discriminant functions. Thus, the bin B is labeled according to the highest m_i in it, and all points in $\Re^n$ in B get the same class label. To classify an input $\mathbf{x}$ (not in $\mathbf{Z}$), the histogram classifier finds the bin where $\mathbf{x}$ belongs and retrieves the bin's class label. Practically, the histogram classifier is a look-up table (usually a big one) whose accuracy depends on the number of bins and the sample size N. It is convenient to define formally the **look-up classifier** as an n-way table, with a finite number of cells, such that for any $\mathbf{x} \in \Re^n$, we can recover the class label from the table cell indexed by the intervals in which the components of $\mathbf{x}$ fall.

Example 3.2.1. Consider the look-up classifier in $\Re^3$ shown in Figure 3.7. The classifier will label $\mathbf{x} = [6, -7, 11]$ in ω_2 because $x_1 \in [5, 10]$, $x_2 \in [-5, -10]$, and $x_3 \in [10, 15]$. (Of course, the cells need not be of equal size.) ■

The number of bins grows exponentially with the dimensionality. If we use M bins per axis, the total number of bins will be M^n. This requires huge data sets because otherwise many bins will be empty or containing too few elements to provide good estimates of the posterior probabilities. This phenomenon is known as the **curse of dimensionality**.

Example 3.2.2. The result from the histogram method with $M \times M = 100$ bins (10 on each axis) for the Normal-mixtures data is shown in Figure 3.8. The bins where the number of elements for both classes tie, or which are empty, are left white while the others are shaded. The gray level indicates the class labels. We calculated the classification error of this design using the leave-one-out method. For each $\mathbf{z}_j \in \mathbf{Z}$ we found its bin, and then recalculated the class label of the bin by updating m_1 or m_2. For example, assume that bin B originally contained $m_1 = 4$ points from ω_1 and $m_2 = 3$ points from ω_2 (so, B is labeled in ω_1). Let $\mathbf{z}_B$ be a point from $\mathbf{Z}$ such that it is located in B, and its class label is $l(\mathbf{z}_B) = \omega_1$. The updated values for the leave-one-out (in this case leave $\mathbf{z}_B$ out) are $m_1 \leftarrow m_1 - 1 = 3$ and $m_2 \leftarrow m_2 = 3$. Now the class label of B is obtained by breaking the tie randomly. Should $\mathbf{z}_B$ be from class ω_2, the label assigned to it is ω_1 because the updated values are $m_1 \leftarrow m_1 = 4$ and $m_2 \leftarrow m_2 - 1 = 2$, i.e., all m_2 elements from ω_2 in B will be misclassified. The misclassification rate of the leave-one-out method is 18.0 %. Using the resubstitution method, i.e., without removing $\mathbf{z}_j$ from the bin, the error rate is 10.8 %.

To illustrate the curse of dimensionality, we used $M = 20$ (400 bins) with the same data set. The result is depicted in Figure 3.9. Now the classification boundary is more precise but a larger part of the feature space remains uncovered by the classifier (the white regions). The error using leave-one-out was 20.4 % and the resubstitution error was 4.4 % (both calculated on the set $\mathbf{Z}$) . This indicates that the classifier is "overtrained". The high error rate is due to the fact that many bins contain only one element. In the leave-one-out method this element is removed and an arbitrary class label is assigned to it.
■

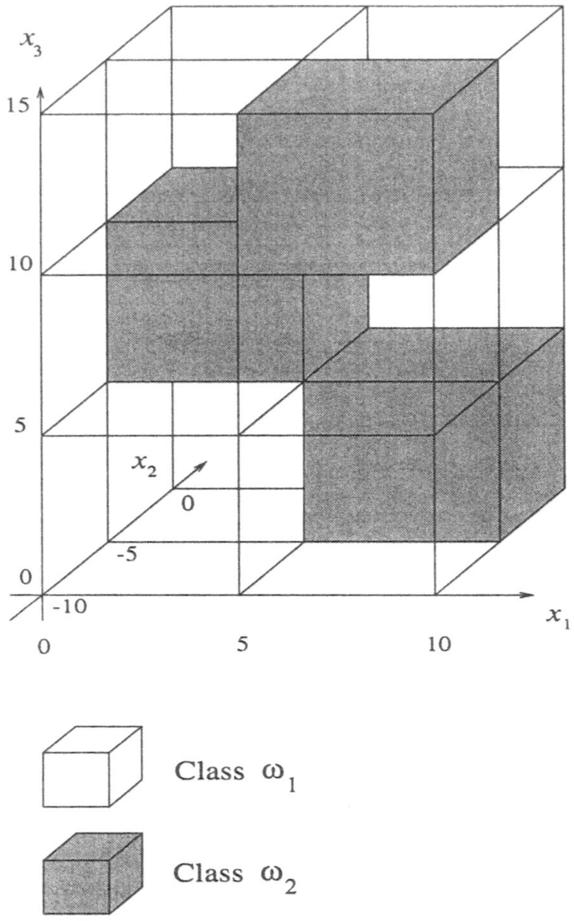

Fig. 3.7. An example of a look-up classifier

The histogram classifier can be viewed as a prototype classifier. We can place a prototype at the geometrical center of each bin. The input $\mathbf{x}$ is considered similar to that prototype, in whose bin it is located. The class label of the bin's prototype becomes the class label of $\mathbf{x}$.

The histogram classifier is both time and memory consuming, and, besides, critically depends on how large our data set is. The number of bins acts as a smoothing factor: the more bins we have, the noisier the approximation of the discriminant function. With large M, it is more likely that large regions of the feature space will not be covered (compare figures 3.8 and 3.9).

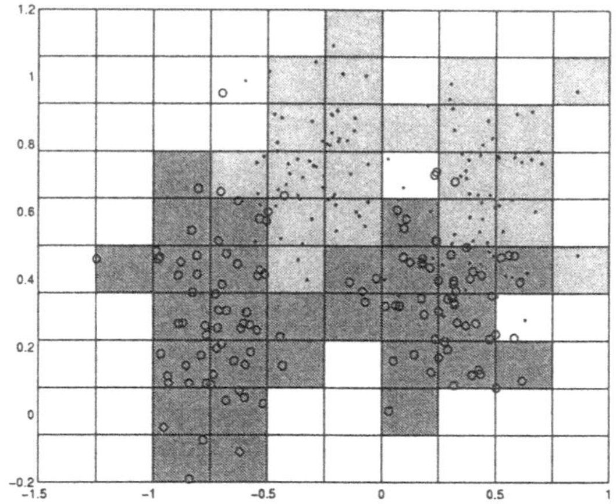

Fig. 3.8. Histogram method with the Normal-mixtures data: M=10, 18.0 % error with leave-one-out and 10.8 % with resubstitution

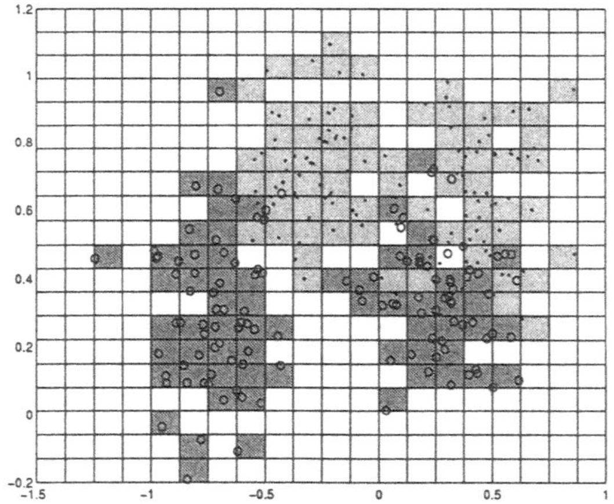

Fig. 3.9. Histogram method with the Normal-mixtures data: M=20, 20.4 % error with leave-one-out and 4.4 % with resubstitution

The method of histograms is hardly ever used in the form described here. The reason to include it here is that many fuzzy classification techniques are drawn upon very similar (sometimes identical) idea.

3.2.3 The k-nearest neighbor rule

One of the most theoretically elegant and yet simple classification techniques is the k-nearest neighbor rule (k-nn)[69, 87, 100]. k-nn is a prototype classifier.

Let $\mathbf{V} = \{\mathbf{v}_1, \ldots, \mathbf{v}_v\}$ be a labeled **reference set** containing v points in $\Re^n$, referred to as **prototypes**. The prototypes are labeled in the c classes, i.e., for any $\mathbf{v}_i \in \mathbf{V}$, we know its class label $l(\mathbf{v}_i) \in \Omega$. Most often $\mathbf{V}$ is the whole of $\mathbf{Z}$ or a subset of it. To classify an input $\mathbf{x}$, the k nearest prototypes are retrieved from $\mathbf{V}$ together with their class labels. The input $\mathbf{x}$ is labeled to most represented class label amongst the k nearest neighbors.

To arrive at this classification method, we fix k and N in (3.20) and allow for a variable V_R. Let R be the hypersphere containing exactly k of the elements of the reference set $\mathbf{V}$. The unconditional p.d.f is approximated as

$$p(\mathbf{x}) \approx \frac{k}{NV_R}. \tag{3.25}$$

Denoting by k_i the number of elements in R from class ω_i, the class-conditional p.d.f for $\omega_i, i = 1, \ldots, c$, approximated in R, is

$$p(\mathbf{x}|\omega_i) \approx \frac{k_i}{N_i V_R}, \tag{3.26}$$

where N_i is the number of elements from class ω_i in $\mathbf{Z}$. Using (3.1) for estimating $P(\omega_i)$'s, the posterior probabilities are obtained as

$$P(\omega_i|\mathbf{x}) = \frac{p(\mathbf{x}|\omega_i)P(\omega_i)}{p(\mathbf{x})} \approx \frac{\frac{k_i}{N_i V_R}\frac{N_i}{N}}{\frac{k}{NV_R}} \tag{3.27}$$

and

$$P(\omega_i|\mathbf{x}) \approx \frac{k_i}{k}. \tag{3.28}$$

Thus, the Bayes classifier using the approximations above will assign $\mathbf{x}$ to the class with the highest posterior probability, i.e., the class most represented amongst the k nearest neighbors of $\mathbf{x}$. The region R and the volume V_R, respectively, are specific for each $\mathbf{x} \in \Re^n$. The k-nn classification rule, however, assigns the class label using only the numbers $k_i, i = 1, \ldots, c$, so the winning label does not depend on V_R.

k-nn is Bayes-optimal when $N \to \infty$ and $V_R \to 0$. The expression $V_R \to 0$ is equivalent to $k \to \infty$ and $k/N \to 0$. That is,

$$\lim_{\substack{N\to\infty \\ k\to\infty \\ \frac{k}{N}\to 0}} P_{k-nn} = P_B, \tag{3.29}$$

where P_B is the Bayes error.

When k is 1 (the **nearest neighbor rule**, denoted 1-nn), and $N \to \infty$, the error rate is bounded from above by twice the Bayes error rate [87]

$$P_{1-nn} \leq 2P_B. \tag{3.30}$$

Notice that "nearest" implies a certain metric over the feature space $\Re^n$. Different metrics define different regions R. Whichever metric is used, the

k-nn rule is applied in the same way. The type of the metric does not change the asymptotic properties of k-nn because the shape of R is not fixed in (3.20).

Example 3.2.3. Figure 3.10 displays the 1-nn and 3-nn rule. The boldface cross (vector to be classified) is linked with arrows to its three nearest neighbors from the 15-point data set. It is assigned to the class of "squares" by the 1-nn rule because such is the label of the closest (circled) neighbor. The 3-nn rule labels the cross as a "snowflake" because such are the labels of the second and the third neighbors (majority: 2 out of 3 votes for class "snowflakes"). The classification regions obtained by the 1-nn rule can be depicted using **Voronoi diagrams** as shown in Figure 3.11. The Voronoi bin V specified by vector $\mathbf{z}_j \in \mathbf{Z}$ is defined as the set of points in $\Re^n$ whose nearest neighbor from $\mathbf{Z}$ is $\mathbf{z}_j$, i.e.,

$$V(\mathbf{z}_j) = \left\{ \mathbf{x} \,\middle|\, \mathbf{x} \in \Re^n, d(\mathbf{z}_j, \mathbf{x}) = \min_{\mathbf{z}_k \in \mathbf{Z}} \; d(\mathbf{z}_k, \mathbf{x}) \right\}, \tag{3.31}$$

where $d(\cdot, \cdot)$ is a distance in $\Re^n$. In the example in Figure 3.11 we used the points from Figure 3.10 as $\mathbf{Z}$ and Euclidean distance. The classification region for class "squares" is shaded. ∎

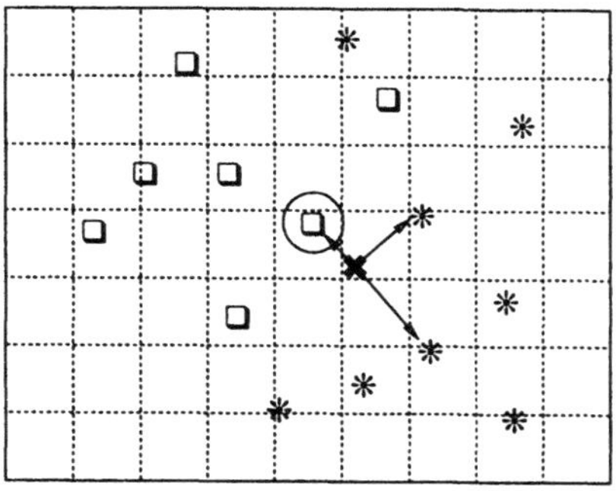

Fig. 3.10. Illustration of the 1-nn and 3-nn classification rules.

Example 3.2.4. The results from applying k-nn to the Normal-mixtures data for odd k (to avoid ties) from 1 to 11 are shown in Table 3.1 below. The leave-one-out method was used, i.e., the vector $\mathbf{z}_j$ to be classified is excluded from $\mathbf{Z}$ and the nearest neighbors are searched for in $\mathbf{Z} - \{\mathbf{z}_j\}$. ∎

There are three basic ways to tune k-nn:

- select the distance metric in $\Re^n$;
- select k;

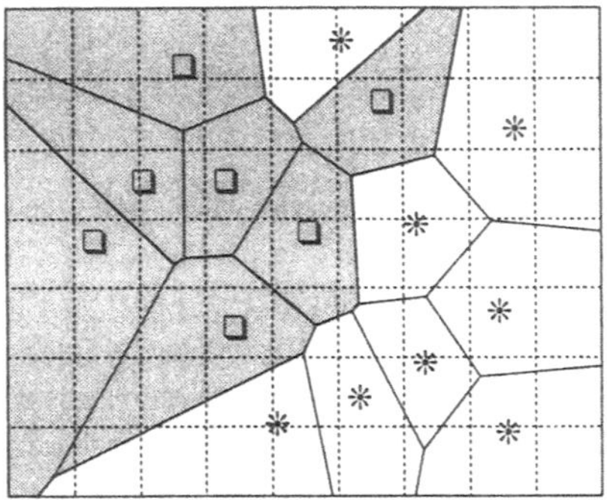

Fig. 3.11. Illustration of Voronoi diagrams.

Table 3.1. k-nn results (leave-one-out) with the Normal-mixtures data

k	1	3	5	7	9	11
errors [%]	14.8	14.4	17.2	14.4	14.4	14.0

- edit $\mathbf{Z}$ to find a "good" reference set $\mathbf{V}$.

Like the histogram method, k-nn is slow and memory-consuming. Unlike the histogram method, however, k-nn covers the whole feature space (e.g., see the Voronoi diagrams). There are many fuzzy versions of k-nn, some of which are discussed in Chapter 7.

3.2.4 Parzen classifier

This model is based again on (3.20) where N and V_R are fixed, and k is found from the data. Let $K(\mathbf{t}), \mathbf{t} = [t_1, \ldots, t_n]^T \in \Re^n$, be a **kernel function** (or *Parzen window*) which peaks at $\mathbf{t} = \mathbf{0}$, is nonnegative, and has integral one over $\Re^n$. The simplest model of a Parzen window is the following function

$$K(\mathbf{t}) = \begin{cases} 1, & \text{if } |t_i| \leq \frac{1}{2},\ \forall i = 1, \ldots, n \\ 0, & \text{otherwise.} \end{cases} \tag{3.32}$$

This function defines a hyperbox in $\Re^n$ with side 1, centered at the origin. For all points $\mathbf{t}$ in the hypercube, $K(\mathbf{t})$ is 1 and for the remaining points in $\Re^n$ is it 0.[3] Let the Parzen window be centered at some $\mathbf{x}$. Using the set $\mathbf{Z}$ for calculating (3.20), and taking into account that $V_R = 1$, the p.d.f at $\mathbf{x}$ can be approximated as

$$p(\mathbf{x}) \approx \frac{k}{N} = \frac{1}{N} \sum_{j=1}^{N} K(\mathbf{Z}_j - \mathbf{x}). \tag{3.33}$$

[3] Various kernel functions are listed in (Fukunaga, 1972)[100].

This formula can be interpreted in a different way if we assume that there are N hyperboxes, each centered on one point from $\mathbf{Z}$. The approximation is then calculated as the proportion of such hyperboxes that contain $\mathbf{x}$. The multidimensional kernel function centered on $\mathbf{z}_j \in \Re^n$ is usually expressed in the form $\frac{1}{h^n}K\left(\frac{\mathbf{x}-\mathbf{z}_j}{h}\right)$, where h is a smoothing parameter and

$$\int_{\Re^n} \frac{1}{h^n} K\left(\frac{\mathbf{x}-\mathbf{z}_j}{h}\right) d\mathbf{x} = 1. \tag{3.34}$$

A common choice of the kernel function is the multidimensional Gaussian kernel [100]:

$$\frac{1}{h^n} K_G\left(\frac{\mathbf{x}-\mathbf{z}_k}{h}\right) = \frac{1}{h^n \sqrt{(2\pi)^n}\sqrt{|S|}} \exp\left\{-\frac{1}{2h^2}(\mathbf{x}-\mathbf{z}_k)^T S^{-1}(\mathbf{x}-\mathbf{z}_k)\right\}. \tag{3.35}$$

Here S is a prespecified covariance matrix determining the n-dimensional shape of the kernel.

The class-conditional p.d.f's are estimated using the sample set $\mathbf{Z}$ [43, 100] by

$$\hat{p}(\mathbf{x}|\omega_i) = \frac{1}{N_i} \sum_{\mathbf{z}_j, l(\mathbf{z}_j)=\omega_i} \frac{1}{h^n} K\left(\frac{\mathbf{x}-\mathbf{z}_j}{h}\right), \tag{3.36}$$

where N_i is the number of elements of $\mathbf{Z}$ from class ω_i. The estimate is asymptotically unbiased if the smoothing parameter h is a function of the number of samples N_i, such that [100]

$$\lim_{N_i \to \infty} h(N_i) = 0. \tag{3.37}$$

Taking (3.1) as the estimates of the prior probabilities we obtain

$$\hat{P}(\omega_i|\mathbf{x}) = \frac{1}{N\, p(\mathbf{x})} \sum_{\mathbf{z}_j, l(\mathbf{z}_j)=\omega_i} \frac{1}{h^n} K\left(\frac{\mathbf{x}-\mathbf{z}_j}{h}\right). \tag{3.38}$$

Using the indicator function (3.2), (3.38) can be rewritten as

$$\hat{P}(\omega_i|\mathbf{x}) = C_1(\mathbf{x}, h, N) \sum_{j=1}^{N} Ind(\mathbf{z}_j, \omega_i)\, K\left(\frac{\mathbf{x}-\mathbf{z}_j}{h}\right). \tag{3.39}$$

where the term $C_1(\mathbf{x}, h, N)$ depends on $\mathbf{x}$, h and N but not on the class label. The approximation of the conditional p.d.f's under the above conditions is asymptotically unbiased. Therefore, the classifier (2.15) using (2.11) with the class-conditional p.d.f's estimated by (3.39) is asymptotically Bayes-optimal. A set of optimal discriminant functions can be obtained from (3.39) by ignoring $C_1(\mathbf{x}, h, N)$, i.e.,

$$g_i(\mathbf{x}) = \sum_{j=1}^{N} Ind(\mathbf{z}_j, \omega_i)\ K\left(\frac{\mathbf{x}-\mathbf{z}_j}{h}\right). \tag{3.40}$$

For the Gaussian kernel (3.35)

$$g_i(\mathbf{x}) = \sum_{j=1}^{N} Ind(\mathbf{z}_j, \omega_i)\ \exp\left\{\frac{1}{2h^2}\left(d_M(\mathbf{x}, \mathbf{z}_j)\right)^2\right\}, \tag{3.41}$$

where

$$\left(d_M(\mathbf{x}, \mathbf{z}_j)\right)^2 = (\mathbf{x}-\mathbf{z}_j)^T S^{-1} (\mathbf{x}-\mathbf{z}_j) \tag{3.42}$$

is the squared **Mahalanobis distance** between $\mathbf{x}$ and $\mathbf{z}_j$ in $\Re^n$.

Parzen classifier can be regarded as a prototype classifier. Here the prototypes are all elements of $\mathbf{Z}$ and each prototype is used in finding the class label of an input $\mathbf{x}$. The prototype label "votes" are weighed according to the distance between $\mathbf{x}$ and the prototype.

Parzen classifier is a beautiful theoretical model which has not been widely applied in its pure form. There are several reasons for this:

- Parzen classifier needs all of $\mathbf{Z}$ as the prototype set;
- the choice of h is difficult: small h leads to "spiky" approximations of the p.d.f's, big h oversmooths the p.d.f's shapes, both increasing the error rate.

A practical recommendation offered by some authors is to try several values of h and select the one producing the smallest error rate. *Reduced* and *weighted* Parzen models are proposed in the literature [21, 136]. The real value of Parzen classifier lies in the fact that it is the statistical counterpart of several important classification methods such as Radial Basis Function networks [67, 270], the probabilistic neural network (PNN) [310], and a number of fuzzy models [234, 313, 339].

3.3 Finding k-nn prototypes

Each of the three nonparametric models introduced above finds and handles the prototypes in a specific way. In this section we give methods to construct the k-nn reference set $\mathbf{V}$. This set, however, is not restricted to the k-nn model only but can be used in other classification paradigms. For example, $\mathbf{V}$ can be used as the initial guess for the locations of the centers of an RBF network or for constructing fuzzy if-then rule-based classifiers.

Let us assume that in the data set $\mathbf{Z}$, there are no identical objects (coinciding points in $\Re^n$) with different class labels. Then the 1-nn classifier with $\mathbf{Z}$ as the reference set, is an "ideal" classifier producing no resubstitution errors. Each object will find itself as the nearest neighbor within $\mathbf{Z}$, thereby getting labeled to the correct class. Finding a smaller number of prototypes is essential for two reasons. The first one is computational: with a smaller number of prototypes, less time and memory resources will be required for the operation of the classifier. Second, in the process of finding $\mathbf{V}$, some "noisy" objects from $\mathbf{Z}$ could be filtered out [94].

Example 3.3.1. Let $\Omega = \{\omega_1, \omega_2\}$. Let $V_{\mathbf{z}_j}$ be the Voronoi cell for $\mathbf{z}_j \in \mathbf{Z}$. Assume that $\mathbf{z}_j$ is labeled in ω_1, i.e., $l(\mathbf{z}_j) = \omega_1$. According to the definition of Voronoi cell, all points in $V_{\mathbf{z}_j}$ will be labeled also in ω_1. Let the true posterior probabilities for $\mathbf{z}_j$ be $P(\omega_1|\mathbf{z}_j) = 0.4$ and $P(\omega_2|\mathbf{z}_j) = 0.6$. Then, although class ω_2 is the favorite for $\mathbf{z}_j$, it is possible that in the data set we might have a realization $\mathbf{z}_j$ from class ω_1. If we assume that $V_{\mathbf{z}_j}$ small, and so the true posterior probabilities for all points $\mathbf{x}$ in the cell satisfy $P(\omega_2|\mathbf{x}) > P(\omega_1|\mathbf{x})$, the correct class label for $V_{\mathbf{z}_j}$ would be ω_2. This makes $\mathbf{z}_j$ a "noisy" object. ■

To reduce the number of prototypes, either a subset of $\mathbf{Z}$ is selected (**prototype selection**) or a set of new prototypes is designed (**prototype extraction**). Prototype selection is mostly related to the 1-nn classifier while prototype extraction is related to a class of neural networks, viz. the vector quantization models, LVQ. In both cases the idea is the same: *Use as few as possible prototypes to achieve as high as possible 1-nn (k-nn) accuracy.* Figure 3.12 shows one possible taxonomy of prototype selection and extraction methods. Listed underneath each leaf of the tree are only the examples which are described algorithmically in the text.

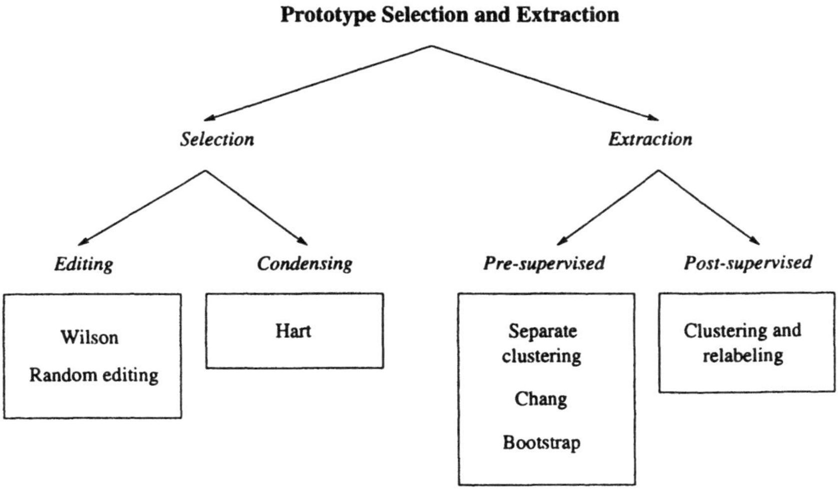

Fig. 3.12. Prototype selection and extraction methods

3.3.1 Edited k-nn (prototype selection)

How to find and label the prototypes? The simplest idea is to select them from **Z**. This is called **editing** of the reference set. There are two basic strategies [69]

- To find a subset $\mathbf{V} \subseteq \mathbf{Z}$ which produces zero *resubstitution* errors on **Z** when used as a reference set with the nearest neighbor (1-nn) classifier. Such subsets are called a **consistent**. The goal is to find the minimal possible consistent subset, i.e., a consistent set with the smallest possible cardinality. Practically, this is a search problem in a high-dimensional search space. We will call techniques for finding consistent subsets **condensing** techniques. Typically, condensing techniques retain points from **Z** that are most likely to be misclassified (e.g., points which are usually close to the classification boundaries).
- To find a small subset $\mathbf{V} \in \mathbf{Z}$ which has low (not necessarily zero) resubstitution error rate and *generalizes* well. This will be called **error-based editing** (in [75] it is called just "editing"). Error-based editing methods tend to eliminate the points close to the boundaries and retain those which belong "most certainly" in their own Bayes classification regions (see Example 3.3.1).

The two strategies can lead to very different reference sets. Due to their design, condensing methods cannot trade accuracy for reduction in the number of prototypes. Such trade off is very desirable because at the expense of one or two misclassifications we might be able to half the number of prototypes that are required. Usually error-based editing methods do not have an explicit mechanism to fix the number of prototypes or penalize reference sets of high cardinality. They often produce more prototypes than are needed. It is reasonable to apply first an error-based and then a condensing method on your data. Alternatively, you may choose editing by random search or guided random search (e.g., genetic algorithms [191, 195, 200] or tabu search [258]) which offer various means to specify the number of prototypes and the amount of accuracy to trade off for reducing the number of prototypes.

Three basic editing algorithms are given in a pseudo-code in Figures 3.13-3.15. Hart's method belongs to the condensing group [127]. It gradually builds the subset **V** starting with $\mathbf{z}_1$ and moving into **V** every element of **Z** which is misclassified by 1-nn using the current **V** as the reference set. The procedure loops through **Z** until there is no misclassified element. This technique tends to retain the points close to the classification boundaries and discard the "inside" points.

Wilson's method [348] is shown in Figure 3.14. Wilson proposes to run k-nn (recommended value $k = 3$) on **Z** and mark for deletion all elements that are misclassified. By deleting the marked elements from **Z**, the reference set **V** to be used with the 1-nn classifier, is obtained. Wilson's method is the basis of the asymptotically optimal editing method called MULTIEDIT

[75]. MULTIEDIT works well on large data sets with cloud-shaped classes. If the overlap of the classes is high, MULTIEDIT could eventually rule out all elements of some of the classes [191, 280].

Shown in Figure 3.15 is the random selection method [200, 308] which is simple and surprisingly successful. We specify the desired number of prototypes $v = |\mathbf{V}|$, $c \leq v < N$, generate T random subset-candidates $\mathbf{V}$ and return the set with the smallest resubstitution error. Better editing results at the expense of longer running time can be achieved by evolutionary algorithms [200, 308] or tabu search [258].

We chose Hart, Wilson and Random selection because these methods are possibly the simplest, have different flavors, and have been the points of departure for many variations.

HART'S CONDENSING METHOD

1. Set $\mathbf{V} = \{\mathbf{z}_1\}$; $flag = true$.
2. While $flag = true$ do
   a) Set $flag = false$.
   b) For $j = 1$ to $|\mathbf{Z}|$ do
      i. Use 1-nn to classify $\mathbf{z}_j$ using $\mathbf{V}$ as the reference set.
      ii. If $\mathbf{z}_j$ is misclassified,
         set $\mathbf{Z} = \mathbf{Z} - \{\mathbf{z}_j\}$
         set $\mathbf{V} = \mathbf{V} \cup \{\mathbf{z}_j\}$
         set $flag = true$
      iii. End % (For j)
   c) End % (While)
3. Return $\mathbf{V}$ with the labels.

Fig. 3.13. Hart's condensing method

WILSON'S EDITING METHOD

1. For $j = 1$ to N do
   a) Use 3-nn to classify $\mathbf{z}_j$ using $\mathbf{Z} - \{\mathbf{z}_j\}$ as the reference set (*leave-one-out*).
   b) If $\mathbf{z}_j$ is misclassified, mark it for deletion.
   c) End % (For j)
2. To obtain $\mathbf{V}$ delete from $\mathbf{Z}$ all marked elements.
3. Return $\mathbf{V}$ with the labels.

Fig. 3.14. Wilson's editing method

The reference sets $\mathbf{V}$ obtained by four editing procedures on Cone-torus data and the Normal-mixtures data are plotted in Figures 3.16 and 3.17. As the fourth method we used Wilson's error-based editing followed by Hart's condensing method (bottom left scatterplots). Table 3.2 displays the the num-

EDITING BY RANDOM SELECTION

1. Choose $v = |\mathbf{V}|$, $c \leq v < N$. Pick the maximal number of steps T. Set $E = 1$.
2. For $i = 1$ to T do
   a) Select at random $\mathbf{V} \subset \mathbf{Z}$, $|\mathbf{V}| = v$.
   b) Calculate the resubstitution error rate $e(\mathbf{V})$ of the 1-nn classifier with $\mathbf{V}$ as the reference set;
   c) If $e(\mathbf{V}) < E$ then store $\mathbf{V}$ and set $E = e(\mathbf{V})$.
   d) End % ($For\, i$)
3. Return the stored $\mathbf{V}$ with the labels.

Fig. 3.15. The RS editing algorithm

ber of prototypes and the training and testing error rates (in %). With both data sets the Random Selection was run for $T = 1000$ and $v = 10$.

Table 3.2. Number of prototypes and resubstitution error rates (in %) of 4 editing techniques

Data set		Hart	Wilson	Wilson + Hart	Random
Cone-torus	v (from 400)	119	336	36	10
	Training error	0	10.75	10.75	16.25
	Testing error	15.50	13.00	13.00	16.25
Normal-mixtures	v (from 250)	67	214	26	10
	Training error	0	10.80	10.40	10.80
	Testing error	15.60	14.40	15.50	10.20

3.3.2 Calculating prototypes from data (prototype extraction)

By *selecting* prototypes from $\mathbf{Z}$ we have a limited choice of points in $\Re^n$. Theoretically, *extracting* prototypes can lead to better results.

Example 3.3.2. Consider the following one-dimensional 3-class set, 2 points in each class:

To recognize correctly all 6 points by the 1-nn rule we need to retain 4 (*selected*) prototypes: any of the clubs, any of the hearts and both diamonds. The 6 points can be recognized correctly by *extracting* only 3 prototypes, e.g., the middle point for each class. ■

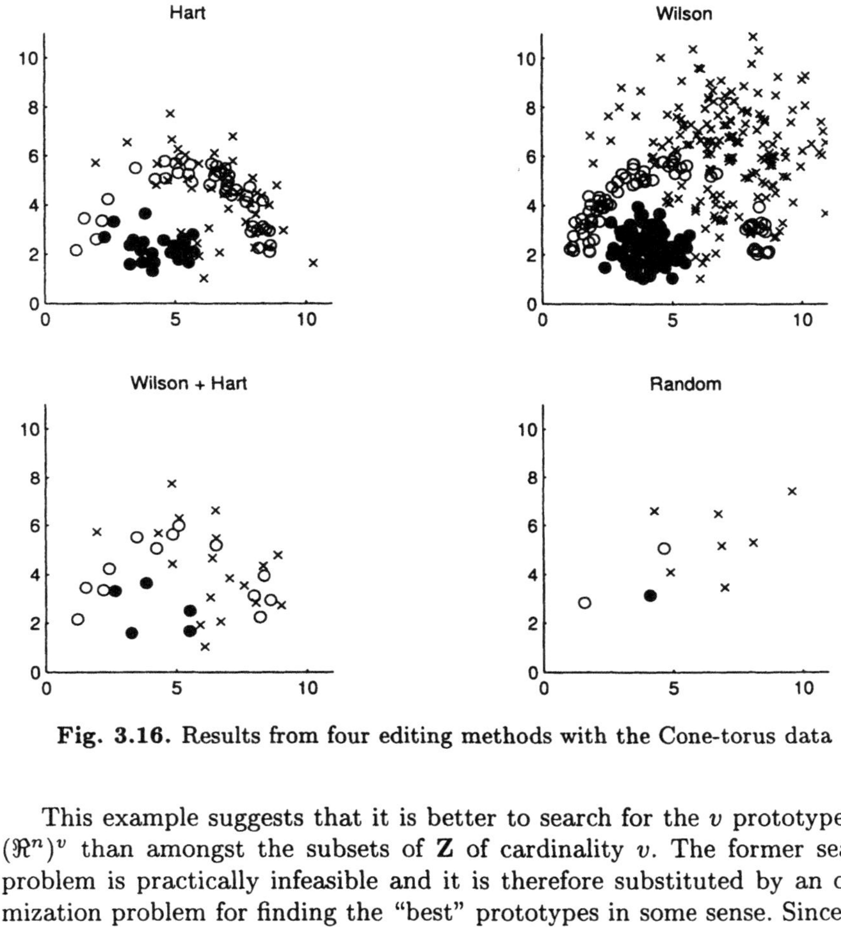

Fig. 3.16. Results from four editing methods with the Cone-torus data

This example suggests that it is better to search for the v prototypes in $(\Re^n)^v$ than amongst the subsets of $\mathbf{Z}$ of cardinality v. The former search problem is practically infeasible and it is therefore substituted by an optimization problem for finding the "best" prototypes in some sense. Since the goal is to achieve the lowest possible error rate with the smallest possible v, the criterion function should be connected to both error rate and the cardinality of $\mathbf{V}$. Usually, v is fixed in advance and the algorithms optimize only the prototype locations. There are numerous strategies and techniques for extracting prototypes, which can be generally grouped as

- *Clustering (and relabeling).* There are two basic approaches: separate clustering of data from each class (pre-supervised approach), and clustering of all of $\mathbf{Z}$ and relabeling the clusters afterwards (post-supervised approach). Pseudo-code algorithms for separate clustering (pre-supervised) and clustering and relabeling (post-supervised) are shown in Figures 3.18 and 3.19, respectively[4].

[4] The relabeling procedure in Figure 3.19 guarantees minimal resubstitution error rate

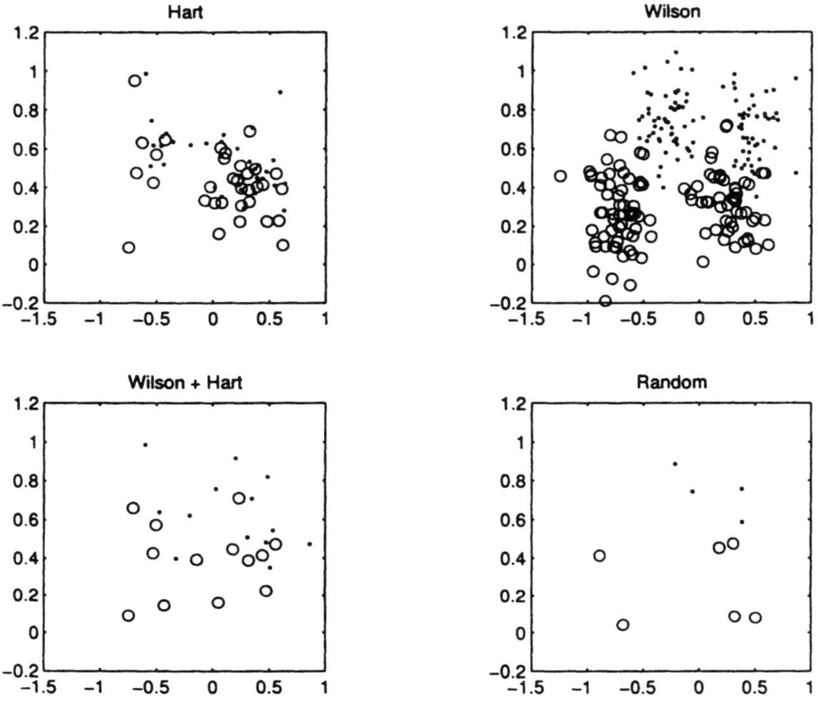

Fig. 3.17. Results from four editing methods with the Normal-mixtures data

- *Mixture modeling.* The pre-supervised and post-supervised approaches can be applied to extract prototypes from data by mixture modeling, usually with Gaussian mixture components [10, 13]. The prototypes correspond to the centers of the Gaussians and express "most typical" objects, either for a certain class (pre-supervised approach) or for all data (post-supervised approach).
- *Competitive learning.* Examples of this group are VQ and LVQ [175, 176] and modifications thereof [78, 106, 257, 295, 351] (LVQ is described in more detail in the neural networks section); competitive clustering algorithms such as dog-rabbit algorithm, and Chang's method. A modification of the original Chang method is proposed by Bezdek et al. [40]. The algorithm (Figure 3.20) finds a *consistent* set of prototypes which is *not a subset* of **Z**. The procedure starts with the whole **Z** as the set of prototypes **V**. The pair of prototypes of the same class label which are closest to each other is identified. The two prototypes are averaged to get a single replacement of the pair. The new set is checked for resubstitution errors. If no errors occur, the merger is accepted. Otherwise, the pre-merger **V** is restored and the pair is marked as ineligible. The search continues with the next (eligible) pair of closest prototypes. Applying Modified Chang to the Cone-torus data

resulted in a set of 74 prototypes (5/38/31 for classes cone/torus/normal, respectively). With the Normal-mixtures data 49 prototypes were found (23/26).

- *Using gradient descent.* Tuning prototype locations by gradient descent is proposed in [72, 317, 359].

SEPARATE CLUSTERING
1. Pick the number of prototypes per class $v_i, i = 1, \ldots, c$. Set $\mathbf{V} = \emptyset$.
2. For $i = 1$ to c do
   a) Cluster the subset of $\mathbf{Z}$ from class ω_i into v_i clusters.
   b) Find the cluster centroids $\mathbf{v}_1, \ldots, \mathbf{v}_{v_i}$ as the average of the vectors in the respective clusters.
   c) Label the new prototypes with class ω_i.
   d) Set $\mathbf{V} = \mathbf{V} \cup \{\mathbf{v}_1, \ldots, \mathbf{v}_{v_i}\}$.
   e) End % (For i)
3. Return $\mathbf{V}$ with the labels.

Fig. 3.18. Separate clustering (pre-supervised approach)

CLUSTERING AND RELABELING
1. Pick the number of prototypes $v = |\mathbf{V}|$.
2. Cluster $\mathbf{Z}$ into v clusters, $C_1, \ldots, C_v$.
3. Find the cluster centroids $\mathbf{v}_1, \ldots, \mathbf{v}_v$ as the average of the vectors in the respective clusters.
4. **Relabel** the prototypes. For cluster C_i, count the number of elements from $\omega_1, \ldots, \omega_c$. Label the prototype $\mathbf{v}_i$ to the class with the highest score. Relabel in this way all prototypes $\mathbf{v}_1, \ldots, \mathbf{v}_v$.
5. Return $\mathbf{V}$ with the labels.

Fig. 3.19. Clustering and relabeling (post-supervised approach)

Hamamoto et al. [126] propose four bootstrap methods for pre-supervised prototype selection. A modification of the simplest, and the most successful of the four (according to [126]) is shown in Figure 3.21. First, v elements of $\mathbf{Z}$ are picked at random, v_i from class ω_i, $i = 1, \ldots, c$. Then, to find the final $\mathbf{V}$, each element is replaced by the mean of its k nearest neighbors from its own class. $k > 1$ is a tuning parameter of the algorithm.

Example 3.3.3. Table 3.3 shows the resubstitution errors with the Cone-torus and Normal-mixtures data produced by the Bootstrap editing method. The best result from $T = 10$ runs is displayed. In all experiments the number of neighbors k was 7.

As expected, the error rate tends to decrease with the number of prototypes increasing. There is no requirement that the number of prototypes per

MODIFIED CHANG

1. Set $\mathbf{V} = \mathbf{Z}$ ($v = N$). Set up an $v \times v$ matrix M as the eligibility mask for pairs of elements of $\mathbf{V}$. Initially all pairs of elements from the same class are eligible.
2. While (there are eligible pairs) do,
   a) Amongst all eligible pairs in $\mathbf{V}$ find the pair with the smallest distance between them (called parents).
   b) Calculate a new candidate-prototype as the average of the two parents and label it in the class of the two parents.
   c) Tentatively replace the two parents by the new candidate-prototype.
   d) Classify $\mathbf{Z}$ using $\mathbf{V}$ as the reference set.
   e) If a misclassification occurs
      i. THEN: Reject the merge: Restore $\mathbf{V}$ (put back the two parents and discard the candidate-prototype). Mark the parent pair as non-eligible.
      ii. ELSE: Accept the merge: Keep the new prototype in $\mathbf{V}$ and update the eligibility matrix M making the new prototype eligible with any other element from its class.
   f) End % (While)
3. Return $\mathbf{V}$ with the labels.

Fig. 3.20. Modified Chang algorithm

BOOTSTRAP PROTOTYPE EXTRACTION

1. Pick the number of iterations T. Choose $v_1, v_2, \ldots, v_c$ ($v = v_1 + \ldots + v_c$). Set $E = 1$.
2. For $i = 1$ to T do
   a) Select at random v_i elements of $\mathbf{Z}$ from class ω_i, $i = 1, \ldots, c$.
   b) Replace each selected element by the average of its k-nearest neighbors *from the same class* from $\mathbf{Z}$.
   c) Calculate the resubstitution error rate $e(\mathbf{V})$ of the 1-nn classifier with $\mathbf{V}$ as the reference set;
   d) If $e(\mathbf{V}) < E$ then store $\mathbf{V}$ and set $E = e(\mathbf{V})$.
   e) End % (i)
3. Return the stored $\mathbf{V}$ with the labels.

Fig. 3.21. Bootstrap prototype extraction

class should be proportional to the prior probability of that class. We can pick any set of numbers $v_1, \ldots, v_c$, ($v_i \leq N_i$), or choose them at random too. ■

3.4 Neural networks

3.4.1 What are they?

Artificial Neural Networks (ANNs or simply NNs) originated from the idea to model mathematically human intellectual abilities by biologically plausible

Table 3.3. Bootstrap editing results: number of prototypes and the resubstitution error rate in % (the best from $T = 10$ runs).

Cone-torus data

Number of prototypes	8(2/2/4)	12(3/3/6)	16(4/4/8)	20(5/5/10)	24(6/6/12)
Error rate (%)	15.50	15.25	14.25	14.25	12.75

Normal-mixtures data (half of the prototypes allocated to each class

Number of prototypes	4	6	8	10	12	14	16	18
Error rate (%)	17.6	12.8	11.2	12.0	12.4	11.2	12.4	11.6

engineering designs. NNs are meant to be massively parallel computational schemes resembling a real brain. Neither of the two components of the original idea has been taken very far: the current models cannot outperform humans on intellectual tasks like face, voice, speech recognition, nor do they bear any striking similarity with the biological prototype – the brain. Nevertheless, NNs are a powerful classification tool with a significant influence on pattern recognition theory and practice.

A neural network, as any other classifier, can be thought of as a **black box**. Its inputs are the feature values ($\mathbf{x} = [x_1, \ldots, x_n]^T \in \Re^n$), and the outputs are c discriminant functions $g_1(\mathbf{x}), \ldots, g_c(\mathbf{x})$. Unlike many classifiers, however, NNs have two very important advantages:

• *Theoretical.* It has been shown that a classifier trained to minimize the squared error

$$E = \frac{1}{2}\sum_{j=1}^{N}\sum_{i=1}^{c}\left(g_i(\mathbf{z}_j) - Ind(\mathbf{z}_j, \omega_i)\right)^2, \tag{3.43}$$

where $Ind(\mathbf{z}_j, \omega_i)$ is the indicator function as in (3.2), gives as its output, $D(\mathbf{x}) = [g_1(\mathbf{x}), \ldots, g_c(\mathbf{x})]^T$, the posterior probabilities for the classes for $N \to \infty$ [279, 289, 334], i.e.,

$$g_i(\mathbf{x}) = P(\omega_i|\mathbf{x}),\ \mathbf{x} \in \Re^n. \tag{3.44}$$

This result was brought to light in connection with NNs but, in fact, it holds true, no matter what the classifier is, provided it can approximate any function with a predefined precision. This **universal approximation** property has been proven for both *Multi-Layered Perceptron* (MLP) and for the *Radial Basis Function* (RBF) networks (for summaries of the literature and proofs refer to [43] and [280]).

• *Practical.* There are good algorithms to train NNs. Although theoretically optimal performance is not guaranteed by the training procedures, the training ability of NNs towards their desirable performance is undoubtedly much richer than that of most of the classifiers considered so far. For example, there is not much that can be done to improve k-nn performance: we can try different types of distance, editing the reference set, varying k, or weighting the neighbors' votes. Such heuristics may or may not improve the classification performance. NNs have much greater training freedom: changing the network architecture, parameter values, initialization, or even switching to another NN model.

Literature on NNs is continuously growing. Many publications – textbooks and monographs [43, 93, 131, 129, 222, 269, 262, 280, 283, 298, 341, 342], paper collections [17], introductory readings [218, 219, 233] – discuss NNs at various theoretical and algorithmic depth. Modeling of the human brain, at either the morphological or functional level, and trying to understand NNs' cognitive capacity are also important research topics [8, 51, 254]. We confine the discussion here to three basic models: MLP, RBF and LVQ, used as pattern classifiers.

3.4.2 Neuron: the processing unit

The processing units in the human brain are neurons of different specialization and functioning. The earliest models of neurons, e.g., the model of McCulloch and Pitts [226], Fukushima's cognitron [101], and others, reprinted in the collection [17], were more similar to the biological neuron. For example, they incorporated both activating and veto-type inhibitory inputs. To avoid confusion artificial neurons are often given other names: "nodes" [277], "units" [43, 280], "neurodes" [222]. How plausible a model of a neuron do we need? Simple models will need a large structure for the whole system to work well (as with weightless neural networks [8]) while with more complex models of neurons a few units will suffice (as with a number of models of fuzzy neurons). In both cases proper algorithms are needed to train the system (structure and/or parameters) to make use of the modeling resources. Complex models without good training algorithms are not of much use. The basic scheme of a processing node is shown in Figure 3.22.

Let $\mathbf{u} = [u_0, \ldots, u_q]^T \in \Re^{q+1}$ be the input vector to the node and $v \in \Re$ be its output. We call $\mathbf{w} = [w_0, \ldots, w_q]^T \in \Re^{q+1}$ a vector of **synaptic weights**. The processing element implements the function

$$v = \phi(\xi); \quad \xi = \sum_{i=0}^{q} w_i u_i, \tag{3.45}$$

where $\phi : \Re \to \Re$ is the **activation function** and ξ is the **net sum**. Typical choices for ϕ are

- The Heaviside (threshold) function

$$\phi(\xi) = \begin{cases} 1, & \text{if } \xi \geq 0, \\ 0, & \text{otherwise.} \end{cases} \tag{3.46}$$

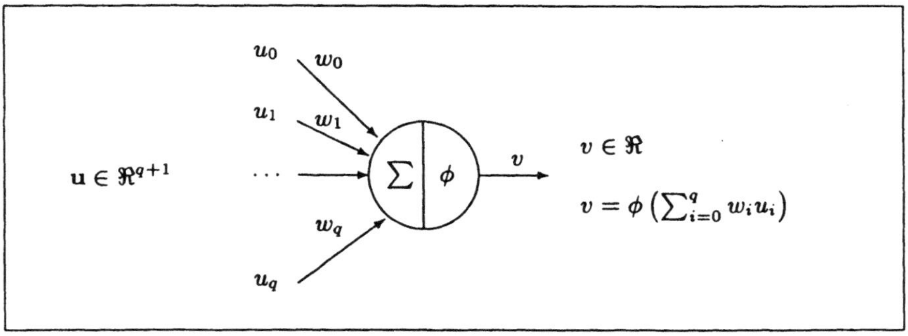

Fig. 3.22. The NN processing unit

- The sigmoid function

$$\phi(\xi) = \frac{1}{1 + \exp(-\xi)}. \tag{3.47}$$

- The identity function

$$\phi(\xi) = \xi. \tag{3.48}$$

The threshold and the sigmoidal functions are drawn in Figure 3.23.

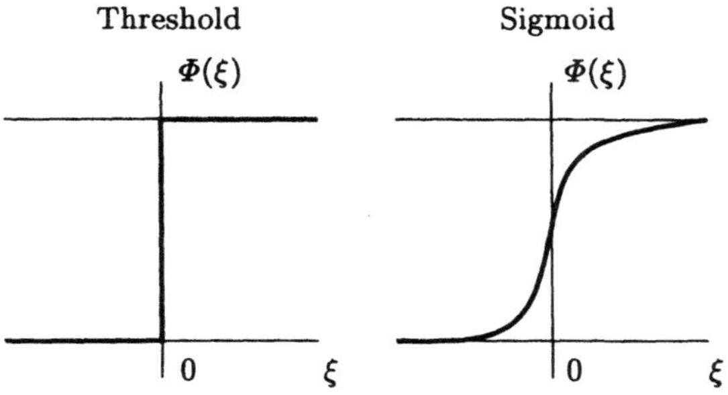

Fig. 3.23. Threshold and sigmoidal activation functions

The sigmoid activation function is the most widely used one because:

- It can model both linear and threshold functions to a desirable precision. Using properly scaled small weights, ϕ is almost linear near the origin, whereas for large weights, ϕ is practically the threshold function.
- The sigmoid function is differentiable, which is important for the NN training algorithms. Moreover, the derivative has the simple form $\phi'(\xi) = \phi(\xi)(1 - \phi(\xi))$.

The weight $-w_0$ is used as a **bias**, and the corresponding input value u_0 is set to 1. Equation (3.45) then becomes

$$v = \phi\left(\zeta - (-w_0)\right) = \phi\left(\sum_{i=1}^{q} w_i u_i - (-w_0)\right), \tag{3.49}$$

where ζ is now the weighted sum of the inputs from 1 to q. Geometrically, the equation

$$\sum_{i=1}^{q} w_i u_i - (-w_0) = 0 \tag{3.50}$$

defines a hyperplane in $\Re^q$. Therefore a node with a threshold activation function (3.46) responds with value +1 to all inputs $[u_1, \ldots, u_q]^T$ on the one side of the hyperplane, and value 0 on the other side.

3.4.3 The Rosenblatt perceptron

An important model was defined by Rosenblatt [284]. It is called "**perceptron**", and is famous for its training algorithm. The perceptron is implemented as (3.45) with a threshold activation function

$$\phi(\xi) = \begin{cases} 1, & \text{if } \xi \geq 0, \\ -1, & \text{otherwise} \end{cases} \tag{3.51}$$

It separates two classes in $\Re^n$ by the linear discriminant function defined by $\xi = 0$. The vectors from one class get output value +1, and from the other, −1. The algorithm starts with random initial weights $\mathbf{w}$ and proceeds by modifying the weights as each sample from $\mathbf{Z}$ is subsequently submitted to the input of the perceptron. The modification takes place *only if* the current vector $\mathbf{z}_j$ is misclassified (appears on the "wrong" side of the hyperplane). The weights are corrected by

$$\mathbf{w} \leftarrow \mathbf{w} - v\eta\mathbf{z}_j, \tag{3.52}$$

where v is the output of the node for $\mathbf{z}_j$ and η is a parameter specifying the **learning rate**. A pseudo-code for the perceptron training algorithm is shown in Figure 3.24.

Beside its simplicity, perceptron training has the following interesting properties:

- If the two classes are ***linearly separable*** in $\Re^n$, the algorithm ***always converges in a finite number of steps*** to a linear discriminant function that gives no resubstitution errors on $\mathbf{Z}$. (This is called "the perceptron convergence theorem").
- If the two classes are not linearly separable in $\Re^n$, the algorithm will never converge – it will loop infinitely through $\mathbf{Z}$. Moreover, there is no guarantee that if we terminate the procedure at some stage the resultant linear function is the one with the smallest possible misclassification count on $\mathbf{Z}$. The fuzzy perceptron model discussed in Chapter 7 remedies this drawback of the perceptron training.

PERCEPTRON TRAINING

1. Initialize the weight vector ($\mathbf{w}$) with small numbers and pick $\eta > 0$.
2. Set $ier = 1$ % (indicator of errors).
3. While $ier = 1$,
   a) Set $ier = 0$
   b) For j=1 to N % (a pass through $\mathbf{Z}$)
      i. Calculate the class label of $\mathbf{z}_j$ according to the perceptron output (3.45) and (3.51).
      ii. If $\mathbf{z}_j$ is misclassified
         A. Set $ier = 1$
         B. Set $\mathbf{w} = \mathbf{w} - v\eta\mathbf{z}_j$ % (correction towards the correct class)
         C. End % (If)
      iii. End % (For j)
   c) End % (While)
4. Return $\mathbf{w}$.

Fig. 3.24. The perceptron training procedure

3.4.4 Multi-layer perceptron (MLP)

By connecting nodes of the type shown in Figure 3.22 we can design an NN structure called the *Multi-Layer Perceptron* (MLP). MLP is a *feed-forward* structure because the output of a certain layer is submitted only to the higher layer, so that no feedback is allowed. A typical feed-forward NN classifier is shown in Figure 3.25. The "black box" is outlined with dashed lines.

Here "layer" means a layer of nodes (not a layer of tunable weights). By default there is an input layer, where $\mathbf{x}$ is submitted to the classifier, and an output layer yielding c discriminant functions $g_1(\mathbf{x}), \ldots, g_c(\mathbf{x})$. The number of hidden layers and the number of nodes at each hidden layer are not limited. We consider the following widely used model:

- The activation function at the input layer is the identity function (3.48).
- There are no lateral connections between the nodes at the same layer (feed-forward structure).
- Non-adjacent layers are not connected directly.
- All nodes at all hidden layers have the same activation function ϕ.

This model is not as constrained as it might look. In fact, most of the theoretical results in NNs are developed exactly for this model:

♣ In the late 80s, based on a simple constructive layout, it was shown that an MLP (as above) with 2 hidden layers of threshold nodes can approximate *any* classification regions in $\Re^n$ with a prespecified precision [218, 221, 138]. Figure 3.26 shows the classification regions that could be formed by an MLP with 1, 2 and 3 layers of threshold nodes.

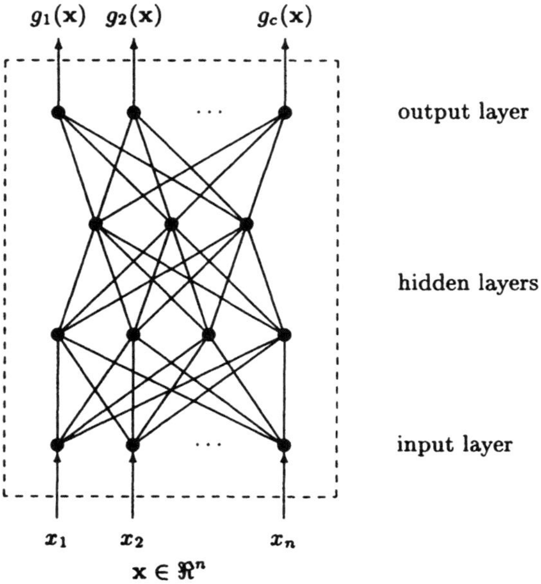

Fig. 3.25. An MLP classifier

♣ It was next proven that even a single-layer perceptron with threshold nodes can approximate *any* function with prespecified precision (for a recap of the literature and proofs see [43, 280, 296]).

The above two results do not tell us *how* to build and train the MLPs and therefore have only theoretical significance. So, if one layer is sufficient, why do we usually use more than one layer? The answer lies in the possibility (or rather impossibility) of designing efficient training algorithms for a certain NN structure. In fact, the resurfacing of NN in the 80s was motivated namely by one such algorithm, the **backpropagation training algorithm**.

3.4.5 Backpropagation training of MLP

We assume that the structure of the NN is already chosen and fixed (the number of hidden layers and the number of nodes at each hidden layer) and that the activation function is differentiable. The problem is to determine the values of the parameters (weights).

Example 3.4.1. How many weights are needed to completely specify the NN in Figure 3.25? For each of the four nodes at the first hidden layer we need $n+1$ parameters (n inputs + 1 bias, see Figure 3.22). For each node at the second hidden layer we need 4+1 parameters (4 inputs from the nodes at the first hidden layer + 1 bias), and for each output node, 3+1 parameters

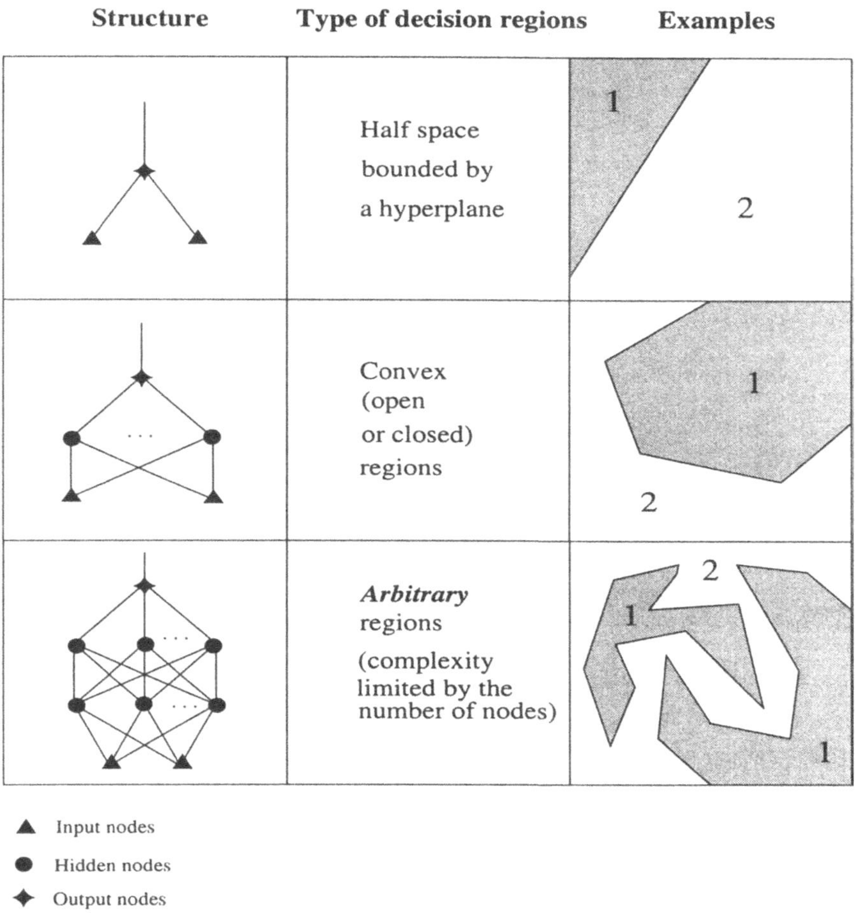

Fig. 3.26. Possible classification regions for an MLP with 1, 2, and 3 layers of threshold nodes. (Note that the "structure" column only indicates the number of hidden layers and not the number of nodes needed to produce the regions in column "examples".)

(3 inputs from the last hidden layer + 1 bias). Then the total number of parameters needed is $4(n+1) + 3(4+1) + c(3+1) = 4n + 4c + 19$. ■

Example 3.4.2. Figure 3.27 shows two equivalent MLP structures. The input is $\mathbf{x} \in \Re^2$ and the output is $g(\mathbf{x})$. All hidden nodes and the output node use sigmoidal activation, and the input nodes are identity functions. Thus, the output can be thought of a discriminant function so that for $g(\mathbf{x}) \geq 0.5$, $\mathbf{x}$ is labeled in ω_1, and for $g(\mathbf{x}) < 0.5$, in ω_2. In the left plot, the bias is represented as an extra weight. A special bias node is added to the structure with no inputs, and output equal to 1. In the right plot, the bias is a parameter of the node and is therefore "hidden". Both structures implement exactly the same function $g(\mathbf{x})$. Table 3.4 shows a possible set of parameter values of this MLP model. Figure 3.28 plots the classification regions obtained from the MLP NN with these parameter values. The network outputs for two points $\mathbf{x}_1 = [2, -1]^T$ and $\mathbf{x}_2 = [-1, 1]^T$ are also shown. ■

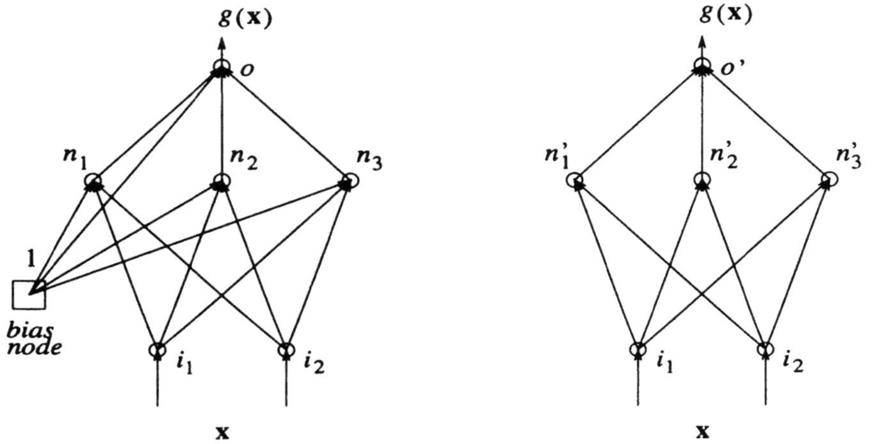

Fig. 3.27. Two identical MLP structures. Left: the bias is taken out as an additional weight. Right: the bias is a hidden parameter of each of the nodes n_1', n_2', n_3' and o'.

Table 3.4. A possible set of parameter values for the MLP model in Example 3.4.2, depicted in Figure 3.27. The bias nodes are indexed with a "b".

Layer	Nodes	Function	Parameters
Input	i_1	identity	N/A
	i_2	identity	N/A
Hidden	$n_1 = n_1'$	sigmoid	$w_{11}^h = 3$; $w_{21}^h = -1$; $w_{b1}^h = 2$
	$n_2 = n_2'$	sigmoid	$w_{12}^h = -1$; $w_{22}^h = 2$; $w_{b2}^h = -2$
	$n_3 = n_3'$	sigmoid	$w_{13}^h = 5$; $w_{23}^h = 1$; $w_{b3}^h = -3$
Output	$o = o'$	sigmoid	$w_{11}^o = 1$; $w_{21}^o = -2$; $w_{31}^o = 1$, $w_b^o = 1$

Let θ be a parameter (e.g., a weight) of the NN and $J(\theta)$ be some error function to be minimized. The gradient descent method updates θ by

$$\theta \leftarrow \theta - \eta \frac{\partial J}{\partial \theta}, \tag{3.53}$$

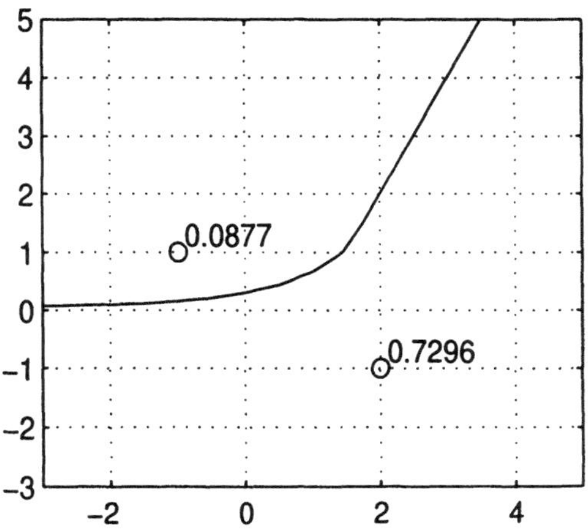

Fig. 3.28. Classification regions for the MLP network in Example 3.4.2.

where $\eta > 0$ is the learning rate. An obvious candidate for $J(\theta)$ is the squared error function E (3.43). Calculating the derivatives of E on all the weights of the MLP is not straightforward. Consider a node somewhere in the NN with net sum ξ, inputs $u_0, \ldots, u_q$ and weights $w_0, \ldots, w_q$. The derivative of E with respect to w_j is

$$\frac{\partial E}{\partial w_j} = \frac{\partial E}{\partial \xi}\frac{\partial \xi}{\partial w_j} = \frac{\partial E}{\partial \xi} u_j. \tag{3.54}$$

We call $\delta = \frac{\partial E}{\partial \xi}$ the **"error"**. Notice that the error is node-specific. Let $\mathbf{x} \in \Re^n$ be the NN input. To calculate the updated weight w_j, we need δ and the inputs u_j of that node for the given $\mathbf{x}$.

Using current weights and starting from the input layer, we can calculate subsequently all the u_js in the network (forward propagation). For the δ's, however, we have to start backwards, i.e., from the output, back to the input (backpropagation). The derivative of E (3.43) with respect to the ith output $g_i(\mathbf{x})$ is

$$\frac{\partial E}{\partial g_i(\mathbf{x})} = g_i(\mathbf{x}) - Ind(\mathbf{x}, \omega_i). \tag{3.55}$$

Let ξ_i^o be the net sum at output node i and $g_i(\mathbf{x}) = \phi(\xi_i^o)$. Using the chain rule,

$$\delta_i^o = \frac{\partial E}{\partial \xi_i^o} = \frac{\partial E}{\partial g_i(\mathbf{x})}\frac{\partial g_i(\mathbf{x})}{\partial \xi_i^o} = (g_i(\mathbf{x}) - Ind(\mathbf{x}, \omega_i))\frac{\partial \phi(\xi_i^o)}{\partial \xi_i^o}. \tag{3.56}$$

For the sigmoidal activation function ϕ (3.47),

$$\delta_i^o = \frac{\partial E}{\partial \xi_i^o} = (g_i(\mathbf{x}) - Ind(\mathbf{x}, \omega_i))g_i(\mathbf{x})(1 - g_i(\mathbf{x})). \tag{3.57}$$

These δ_i^o's, $i = 1, \ldots, c$, are the errors of the c nodes at the output layer and are used to update the weights from the last hidden to the output layer. The inputs u_i are the outputs of the nodes at the last hidden layer which are calculated through the forward propagation. Then we have all components to calculate the derivative of E with respect to w_{ik}^o, where w_{ik}^o is the weight of the connection between kth node at the last hidden layer and the ith output.

Consider now the kth node at the last hidden layer. Let w_{jk}^h denote the weight of the connection coming from the jth node at the previous (hidden or input) layer. Let ξ_k^h be the net sum and $v_k^h = \phi(\xi_k^h)$ be the output of the kth node at the last hidden layer. The error term for the kth node, $\delta_k^h = \frac{\partial E}{\partial \xi_k^h}$, needed for updating all incoming weights from the previous layer, is obtained as

$$\delta_k^h = \frac{\partial E}{\partial \xi_k^h} = \frac{\partial E}{\partial v_k^h}\frac{\partial v_k^h}{\partial \xi_k^h} = \frac{\partial E}{\partial v_k^h}\frac{\partial \phi(\xi_k^h)}{\partial \xi_k^h}. \tag{3.58}$$

The first term can be obtained via the chain rule again. Noticing that E depends on v_k^h through the net sums at the output nodes, $\xi_1, \ldots, \xi_c$, and each net sum participates in a separate summation term of E,

$$\frac{\partial E}{\partial v_k^h} = \sum_{i=1}^{c} \frac{\partial E}{\partial \xi_i}\frac{\partial \xi_i}{\partial v_k^h} = \sum_{i=1}^{c} \delta_i^o w_{ki}^o. \tag{3.59}$$

The errors $\delta_i^o = \frac{\partial E}{\partial \xi_i^o}$ are already obtained at the output layer. Using (3.58), the error at our kth node becomes

$$\delta_k^h = \frac{\partial E}{\partial \xi_k^h} = \left(\sum_{i=1}^{c} \delta_i^o\, w_{ki}^o\right)\frac{\partial \phi(\xi_k^h)}{\partial \xi_k^h}. \tag{3.60}$$

The above equation illustrates the ***backpropagation*** of the error: to calculate the error term for a node at a certain hidden layer we need the errors calculated already at the higher adjacent layer. For a sigmoidal ϕ, (3.60) becomes

$$\delta_k^h = \frac{\partial E}{\partial \xi_k^h} = v_k^h(1 - v_k^h)\sum_{i=1}^{c} \delta_i^o\, w_{ki}^o. \tag{3.61}$$

Propagating the above formulas backwards through the NN, we calculate the derivatives of E for a single input $\mathbf{x}$. There are two ways to implement the training procedure: batch and on-line. In the ***batch*** version, the updating (3.53) takes place once after a pass through the whole $\mathbf{Z}$, called an **epoch**. Therefore the derivatives of E for the whole $\mathbf{Z}$ are obtained by summing up the derivatives for all $\mathbf{z}_j \in \mathbf{Z}$, individually for every weight to be updated. Then we apply (3.53) to every weight of the NN. In the on-line implementation the weights are updated for each $\mathbf{z}_j$ taken as an input to the NN, with the derivatives calculated as above.

Example 3.4.3. Consider again the MLP NN from Example 3.4.2. Take the weights in Table 3.4 as the current approximation that is to be updated. Assume that $\mathbf{z}_j = [-1, 1]^T$ with label $l(\mathbf{z}_j = \omega_2$ is submitted to the algorithm. To simplify notation, the output of the nodes will be denoted with the node identifiers. The forward propagation pass gives the following node outputs $n_1 = 0.8808$, $n_2 = 0.2689$, $n_3 = 0.9991$, and $o = 0.08877$. The input nodes $i_1 = 2$ and $i_2 = -1$ reproduce $\mathbf{z}_j$. Let us calculate the updated value of w_{21}^h using (3.53), for learning rate $\eta = 0.4$. With the forward pass completed, to find $\frac{\partial E}{\partial w_{21}^h}$, we need the error term δ_1^h and the input u associated with this weight (see (3.54)). The input u is the output of the second node at the previous (input) layer, i.e., $u = 1$. First, δ^o has to be calculated. Using (3.57) which applies for sigmoidal outputs,

$$\delta^o = o^2(1 - o) = 0.0877^2 \times (1 - 0.0877) \approx 0.0070. \tag{3.62}$$

In this calculation we used the fact that the target value for $\mathbf{z}_j$ was 0, because $l(\mathbf{z}_j = \omega_2$. The error term needed for the weight w_{21}^h is calculated through (3.61), i.e.,

$$\delta_1^h = n_1(1 - n_1)\delta^o = 0.8808 \times (1 - 0.8808) \times 0.0070 \approx 0.00073. \tag{3.63}$$

Finally, the update is

$$w_{21}^h \leftarrow w_{21}^h - \eta\delta_1^h u = -1 - 0.4 \times 0.00073 \times 1 \approx -1.0003. \quad \blacksquare \tag{3.64}$$

The on-line backpropagation algorithm is given in Figure 3.25. There are a number of interesting modifications of backpropagation training aiming at higher effectiveness (e.g., faster convergence, higher stability) [222].

3.4.6 Radial basis function (RBF) networks

Radial Basis Function (RBF) networks use a single hidden layer with *local* processing elements. Instead of the net sum as in (3.45), the activation function for such local nodes takes as its argument a certain distance between the input vector $\mathbf{x} \in \Re^q$ and a *center*, a vector $\mathbf{v} \in \Re^q$, specific for the node. The activation function $\psi(\xi)$ is usually defined as a nonincreasing function $\psi : \Re^+ \rightarrow \Re$ which takes its maximal value for $\xi = 0$, i.e., when $\mathbf{x}$ coincides with $\mathbf{v}$, and declines with the distance between the two points in $\Re^q$. Let M be the number of the hidden nodes. The hidden layer outputs are

$$v_j = \psi(||\mathbf{x} - \mathbf{v}_j||),\ j = 1, \ldots, M, \tag{3.66}$$

and the output nodes are the same as those of the MLP. We can write the formal dependence of the output of the RBF network, $g_1(\mathbf{x}), \ldots, g_c(\mathbf{x})$, on its input $\mathbf{x}$

BACKPROPAGATION MLP TRAINING

1. Choose an MLP structure: pick the number of hidden layers, the number of nodes at each layer and the activation functions.
2. Initialize the training procedure: pick small random values for all weights (including biases) of the NN. Pick the learning rate $\eta > 0$, the maximal number of epochs T and the stopping constant $\epsilon > 0$.
3. Set $E_{old} = \infty$ and calculate E_{new} using the current weights and equation (3.43). Set $t = 1$.
4. While ($||E_{old} - E_{new}|| > \epsilon$ and $t \leq T$) do
   a) Set $E_{old} = E_{new}$, $t \leftarrow t + 1$.
   b) Calculate the output of every node of the NN with the current weights (forward propagation).
   c) Calculate the error term δ_i at each node at the output layer by (3.57).
   d) Calculate recursively all error terms at the nodes of the hidden layers (δ_k^h) using (3.60) (backward propagation).
   e) For each hidden and output node update the weights by
   $$w_{i,new} = w_{i,old} - \eta \delta u, \tag{3.65}$$
   using the respective δ and u.
   f) Calculate E_{new} using the new weights and equation (3.43).
5. End % (While)

Fig. 3.29. Backpropagation MLP training

$$g_i(\mathbf{x}) = \phi\left(\sum_{j=1}^{M} w_{ji}\psi(||\mathbf{x} - \mathbf{v}_j||) + w_{i0}\right). \tag{3.67}$$

Notice the similarity between the definition of ψ and the kernel functions defined for the Parzen classifier. RBF design involves:

- Choosing the type of the output function ϕ and kernel function ψ. Chen et al. [59] claim that the choice of the kernel function is not very important for the network performance. As with the Parzen classifier, the Gaussian (3.35) is the most popular choice for ψ.
- Choosing the number and the initial location of the centers of the hidden nodes.
- Tuning the parameters of the network (the centers, the parameters of ψ, if any, and the weights from the hidden to the output layer).

In the simplest case we can use only one "spreadout" parameter (width) for all hidden layer nodes, e.g.,

$$v_j = \exp\left\{-\frac{||\mathbf{x} - \mathbf{v}_j||^2}{2s^2}\right\}, \quad s > 0. \tag{3.68}$$

For a better flexibility we can use individual widths $s_j, j = 1, \ldots, M$. In the most general case we have a full individual covariance matrix S_j for each node, i.e.,

$$v_j = \exp\left\{-\frac{1}{2}(\mathbf{x}-\mathbf{v}_j)^T S_j^{-1}(\mathbf{x}-\mathbf{v}_j)\right\}. \tag{3.69}$$

Substituting (3.69) in (3.67), and taking ϕ as the identity function, we obtain the most popular RBF configuration whose input-output relationship is given by

$$g_i(\mathbf{x}) = \sum_{j=1}^{M} w_{ij} \exp\left\{-\frac{1}{2}(\mathbf{x}-\mathbf{v}_j)^T S_j^{-1}(\mathbf{x}-\mathbf{v}_j)\right\} + w_{i0}. \tag{3.70}$$

The prototypes (centers) can be selected or extracted in advance as the first stage of**a two-stage training**, and kept fixed during the second stage at which S_j's (if necessary) and w_{ji} are tuned. There are many ways to initialize the prototypes [131]: we can place them onto the vertices of a large grid in $\Re^n$, choose them at random from $\mathbf{Z}$, select them by editing $\mathbf{Z}$ (see section 3.3), use some addition/deletion scheme, find them as cluster centroids in $\mathbf{Z}$ (as the original model of Moody and Darken [241]), evolve or select them by genetic algorithms [196], etc. Another training option is to tune all the parameters of the RBF network together in a **single training stage**. Tarassenko and Roberts propose backpropagation training [317]. One of the most popular training schemes is the *orthogonal least squares* (OLS) algorithm proposed by Chen, Cowan and Grant (1991)[59]. Sherstinsky and Picard (1996)[303] showed that the OLS does not guarantee the smallest possible network for a given approximation error. Whitehead [346] and Whitehead and Choate [347] use specially adjusted genetic algorithms to train RBF networks.

A training method for RBF networks based on **Support Vector Machines** (SVM), is presented by Schölkopf et al. [297]. The data set $\mathbf{Z}$ is mapped non-linearly into a new high-dimensional feature space and a hyperplane is constructed which separates the classes in the new space. The parameters of the hyperplane are derived using (the images of) those elements of $\mathbf{Z}$ which are closest to the decision boundary. These elements are called *support vectors* and are retained as the centers $\mathbf{v}_j$ in the original feature space. The experiments reported in [297] verify that SVM training outperformed some classical two-stage training models for RBF networks. We can draw an analogy between the selection of support vectors and the condensing methods for 1-nn editing. Both SVM and the condensing methods (e.g., Hart) retain elements which are close to the classification boundary.

In contrast to MLP and LVQ models, RBF networks have a well defined statistical counterpart. They can be viewed as nonparametric classifiers with kernel-type of p.d.f's estimation [67, 185, 326, 310, 363], which gives a background for analysis of their asymptotical properties RBF networks have also been proven to be universal approximators [128, 184, 264, 265], meaning that for any (true) discriminant function, there is a set of centers and widths such that the respective RBF network can approximate the function on a compact set of points with arbitrary precision.

3.4.7 Learning vector quantization (LVQ)

LVQ classifiers compute a set of prototypes $\mathbf{V} = \{\mathbf{v}_1, \ldots, \mathbf{v}_v\}$ by sequential competitive learning [131, 176]. These prototypes are then used as the reference set for the 1-nn classifier. LVQ is an intuitive method and rigorous analytical studies showing what exactly it is that competitive learning optimizes, are scarce [77]. Nevertheless the model has been widely used because it is simple and practically successful. The training is based on **Hebbian learning** which is a plausible biological learning model. According to it, the connection between neurons that fire simultaneously is strengthened whilst the connection between neurons that act differently is loosened. Figure 3.30 shows the pseudo-code of the "classical"member of the LVQ family called **LVQ1** [77, 175, 176].

The number of prototypes v can be either specified in advance or adjusted during training. The algorithm starts with some initialization of the prototypes ($\mathbf{V}_0$), most often by assigning random values or by randomly taking v vectors from $\mathbf{Z}$. It makes sequential passes through $\mathbf{Z}$. To remove the influence of the order of submitting the elements of $\mathbf{Z}$, the data set can be shuffled before each pass. The algorithm checks every $\mathbf{z}_j$ with 1-nn using the current $\mathbf{V}$ as the reference set. If the closest prototype (winning the "closeness" competition), say, $\mathbf{v}_i$ is from the same class as $\mathbf{z}_j$, the prototype is moved towards $\mathbf{z}_j$ in $\Re^n$ (reward). Otherwise, it is moved away (punishment). The new value of $\mathbf{v}_i$ (the winner) at step t, denoted $\mathbf{v}_{i,t}$, is calculated as

$$\mathbf{v}_{i,t} = \mathbf{v}_{i,t-1} + \alpha_{ij,t}(\mathbf{z}_j - \mathbf{v}_{i,t-1}), \quad i = 1, \ldots, c, \tag{3.71}$$

where $a_{ij,t}$ is the **learning rate distribution**. This parameter determines by how much $\mathbf{v}_i$ should be modified at step t. The learning rate distribution can be a function of the three parameters i (prototype), j (datum) and t (iteration). Most often α depends only on t. Starting with an initial value in the interval (0,1], a typical formula to decrement α with the iterations is

$$\alpha_t = \eta \alpha_{t-1}, \tag{3.72}$$

where $\eta \in (0, 1)$ is a parameter. By this, the update of the prototypes is greater at the beginning and gradually declines towards termination of the procedure.

If the winner $\mathbf{v}_{i,t}$ is not from the same class as $\mathbf{z}_j$, it is "punished" using

$$\mathbf{v}_{i,t} = \mathbf{v}_{i,t-1} - \alpha_{ij,t}(\mathbf{z}_j - \mathbf{v}_{i,t-1}), \quad i = 1, \ldots, c, \tag{3.73}$$

The algorithm terminates either if the maximum iteration number is reached or if there have been no misclassifications on $\mathbf{Z}$. Iteration is sometimes used to denote a submission of an element from $\mathbf{Z}$. Here we call an iteration a pass through the whole $\mathbf{Z}$ (corresponding to the term *epoch* in backpropagation MLP training).

LVQ

1. Pick the number of prototypes v and the maximal number of iterations T.
2. Initialize $\mathbf{V} = \mathbf{V}_0$. Set $\alpha = \alpha_{ini}$.
3. Set $ier = 1$ (indicator of errors) and $t = 1$.
4. While ($ier = 1$ and $t \leq T$),
   a) Set $ier = 0$
   b) For j=1 to N (a pass through $\mathbf{Z}$)
      i. Use $\mathbf{V}$ with 1-nn to calculate the class label of $\mathbf{z}_j$.
      ii. If $\mathbf{z}_j$ is misclassified
         Then: "Punish" the winning prototype by (3.73).
         Else: "Reward" the winning prototype by (3.71).
      iii. Update α by (3.72).
      iv. End % (For j).
   c) Classify $\mathbf{Z}$ by 1-nn using the current $\mathbf{V}$ as the reference set.
   d) If a misclassification occurs, set $ier = 1$, set $\alpha = \alpha_{ini}$ (optional), and shuffle $\mathbf{Z}$ (optional).
   e) End % (While)
5. Return$\mathbf{V}$ with the class labels.

Fig. 3.30. LVQ training algorithm

There are many LVQ variants depending on which prototypes are updated and whether an action is taken on every step or only when a misclassification occurs.

Decision surface mapping (DSM) [106] is a variation of LVQ. Instead of updating a prototype at each step, DSM takes action only when $\mathbf{z}_j$ is misclassified by the nearest prototype. Then the winner is "punished" as in (3.73). At the same step, the nearest prototype from the correct class is found and is "rewarded" as in (3.71).

LVQ with training count (LVQTC) is proposed in [257]. During training, each node "remembers" how many times it has been modified (both rewarded or punished). Furthermore, the node also "remembers" which of the input vectors (with their class labels) have been used to modify it and which class label has caused the modification. At the end of each epoch, the network is reinitialized: nodes which are seldom picked for an update are pruned and nodes which have often been punished many times by vectors of a same class are replaced and relabeled to that class. All counters are reset to 0 and the learning rate is decreased. In this way, the NN can shrink the hidden layer adaptively during training. The training counters for each node can be used further in the classification process to account for the certainty of the classification decision.

3.4.8 An experimental example

Which of the NN models is the best one? As with all other classifiers, there is no clear preference. Different networks may prove useful for different problems. The three NN models (MLP, RBF and LVQ) have been applied to the four data sets: Cone-torus, Normal mixtures, Phoneme and Satimage. With the first two sets the designated subsets were used for training and for testing. With the Phoneme and Satimage data, the first 500 points were used for training and the remaining (4904 for Phoneme and 5935 for Satimage) points were used for testing. The results are presented in the Appendix. Figures 3.31 to 3.33 plot the results from the NN experiments, one figure per data set. The Satimage data was excluded from the plots because of the failure of the particular MLP configuration (commented below). The classification error (training and testing) are plotted against the number of hidden nodes (or prototypes for LVQ). Thick lines show the testing error rate. For comparison, classification results with the four data sets and various methods described in Sections 3.1 and 3.2 are displayed in the Appendix, in Tables A.1, A.11, A.21, and A.26, respectively.

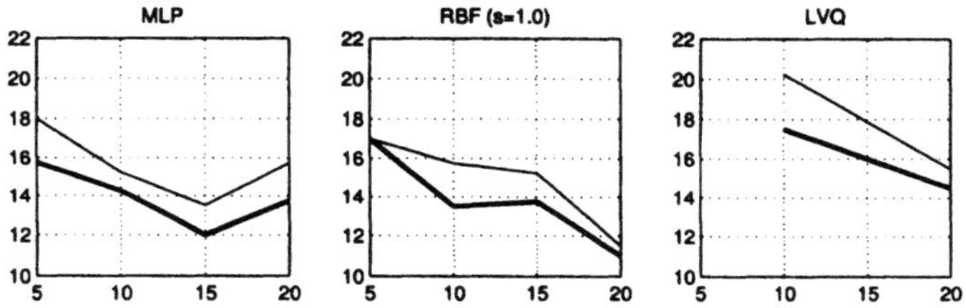

Fig. 3.31. Training and testing error rates with MLP, RBF and LVQ1 on the **Cone-torus** data.

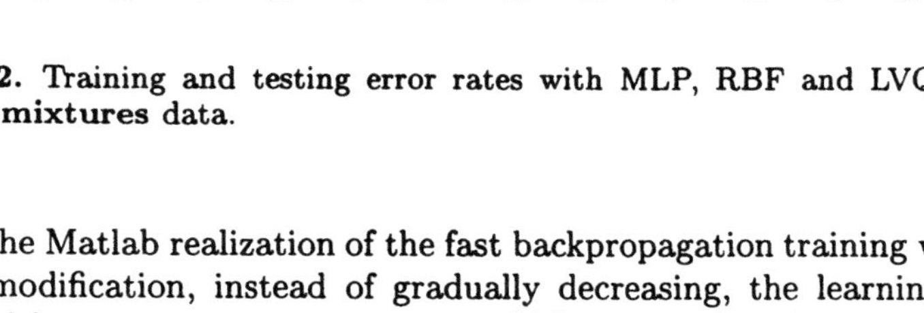

Fig. 3.32. Training and testing error rates with MLP, RBF and LVQ1 on the **Normal-mixtures** data.

MLP. The Matlab realization of the fast backpropagation training was used. In this modification, instead of gradually decreasing, the learning rate is controlled by a *momentum* parameter which is introduced to speed up the convergence.[5] The default procedures and constants were used for parameter

[5] A variety of enhanced backpropagation algorithms are given in [222].

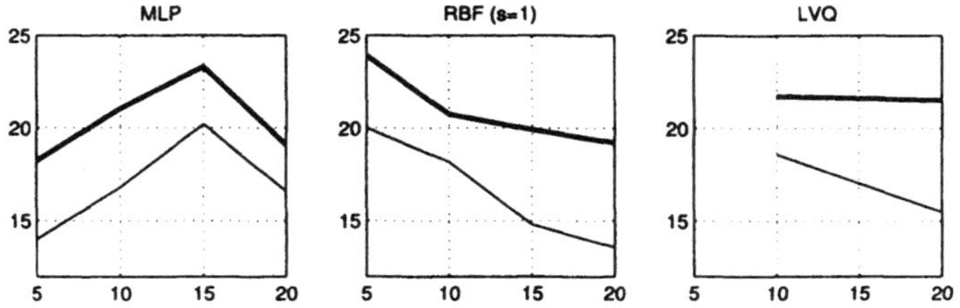

Fig. 3.33. Training and testing error rates with MLP, RBF and LVQ1 on the **Phoneme** data.

initialization. A simple MLP configuration was tried: one hidden layers with 5, 10, 15 or 20 nodes. For the first three data sets the results (training and testing error rates) were within the expected range but for the Satimage data the MLP was unable to train properly and classification error varied around 70 (!) %. The reason for this could be that the network had to approximate 6 outputs (one per class) based on 4 features, while the dimensionalities of the other three problems were smaller. Therefore, to reach acceptable error rates with the Satimage data, the hidden layer was set with 65, 70, 75 and 80 nodes. The first plots in the four Figures display the MLP training and testing error rates.

RBF. The RBF OLS procedure from the Neural Network Toolbox of Matlab was used. The decisive parameter for the success of the network is the width s of the Gaussian activation function (3.68). We experimented with a few values for s (the detailed results are presented in the Appendix) and chose the value for which the least *training* error was obtained. The s values (specific for each data set) are shown in the parentheses of the RBF subplots.

LVQ1. The algorithm in Figure 3.30 was encoded in Matlab. The initial learning rate was $\alpha = 0.3$, the constant $\eta = 0.8$, the learning rate was reinitialized to 0.3 after each pass through $\mathbf{Z}$, T was 100. Experiments for 10 and 20 prototypes were carried out and the best *training* result together with the corresponding test results are shown in the third subplots of Figures 3.31 to 3.33.

The conjecture from the experimental study presented in [136] was that "good statistical classifiers included the k-nearest neighbor (k-nn) and good neural network classifiers included learning vector quantization (LVQ) and radial basis function (RBF) classifiers." Our experiments did not indicate a clear winner between the three networks. The "best" network is specific for each data set. The arrangement and the test error rates (in %) corresponding to the best training rates are

Cone-torus:	RBF (11.00)	$<$	MLP (12.00)	$<$	LVQ1 (14.50)
Normal-mixtures:	LVQ1 (9.15)	$<$	MLP (9.2)	$=$	RBF (9.2)
Phoneme:	MLP (18.23)	$<$	RBF (19.23)	$<$	LVQ1 (21.53)
Satimage:	RBF (14.69)	$<$	MLP (15.32)	$<$	LVQ1 (16.68)

Generally, the test error follows the pattern of the training error and better performance can be expected with more hidden neurons. Training MLP and LVQ1 takes longer than training RBF by OLS. Backpropagation is sensitive to the initialization and can be trapped in local extrema. The experiments with the Satimage data showed that with the same number of hidden nodes, the same number of iterations, and different initial guess for the weights, the training might converge to substantially different error rates. LVQ1 also depends on the initialization and the data arrangement in $\mathbf{Z}$. RBF networks with OLS training could be nominated as the favorite amongst the three designs because together with being the fastest, it also provides consistently good results.

None of the three training algorithms optimize directly the error rate. Backpropagation training and OLS minimize the discrepancy between the target class labels and the output of the network, and LVQ1 is even farther from the error rate [77]. However, NN rate high amongst classification models and are widely used.

4. Fuzzy sets

> ...The Owl put her ear to Buratino's chest: 'The patient is *rather dead than alive*', - she whispered and turned her head back at 180 degrees. The Frog stroked Buratino with her moist flipper. In deep thought, with her goggle eyes looking in opposite directions, the Frog clapped with her big lips: 'The patient is *rather alive than dead*'.
>
> A. Tolstoy
> "The Golden Key or the Adventures of Buratino"
> Moscow, DL, 1997, p. 41 (In Russian)

4.1 Fuzzy logic, an oxymoron?

In Bangor, North Wales, it is often drizzling or raining. At any time you will find in the streets people with umbrellas and without umbrellas, varying in number. For some of them it is raining, and yet for some others, it is not raining – both at the same time. So, 'raining' is a matter of judgement. We can assign a degree (of truth) to the statement "It is raining", and funnily enough, both the proposition and its negation can hold true *to a certain degree* at the same time.

Some people may argue that the *degree of raining* that I assign to a certain instance of North Wales weather is the same thing as the probability that I open up my umbrella in this weather (call this "individual opinion"). We can also measure the degree of raining by the proportion of people in the street with open umbrellas ("public opinion"). There are many ways to estimate such a degree. Whether or not we take up the probabilistic interpretation, we cannot deny that there are matters which are inherently non-binary, and can hold partly true and partly false at the same time – even life and death, as the above motto from about. We can bring thousands of examples of notions, characteristics, categories, statements, in which the transition between truth and non-truth is not clear-cut.

Fuzzy sets were proposed by Lotfi Zadeh in 1965 [366] as a numerical means to handle the uncertainty and vagueness inherent to human perception, speech, thinking, decision making, etc. The most straightforward example is the *linguistic* uncertainty of natural language [367]. Despite the huge body of fuzzy set literature, Dubois and Prade [85] find and highlight the lack of a unique and well understood semantic of fuzzy set theory. The term "fuzziness" can be used to express: uncertainty, typicality, severity, importance, possibility, compatibility, fitness, similarity, belief; and more: degree of involvement, damage, beauty, desirability, fatigue, expense, etc. All these can be conveniently (although not very crisply) grouped into three basic semantic categories of degrees of membership: *similarity, preference*, and *uncertainty* [85], and quantified in a uniform "fuzzy set" way. Maybe this variety of interpretations and degrees of freedom in choosing and implementing fuzzy models is the blessing and the curse of fuzzy set theory. There have been, and will always be those who will call fuzzy logic a misnomer, useless mathematical abstraction, and instigate extensive discussions in the literature (see [36, 135, 253, 368]). And, yet, there have been, and will always be those who bring about the applications of fuzzy sets to space shuttle control, washing machines, cameras, helicopters, electronic noses, magnetic resonance image segmentation, signature verification, etc., and who agree with Paul Wasserman that [342]:

> "Ultimately, mathematical theories are judged by their consistency, beauty, and utility. Fuzzy theory passes examinations on all accounts."

Fuzzy theory is not a religion or a taboo, it is a *theory* and we will look at how far and how successful it has been and can be for classifier design.

4.2 Basic definitions

4.2.1 Fuzzy set, membership function

Let $U = \{u_1, \ldots, u_n\}$ be a **set** . U will be called the **universal** set. To distinguish between ordinary and fuzzy sets, ordinary sets are often called in the fuzzy set literature **crisp** sets. Each (crisp) subset A of U can be described by a characteristic function or a **membership function**

$$\mu_A : U \to \{0,1\}, \quad u \in U, \tag{4.1}$$

where $\mu_A(u)$ is one if $u \in A$, and zero, otherwise. A **fuzzy set** B on U (also called a **fuzzy subset** of U) is described by its membership function

$$\mu_B : U \to [0,1] \tag{4.2}$$

where $\mu_B(u)$ expresses the degree in which the element u *belongs* in B, and is called a **degree of membership**.[1] Since the membership function deter-

[1] To prevent over-notation, all type of subsets of U (crisp or fuzzy) will be denoted by capital letters, e.g., A, B. The type of any particular set will be explicitly declared.

mines uniquely the fuzzy set, we will use the two notions μ_B and the fuzzy set B, interchangeably. We note that crisp sets is a special case of fuzzy sets such that the degrees of membership of the elements take only values 0 and 1.

Usually the membership function of a fuzzy set corresponding to a linguistic term, such as *small, tall, young, fast*, is designed to peak at the most typical value(s) for this term. Some widely used types of membership functions: triangular, trapezoidal, and Gaussian, are depicted in Figure 4.1, and calculated by the formulas in Table 4.1.

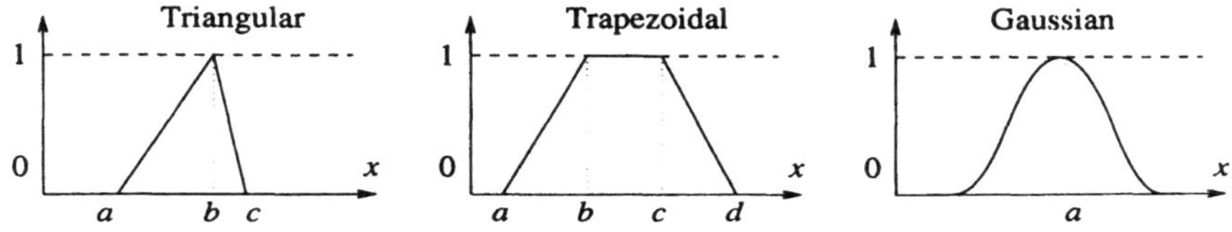

Fig. 4.1. Triangular, trapezoidal and Gaussian membership functions

Table 4.1. Membership function formulas

Triangular:	$\mu(x) = \begin{cases} \frac{x-a}{b-a}, & \text{if } x \in (a,b], \\ \frac{c-x}{c-b}, & \text{if } x \in (b,c], \\ 0, & \text{otherwise.} \end{cases}$
Trapezoidal:	$\mu(x) = \begin{cases} \frac{x-a}{b-a}, & \text{if } x \in (a,b], \\ 1, & \text{if } x \in (b,c], \\ \frac{d-x}{d-c}, & \text{if } x \in (c,d], \\ 0, & \text{otherwise.} \end{cases}$
Gaussian:	$\mu(x) = \exp\left\{-\frac{(x-a)^2}{2\sigma^2}\right\}.$

Example 4.2.1. A **fuzzy number** "about 5" can be designed on $U = \Re$, following some logical guidelines:

- $\mu_{about\ 5}(5) = 1$,
- the membership function should be symmetrical about 5,
- $\mu_{about\ 5}(u)$ should decrease as u goes away from 5, $u \in U$.

Any membership function that satisfies these three requirements can be used. For example, $\mu_{about\ 5}$ could be

$$\mu_{about\ 5}(u) = \begin{cases} 1 - \frac{|u-5|}{2}, & 3 \le u \le 7, \\ 0, & \text{otherwise.} \end{cases} \tag{4.3}$$

Compare this equation with the membership function of the crisp set $A = 5$, $A \subset U$,

$$\mu_5(u) = \begin{cases} 1, & u = 5, \\ 0, & \text{otherwise.} \end{cases} \tag{4.4}$$

Pointing at the uniqueness of crisp membership functions and the infinite variety of fuzzy membership functions for the same concept, Jim Bezdek says [35]:

> "Uniqueness is sacrificed (and mathematicians howl), but flexibility is increased (and engineers smile)."

■

Practically, except for some vague semantic guidelines, there are no restrictions in designing a membership function. This fact causes a major problem in comparing fuzzy and non-fuzzy techniques. If a fixed non-fuzzy technique gives better results in terms of the criterion used for the comparison, there will always be the assumption that another design of the membership functions could bring the reverse result.

Let A be a fuzzy set on U. Below are three different notations for A (notice that '/' is not used for division but shows a link

$$A = \{u_1/\mu_A(u_1), \ldots, u_n/\mu_A(u_n)\}; \tag{4.5}$$

$$A = u_1/\mu_A(u_1) + \ldots + u_n/\mu_A(u_n); \tag{4.6}$$

$$A = \{(u_1, \mu_A(u_1)), \ldots, (u_n, \mu_A(u_n))\} \tag{4.7}$$

Whichever way we choose, notice that a fuzzy set A on U is completely specified iff for any $u \in U$, the degree of membership $\mu_A(u)$ can be calculated or directly retrieved from the description.

Example 4.2.2. Let U be my list of holiday places on the Black Sea for this summer, e.g., U = {Sozopol, Ravda, Varna, Golden Sands, Duni, Kiten}. Let the fuzzy set A correspond to "a place of art". Then A can be defined as: A = {(Sozopol, 0.9), (Ravda, 0.1), (Varna, 0.7), (Golden Sands, 0.6), (Duni, 0.0), (Kiten, 0.1)}. The values of the membership function (degrees of membership) in this example have been assigned subjectively. Different values can be assigned by checking and evaluating the art calendar of each of these resorts. ■

4.2.2 Support, core, height, level-set

Let A be a fuzzy set on U, with membership function $\mu_A : U \to [0,1]$.

The **support** of a fuzzy set A on U is the (crisp) set *supp*(A) obtained as

$$supp(A) = \{u \mid u \in U,\ \mu_A(u) > 0\} \tag{4.8}$$

The **core** of a fuzzy set A on U is the (crisp) set *core*(A) obtained as

$$core(A) = \{u \mid u \in U,\ \mu_A(u) = 1\} \tag{4.9}$$

The core can be thought of as the most representative elements of A. The **height** of a fuzzy set A on U is obtained as

$$height(A) = \sup_{u \in U} \mu_A(u), \tag{4.10}$$

which becomes

$$height(A) = \max_{u \in U} \mu_A(u), \tag{4.11}$$

for a finite U.

Fuzzy sets with height 1 are called **normal**, and these with height less than one are called **subnormal**.

A fuzzy set with a support of cardinality one is called a **singleton**. Singletons can be normal or subnormal.

The fuzzy set A in example 4.2.2 has *height*$(A) = 0.9$ (i.e., it is subnormal), *core*$(A) = \emptyset$ and *supp*$(A) =$ {Sozopol, Ravda, Varna, Golden Sands, Kiten}.

The **α-level set** (**α-cut**) of a fuzzy set A on U is the (crisp) set

$$A_\alpha = \{u \mid u \in U,\ \mu_A(u) \geq \alpha\} \tag{4.12}$$

The α-level sets of a fuzzy set A are *nested*, i.e.,

$$\alpha_1 \geq \alpha_2 \iff A_{\alpha_1} \subseteq A_{\alpha_2}. \tag{4.13}$$

Any fuzzy set can be represented using its level sets. This representation is called also the **α-cut decomposition (representation)** of the fuzzy set and is formulated as a theorem. The theorem states that for any fuzzy set A on U,

$$\mu_A(u) = \sup_{\alpha \in [0,1]} \min\{\alpha, \mu_{A_\alpha}(u)\}, \quad \forall u \in U. \tag{4.14}$$

Notice that $\mu_{A_\alpha}(u)$ is a membership function of a crisp set and assumes either 0 or 1, as in (4.1).

4.2.3 Cardinality, complement, measures of fuzziness

Let $U = \{u_1, \dots, u_n\}$ be a universal set. The **cardinality** of a fuzzy set A on U is

$$|A| = \sum_{i=1}^{n} \mu_A(u_i), \quad u_i \in U. \tag{4.15}$$

The **relative cardinality** of a fuzzy set A on U is

$$\| A \| = \frac{|A|}{|U|} = \frac{1}{n} \sum_{i=1}^{n} \mu_A(u_i), \quad u_i \in U. \tag{4.16}$$

The **complement** of a fuzzy set A on U ($\mu_{\bar{A}} : U \to [0,1]$), is a fuzzy set $\bar{A}$ on U, defined by the function $h : [0,1] \to [0,1]$ via the following set of axioms:

- (i): h is a function of one argument in [0,1], taking values in [0,1] ($\mu_{\bar{A}}(u)$ depends only on $\mu_A(u)$, $u \in U$).
- (ii): $h(0) = 1$ and $h(1) = 0$;
- (iii): h is continuous and strictly monotonically decreasing;
- (iv): h is involutive, i.e., $h(h(a)) = a, a \in [0,1]$.

These four axioms do not determine a unique h. Adding to this set

- (v): $a + b = 1, \;\; a, b \in [0,1] \iff h(a) + h(b) = 1$,

specifies a unique h as

$$h(a) = 1 - a, \quad \text{i.e.,} \quad \mu_{\bar{A}}(u) = 1 - \mu_A(u). \tag{4.17}$$

Equation (4.17) is the complement defined originally by Zadeh, and will be called, the **standard complement**. Sugeno defined another complement, called **λ-complement**, denoted by $\bar{A}^{\lambda}$,

$$\mu_{\bar{A}^{\lambda}}(u) = \frac{1 - \mu_A(u)}{1 + \lambda \mu_A(u)}, \quad \lambda \in (-1, \infty), \tag{4.18}$$

which satisfies axioms (i) to (iv).

The **closest crisp set** of a fuzzy set A on U is the (crisp) set A' such that

$$\mu_{A'}(u) = \begin{cases} 0, & \text{if } \mu_A(u) \le 0.5, \\ 1, & \text{otherwise,} \end{cases} \quad , u \in U. \tag{4.19}$$

A **sharpened version** of a fuzzy set A on U is any fuzzy set A^*, such that

$$\begin{cases} \mu_{A^*}(u) < \mu_A(u), & \text{if } \mu_A(u) < 0.5, \\ \mu_{A^*}(u) = \mu_A(u) = 0.5, & \text{if } \mu_A(u) = 0.5, \\ \mu_{A^*}(u) > \mu_A(u), & \text{if } \mu_A(u) > 0.5 \end{cases} \tag{4.20}$$

To find *how fuzzy* a fuzzy set A on U is, a **measure of fuzziness** $H(A)$ can be designed based on the following set of properties (cf. [259])

- *Sharpness P1.* $H(A)$ takes its minimal value if and only if $\mu_A(u) \in \{0, 1\}$ for all $u \in U$, i.e., if and only if A is a crisp subset of U.
- *Maximality P2.* $H(A)$ takes its maximal value if and only if $\mu_A(u) = 0.5$ for all $u \in U$, i.e., for the "fuzziest" set on U.
- *Resolution P3.* For any sharpened version A^* of A on U, $H(A) \geq H(A^*)$,
- *Symmetry P4.* $H(A) = H(\bar{A})$, where $\bar{A}$ is the complement of A.

Kaufmann [161] proposed two measures based respectively on the Hamming and Euclidean distance between A and its closest crisp set A'

$$H_{Hamming}(A) = \nu(A) = \frac{2}{n} d_H(A, A') = \frac{2}{n} \sum_{i=1}^{n} |\mu_A(u_i) - \mu_{A'}(u_i)| \tag{4.21}$$

and

$$\begin{aligned} H_{Eucludean}(A) = \eta(A) &= \frac{2}{\sqrt{n}} d_E(A, A') \\ &= 2\sqrt{\frac{1}{n} \sum_{i=1}^{n} (\mu_A(u_i) - \mu_{A'}(u_i))^2} \end{aligned} \tag{4.22}$$

Using the Shannon entropy, which is a measure of uncertainty and information formulated in terms of probability theory, De Luca and Termini [73] defined an entropy-based measure of fuzziness

$$H_{Entropy}(A) = -K \sum_{i=1}^{n} \left[\mu_A(u_i) \log(\mu_A(u_i)) + \mu_{\bar{A}}(u_i) \log(\mu_{\bar{A}}(u_i))\right], \tag{4.23}$$

where K is a scaling coefficient.

At many places in this chapter we consider only a finite U. The definitions can be generalized for infinite U, equipped with a measure (see [174]).

4.3 Operations on fuzzy sets

4.3.1 Intersections and unions, *t*-norms and *t*-conorms

Let A and B be fuzzy sets on $U = \{u_1, \ldots, u_n\}$. The **intersection** of A and B is a fuzzy set, $A \cap B$, defined as

$$\mu_{A\cap B}(u) = \min\{\mu_A(u), \mu_B(u)\}, \; \forall u \in U. \tag{4.24}$$

The **union** of A and B is a fuzzy set, $A \cup B$, defined as

$$\mu_{A\cup B}(u) = \max\{\mu_A(u), \mu_B(u)\}, \; \forall u \in U. \tag{4.25}$$

We can easily verify that the above definitions coincide with the conventional set theoretic intersection and union if the operands are crisp sets.

An example of two fuzzy sets on $U = \Re$ (displayed by their membership functions) and their intersection and union is shown in Figure 4.2. Handy algebraic expressions for the minimum and the maximum are

$$\min\{a,b\} = \frac{1}{2}\,\{a + b - |a - b|\} \tag{4.26}$$

and

$$\max\{a,b\} = \frac{1}{2}\,\{a + b + |a - b|\}. \tag{4.27}$$

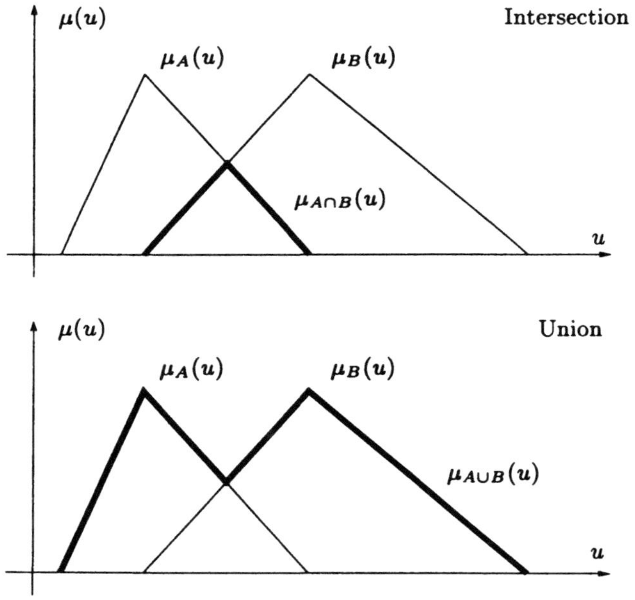

Fig. 4.2. Fuzzy sets: A, B, $A \cap B$, and $A \cup B$ (the classical "min" and "max" definitions)

Example 4.3.1. Recall example 4.2.2 about the holiday places. $U = \{$Sozopol, Ravda, Varna, Golden Sands, Duni, Kiten$\}$. Let again A be "a place of art" $A = \{$ (Sozopol, 0.9), (Ravda, 0.1), (Varna, 0.7), (Golden Sands, 0.6), (Duni, 0.0), (Kiten, 0.1)$\}$. Define a fuzzy set B on U, corresponding to the "beauty of the landscape". $B = \{$ (Sozopol, 0.7), (Ravda, 0.4), (Varna, 0.5), (Golden Sands, 0.6), (Duni, 0.9), (Kiten, 0.4)$\}$. Then the fuzzy set on U corresponding to "a place of arts *or* beauty of the landscape" is $A \cup B = \{$ (Sozopol, 0.9), (Ravda, 0.4), (Varna, 0.7), (Golden Sands, 0.6), (Duni, 0.9), (Kiten, 0.4)$\}$, and the fuzzy set on U corresponding to "a place of arts *and* beauty of the landscape" is $A \cap B = \{$ (Sozopol, 0.7), (Ravda, 0.1), (Varna, 0.5), (Golden Sands, 0.6), (Duni, 0.0), (Kiten, 0.1)$\}$. ■

It can be easily verified that each of the intersection and union operations is

1. **Commutative**: $A \cap B = B \cap A$ and $A \cup B = B \cup A$;
2. **Associative**: $A \cap (B \cap C) = (A \cap B) \cap C$ and $A \cup (B \cup C) = (A \cup B) \cup C$;
3. **Idempotent**: $A \cap A = A$ and $A \cup A = A$;

The two operations are mutually **distributive**

$$A \cap (B \cup C) = (A \cap B) \cup (A \cap C),$$

and

$$A \cup (B \cap C) = (A \cup B) \cap (A \cup C),$$

and also satisfy **De Morgan's Law**:

$$\overline{A \cap B} = \bar{A} \cup \bar{B},$$

and

$$\overline{A \cup B} = \bar{A} \cap \bar{B}.$$

What instigates a series of fruitless debates in the literature is the fact that in fuzzy logic, unlike in Boolean logic, the following properties generally *do not hold*:

The **law of excluded middle**

$$A \cup \bar{A} = U \tag{4.28}$$

and the **noncontradiction principle**

$$A \cap \bar{A} = \emptyset. \tag{4.29}$$

Instead of minimum and maximum, many other operations can be used on $\mu_A(u)$ and $\mu_B(u)$, for $u \in U$. If A and B are crisp subsets of U, these operations lead to the conventional intersection and union. The intersection-type operations on two fuzzy sets are implemented by ***t*-norms**, and the union-type operations, by **t-conorms**, sometimes called s-norms. These are two-place operations, i.e.,

$$t : [0,1] \times [0,1] \to [0,1]; \quad \text{and} \quad s : [0,1] \times [0,1] \to [0,1], \tag{4.30}$$

designed according to a set of axioms [356]. For the t-norms

1. **Commutativity**: $t(a,b) = t(b,a)$.
2. **Associativity**: $t(a,t(b,c)) = t(t(a,b),c)$.
3. **Monotonicity** on both arguments: if $a \geq c$ and $b \geq d$ then $t(a,b) \geq t(c,d)$.
4. **One identity**: $t(a,1) = a$.

The Minimum is a t-norm. It is the largest possible t-norm which comes from its **idempotency**: $t(a,a) = a$. If we add the idempotency as the fifth axiom, the only t-norm that satisfies all five axioms will be the minimum. The counterpart of the idempotency property is called the **archimedean** property [83], i.e.,

$$t(a,a) < a, \quad \forall a \in (0,1).^2 \tag{4.31}$$

Operations which satisfy the archimedean property are called **strict** operations. A similar set of axioms (only axiom 4 is different) is postulated for the t-conorms (s-norms)

1. **Commutativity**: $s(a,b) = s(b,a)$.
2. **Associativity**: $s(a,s(b,c)) = s(s(a,b),c)$.
3. **Monotonicity** on both arguments: if $a \geq c$ and $b \geq d$ then $s(a,b) \geq s(c,d)$.
4. **Zero identity**: $s(a,0) = a$.

Maximum is a t-conorm, and, besides, it is the smallest possible t-conorm because of its idempotency: $s(a,a) = a$. Again, if we add the idempotency as the fifth axiom, the only t-conorm that satisfies all five axioms will be the maximum. The archimedean property for the t-conorms is

$$s(a,a) > a, \quad \forall a \in (0,1). \tag{4.32}$$

Using the standard complement (4.17), a t-norm and a t-conorm are called **dual** if [83]

$$t(a,b) = 1 - s(1-a, 1-b), \tag{4.33}$$

which is identical to

$$s(a,b) = 1 - t(1-a, 1-b). \tag{4.34}$$

[2] Notice that a takes values in the *open* interval (0,1)

Generally, if h is a complement function satisfying axioms (i) to (iv) on page 84, t and s are h-dual if

$$t(a,b) = h[s(h(a), h(b))]. \tag{4.35}$$

The equivalent expression is obtained from (4.35) by

$$h(t(h(a), h(b)) = h(h(s(h(h(a)), h(h(b))), \tag{4.36}$$

and since h is involutive (axiom iv),

$$h(t(h(a), h(b)) = s(a,b). \tag{4.37}$$

Table 4.2 gives some basic t-norms and their dual t-conorms

Table 4.2. Three pairs of widely used t-norms and t-conorms

t-norm	t-conorm	Name
$\min\{a,b\}$	$\max\{a,b\}$	min/max
ab	$a+b-ab$	product/probabilistic sum
$\max\{0, a+b-1\}$	$\min\{1, a+b\}$	bounded difference/bounded sum
$\begin{cases} a, & \text{if } b=1, \\ b, & \text{if } a=1, \\ 0, & \text{otherwise} \end{cases}$	$\begin{cases} a, & \text{if } b=0, \\ b, & \text{if } a=0, \\ 1, & \text{otherwise} \end{cases}$	drastic product/drastic sum

Table 4.3 summarizes some parametric families of (dual) union (U) and intersection (I) type fuzzy operators. It is similar to that in [356], originally from [174], with a few changes and additions, cf. [83]. It The references are cited after [83].

Figure 4.3 shows the values of some intersection and union operations for $a = 0.4$ and $b = 0.7$.

Leaving aside the algebraic beauty of the various union and intersection operations, it is difficult to recommend any particular operation for practical purposes. For the parametric families, by varying the parameter, the operations can be made more or less "pessimistic", covering the whole range under the minimum (for the intersection) and above the maximum (for the union).

A class of fuzzy set operations which has raised debates in the fuzzy set literature is the **fuzzy implication** [32, 327]. Implication $A \rightarrow B$ (A implies B) is a necessary component of any if-then system to connect the antecedent with the consequent parts of the if-then rules. Fuzzy implication is defined over $[0,1] \times [0,1]$ and, unlike most intersection and union operators, it *does*

Table 4.3. Parametric families of t-norms and t-conorms

Author	Formula	U/I	Range
Schweizer & Sklar [299]	$1-\left(\max\{0,(1-a)^{-p}+(1-b)^{-p}-1\}\right)^{-\frac{1}{p}}$	U	$p \in \Re$
	$\left(\max\{0,a^{-p}+b^{-p}-1\}\right)^{-\frac{1}{p}}$	I	
Sugeno [312]	$\min\{1, a+b+\lambda ab\}$	U	$\lambda > -1$
	$\max\{0,(a+b-1)(1+\lambda)-\lambda ab\}$	I	
Hamacher [124]	$\dfrac{a+b-(2-\gamma)ab}{1-(1-\gamma)ab}$	U	$\gamma \in [0,\infty)$
	$\dfrac{ab}{\gamma+(1-\gamma)(a+b-ab)}$	I	
Frank [98]	$1-\log_S\left[1+\dfrac{(S^{1-a}-1)(S^{1-b}-1)}{S-1}\right]$	U	$S \in [0,\infty)$
	$\log_S\left[1+\dfrac{(S^{a}-1)(S^{b}-1)}{S-1}\right]$	I	
Yager [353]	$\min\left\{1,(a^w+b^w)^{\frac{1}{w}}\right\}$	U	$w \in [0,\infty)$
	$\max\left\{0,1-((1-a)^w+(1-b)^w)^{\frac{1}{w}}\right\}$	I	
Dubois & Prade [82]	$\dfrac{a+b-ab-\min\{a,b,1-\alpha\}}{\max\{1-a,1-b,\alpha\}}$	U	$\alpha \in [0,1]$
	$\dfrac{ab}{\max\{a,b,\alpha\}}$	I	

not always coincide with its nonfuzzy counterpart when both arguments are binary (True/False, as for crisp sets). Most frequently, minimum is used as the fuzzy implication. Compare the truth tables (Table 4.4) of the implication

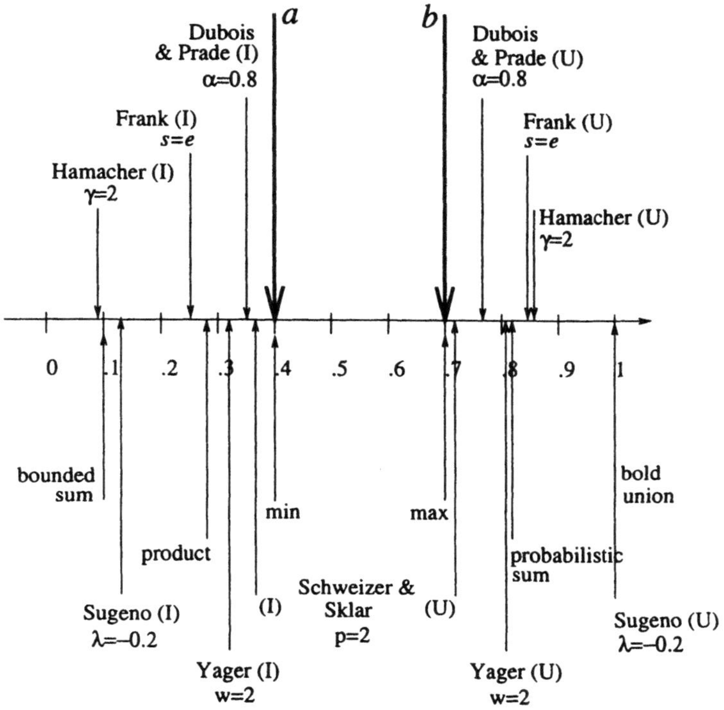

Fig. 4.3. Intersection and union operations for $a = 0.4$ and $b = 0.7$

as defined in Boolean logic, and the minimum operation used as substitute in fuzzy logic when the inputs are crisp, i.e., $a, b \in \{0, 1\}$. A set of fuzzy implication operations is given in Table 4.5, reproduced from [32]. The last column indicated whether the fuzzy implication verifies the Boolean implication truth table 4.4 . The fact that fuzzy and Boolean implications do not coincide is worth mentioning but it does not invalidate the use of minimum operation at the place of THEN. To avoid confusion, instead of "implication" we can talk about "association" [66].

Table 4.4. Truth tables for Boolean implication and minimum

a	b	implication $(a \to b)$	minimum $(\min\{a, b\})$
0	0	1	0
0	1	1	0
1	0	0	0
1	1	1	1

Table 4.5. Fuzzy implication operators on $a, b \in [0, 1]$

Implication operator	Formal expression	Boolean
Mamdani	$\min\{a, b\}$	N
Larsen	$a.b$	N
Lukasiewicz	$\min\{1, 1-a+b\}$	Y
Kleen-Dienes	$\max\{1-a, b\}$	Y
Bounded product	$\max\{0, a+b-1\}$	N
Zadeh	$\max\{\min\{a, b\}, 1-a\}$	Y
Standard	1, if $a \le b$, 0, if $a > b$	Y
Drastic product	(see Table 4.2)	N
Gougen	1, if $a \le b$, b/a, if $a > b$	Y
Godelian	1, if $a \le b$, b, if $a > b$	Y

Note that fuzzy implications, t-norms and t-conorms between two fuzzy sets A and B on U are defined on $[0,1] \times [0,1]$, i.e., on a pair of degrees of membership $\mu_A(u)$ and $\mu_B(u)$. Therefore, to obtain the resultant fuzzy set, the operation is applied to $\mu_A(u)$ and $\mu_B(u)$ for every $u \in U$.

4.3.2 Aggregation operations

Fuzzy intersections and unions are alternatives of the set-theoretic operations for crisp sets. Being a richer model, however, fuzzy sets can be combined by other formulas, which, together with the set-theoretic operations, are called **fuzzy aggregation connectives** [44, 83]. Intersection and union are not good enough for problems where the fuzzy sets represent properties which can compensate for each other.

Example 4.3.2. In the holiday example 4.3.1, there can be a place that is a good compromise between the two criteria expressed by the fuzzy sets A and B. A weight 0.6 can be assigned to the first criterion: art, and 0.4 to the second criterion: beauty of the landscape. Then the compromise can be found by designing a fuzzy set C using $\mu_C(u) = 0.6\mu_A(u) + 0.4\mu_B(u)$, i.e.,

$$C = \{(\text{Sozopol}, 0.82), (\text{Ravda}, 0.22), (\text{Varna}, 0.62), (\text{Golden Sands}, 0.60),$$
$$(\text{Duni}, 0.36), (\text{Kiten}, 0.22)\}.$$

■

These compensatory connectives are not needed in classical set theory [83], and therefore do not exist for crisp sets.

Mean and median operations. Mean and median operations fill in the space between the minimum and the maximum (e.g., between a and b in Figure 4.3). For now we will consider only two-place operations $m : [0,1] \times [0,1] \to [0,1]$, based on the following set of axioms

1. **Commutativity**: $m(a,b) = m(b,a)$.
2. **Monotonicity and continuity** on both arguments.
3. **Range**: $\min\{a,b\} \le m(a,b) \le \max\{a,b\}$, $m \ne \min$, $m \ne \max$.

From axiom 3 it follows that mean operators are **idempotent**, i.e., $m(a,a) = a$. Dubois and Prade [83] assert that idempotency and associativity are seldom consistent, and define the **median** as the only mean operation that is associative (and idempotent by definition). For $a, b \in [0,1]$, $a \le b$,

$$\mathrm{med}_\alpha(a,b) = \begin{cases} b, & \text{if } a \le b \le \alpha, \\ \alpha, & \text{if } a \le \alpha \le b, \\ a, & \text{if } \alpha \le a \le b. \end{cases} \tag{4.38}$$

A class of mean operations that encompasses a number of traditionally known means is [83]

$$m_\alpha(a,b) = \left(\frac{1}{2}(a^\alpha + b^\alpha)\right)^{\frac{1}{\alpha}}. \tag{4.39}$$

Classical operations are retrieved from m_α for some values of α, as shown in Table 4.6[3].

The non-parametric two-place operations on a and b have the following order

$$\begin{aligned}
&\text{drastic product}(a,b) \le \\
&\max\{0, a+b-1\} \le \\
&ab \le \\
&\min\{a,b\} \le \\
&\frac{2ab}{a+b} \le \\
&\sqrt{(ab)} \le \\
&\frac{a+b}{2} \le \\
&\max\{a,b\} \le \\
&a+b-ab \le \\
&\min\{1, a+b\} \le \\
&\ \text{drastic sum}(a,b).
\end{aligned} \tag{4.40}$$

[3] We note that, although retrieved from (4.39), the minimum and the maximum do not belong to the mean family by definition.

Table 4.6. Operations retrieved from (4.39)

α	$m_\alpha(a,b)$	Name
$-\infty$	$\min\{a,b\}$	minimum
-1	$\frac{2ab}{a+b}$,	harmonic mean
0	$\sqrt{ab}$	geometric mean
1	$\frac{a+b}{2}$,	arithmetic mean
∞	$\max\{a,b\}$	maximum

The parametric families of operations cannot be ordered because each family covers a range of values.

Bloch uses the name **constant behavior operators** [44] for operations of either of the below types

- **Conjunctive**, whose results is always no greater than the minimum, e.g., the t-norms: $t(a,b) \leq \min\{a,b\}$;
- **Disjunctive**, whose result is always no smaller than the maximum, e.g, the t-conorms: $s(a,b) \geq \max\{a,b\}$;
- **Compromise** , whose result is between the minimum and the maximum, e.g, the means: $\min\{a,b\} \leq m(a,b) \leq \max\{a,b\}$,

irrespectively of what their arguments are. The aggregation operations introduced so far have constant behavior. If we drop the limit condition restricting the mean values between $\min\{a,b\}$ and $\max\{a,b\}$ we can open up space for the **variable behavior** operations [44]. An example of such operations are the **symmetric sums** $ss(a,b) : [0,1] \times [0,1] \to [0,1]$ defined by the following set of axioms

1. **Commutativity.**
2. **Monotonicity and continuity** on each argument.
3. **Limit conditions**: $ss(0,0) = 0, ss(1,1) = 1$.
4. **Auto-duality** with respect to the standard complement (4.17):

$$1 - ss(a,b) = ss(1-a, 1-b)$$

Symmetric sums can exhibit conjunctive, disjunctive and compromise behavior depending on the values of their arguments.

A general equation that can be used to construct symmetric sums is [44, 83]

$$ss(a,b) = \frac{g(a,b)}{g(a,b) + g(1-a, 1-b)}, \tag{4.41}$$

where g is called a *generator function*. It is continuous, increasing, positive definite, and satisfying $g(0,0) = 0$. Examples of symmetric sums with their generator functions are given in Table 4.7. Constant-behavior and variable-behavior operations will be revisited further on in this chapter.

Table 4.7. Examples of symmetric sums

$ss(a,b)$	Generator $g(a,b)$	Comment
$ss_{\min}(a,b) = \dfrac{\min\{a,b\}}{1-\lvert a-b\rvert}$	$g(a,b) = \min\{a,b\}$	mean-type
$ss_0(a,b) = \dfrac{ab}{1-a-b+2ab}$	$g(a,b) = ab$	associative
$ss_+(a,b) = \dfrac{a+b-ab}{1+a+b-2ab}$	$g(a,b) = a+b-ab$	non-associative
$ss_{\max}(a,b) = \dfrac{\max\{a,b\}}{1+\lvert a-b\rvert}$	$g(a,b) = \max\{a,b\}$	mean-type

4.3.3 Aggregation of more than two fuzzy sets

Up to now we have considered only two-place operations, i.e., operations on *two* fuzzy sets A and B on U. More often, there are a greater number of fuzzy sets to be aggregated.

Example 4.3.3. Let $U = \{u_1, \ldots, u_n\}$ be the set of participants in the semi-final round of a beauty contest. The jury consists of L experts, $E_1, \ldots, E_L$, who have to select the finalists. Assume that there is no limit on the number of finalists out of n. Each u_i should be put in one of the two classes $\Omega = \{pass,\ drop\}$. Each member of the jury expresses their support for u_i going to the final round (*pass*) in a scale from 0 to 10. The two classes are mutually exclusive, therefore we could assume the following model: if the degree is above 5, u_i is labeled in class *pass*. Let A_j be a fuzzy set on U corresponding to the opinion of expert E_j, $j \in \{1, \ldots, L\}$, with membership function for class *pass*

$$\mu_{A_j}(u_i) = \frac{E_j\text{'s support for candidate } u_i}{10}.$$

Then for each candidate u_i there will be a set of L degrees of membership $\{\mu_{A_1}(u_i), \ldots, \mu_{A_L}(u_i)\}$. To find the final degree of membership we need to aggregate all L values, so the two-place operations considered so far are insufficient. ■

An L-**place aggregation** operation $\mathcal{A}$ is defined as

$$\mathcal{A} : [0,1]^L \to [0,1], \text{ i.e., } \mathcal{A}(a_1, \ldots, a_L) \in [0,1], \quad a_i \in [0,1]. \tag{4.42}$$

A natural set of properties for L-place aggregation operations is encoded in the following set of axioms

1. **Commutativity**: $\mathcal{A}(a_1, \ldots, a_L) = \mathcal{A}(a_{i_1}, \ldots, a_{i_L})$, for any permutation $i_1, \ldots, i_L$ of $1, \ldots, L$.
2. **Monotonicity** on each argument:

$$b_i \geq a_i,\ i = 1, \ldots, L \ \Rightarrow \ \mathcal{A}(b_1, \ldots, b_L) \geq \mathcal{A}(a_1, \ldots, a_L).$$

3. **Limit conditions**: $\mathcal{A}(0, \ldots, 0) = 0$ and $\mathcal{A}(1, \ldots, 1) = 1$

As with the two-place operations, $\mathcal{A}$ can be characterized as conjunctive, disjunctive or compromise, i.e., for any $a_1, \ldots, a_L,\ a_i \in [0,1]$,

- Conjunctive: $\mathcal{A}(a_1, \ldots, a_L) \leq \min\{a_1, \ldots, a_L\}$;
- Disjunctive: $\mathcal{A}(a_1, \ldots, a_L) \geq \max\{a_1, \ldots, a_L\}$;
- Compromise: $\min\{a_1, \ldots, a_L\} \leq \mathcal{A}(a_1, \ldots, a_L) \leq \max\{a_1, \ldots, a_L\}$;

Not all of the 2-place operations considered so far can be extended straightforwardly for the L-dimensional case specified by the three axioms. Associativity is sufficient for such an extention to be possible [83]. Below is a list of some simple L-place aggregation operations

- Minimum

$$\mathcal{A}(a_1, \ldots, a_L) = \min\{a_1, \ldots, a_L\}; \tag{4.43}$$

- Maximum

$$\mathcal{A}(a_1, \ldots, a_L) = \max\{a_1, \ldots, a_L\}; \tag{4.44}$$

- Product

$$\mathcal{A}(a_1, \ldots, a_L) = \prod_{i=1}^{L} a_i; \tag{4.45}$$

- Average

$$\mathcal{A}(a_1, \ldots, a_L) = \frac{1}{L}\sum_{i=1}^{L} a_i; \tag{4.46}$$

- Generalized mean

$$\mathcal{A}_\alpha(a_1,\ldots,a_L) = \left(\frac{1}{L}\sum_{i=1}^{L} a_i^\alpha\right)^{\frac{1}{\alpha}} ; \tag{4.47}$$

with the following special cases (see Table 4.6)[4]

$$\alpha \to -\infty \;\Rightarrow\; \mathcal{A}_\alpha = \min_i\{a_i\} \tag{4.48}$$

$$\alpha = -1 \;\Rightarrow\; \mathcal{A}_\alpha = \left(\frac{1}{L}\sum_{i=1}^{L}\frac{1}{a_i}\right)^{-1} \quad \text{(harmonic mean)} \tag{4.49}$$

$$\alpha = 0 \;\Rightarrow\; \mathcal{A}_\alpha = (a_1\ldots a_L)^{1/L} \quad \text{(geometric mean)} \tag{4.50}$$

$$\alpha = 1 \;\Rightarrow\; \mathcal{A}_\alpha = \frac{1}{L}\sum_{i=1}^{L}\frac{1}{a_i} \quad \text{(arithmetic mean)} \tag{4.51}$$

$$\alpha \to \infty \;\Rightarrow\; \mathcal{A}_\alpha = \max_i\{a_i\} \tag{4.52}$$

4.3.4 Ordered weighted averaging (OWA)

An interesting class of parametric mean connectives are Yager's **Ordered Weighted Averaging (OWA)** operations [355, 357].

Let $\mathbf{b} = [b_1,\ldots,b_L]^T \in [0,1]^L$ be a vector with coefficients and

$$\sum_{k=1}^{L} b_k = 1.$$

The aggregation operation is implemented as the dot product of $\mathbf{b}$ and the vector $[a_{i_1},\ldots,a_{i_L}]^T$, where $i_1,\ldots,i_L$ is a permutation of the indices $1,\ldots,L$, such that $a_{i_1} \geq a_{i_2} \geq \ldots \geq a_{i_L}$. That is,

$$\mathcal{A}_{\mathbf{b}}^{OWA}(a_1,\ldots,a_L) = \sum_{k=1}^{L} a_{i_k} b_k. \tag{4.53}$$

It can be verified that OWA operators are commutative, monotonic and idempotent.

Example 4.3.4. When a jury has to judge a sport performance (e.g., in gymnastics, acrobatics, ice-skating), to avoid, or at least reduce, subjective bias, usually the highest and the lowest marks are dropped, and the remaining $L-2$ marks are averaged. Now assume that in example 4.3.3, the participant u in the beauty contest has obtained the marks shown in Table 4.8 from the $L = 5$ members of the jury.

[4] Note that the min and the max operators should not be classed as *means* because of Axiom 3 on page 93

Table 4.8. Marks for competitor u from the 5 members of the jury

#	Member	Mark
1	The photographer of a popular magazine	6
2	The chief-manager of the model agency which u has a contract with	7
3	This year's world champion in aerobics – boyfriend of one of u's rivals from U	2
4	The Chairman of the contest	6
5	The last year winner of the beauty contest	6

The degrees of membership assigned by the experts are $[.6, .7, .2, .6, .6]^T$. To implement the **competition jury** model, we use OWA aggregation with $\mathbf{b} = [0, 1/3, 1/3, 1/3, 0]^T$. This yields

$$\mathcal{A}_{\mathbf{b}}^{OWA}(0.6, 0.7, 0.2, 0.6, 0.6) =$$

$$\left[0, \frac{1}{3}, \frac{1}{3}, \frac{1}{3}, 0\right] [0.7, 0.6, 0.6, 0.6, 0.2]^T = (0.6 + 0.6 + 0.6)/3 = 0.6,$$

which looks a more realistic overall mark than the average 0.54. ■

By selecting a specific $\mathbf{b}$, a number of operations can be modeled, and further operations can be created

- Minimum: $\mathbf{b} = [0, 0, \ldots, 0, 1]^T$.
- Maximum: $\mathbf{b} = [1, 0, \ldots, 0, 0]^T$.
- Average: $\mathbf{b} = [1/L, 1/L, \ldots, 1/L]^T$.
- Competition jury: $\mathbf{b} = [0, 1/(L-2), \ldots, 1/(L-2), 0]^T$.

The coefficient vector $\mathbf{b}$ can be either designed in advance or found algorithmically from data. Yager and Filev [356] show how OWA coefficient vector can be designed to model linguistic quantifiers such as *almost all, few, many, most, nearly half*, etc.

4.3.5 Fuzzy integral

The fuzzy integral can be used as an aggregation connective [112, 164] and will be introduced here to the extent that is needed for this interpretation.

Let $E = \{E_1, \ldots, E_L\}$ be a crisp set. Three useful pattern recognition interpretations of E are

- E is a set of "experts" as in the competition examples.
- E is a set of features, or a set of feature subsets [115, 118, 117, 111, 113, 116, 114]. We assume that each element of E is used to calculate a degree of membership for an object $u \in U$ with respect to a fixed class $\omega_i \in \Omega, i = 1, \ldots, c$. For the moment we will not discuss the way these degrees

are obtained and how they are related to each other. Here we are interested in how we can *combine* these degrees to obtain a final (aggregated) value showing the support for the hypothesis u comes from class ω_i.

- E is a set of classifier outputs [62, 63, 102, 201, 335]. Practically the interpretation is the same as in the previous item. The difference is that the E_js are not necessarily sets of features, but can be any classifiers designed on the same or on different feature sets.

Let $\mathcal{P}(E)$ be the power set of E. A **fuzzy measure** on E is the set function[5] g.

$$g : \mathcal{P}(E) \rightarrow [0, 1], \tag{4.54}$$

such that

1. $g(\emptyset) = 0, \quad g(E) = 1;$
2. For any A and B, crisp subsets of E, $A \subset B \Rightarrow g(A) \leq g(B)$.

The function g is a **probabilistic measure** if the second property is replaced by the stronger requirement: For any A and B, crisp subsets of E, such that $A \cap B = \emptyset$,

$$g(A \cup B) = g(A) + g(B). \tag{4.55}$$

g is called a **λ-fuzzy measure** if for any A and B, crisp subsets of E, such that $A \cap B = \emptyset$,

$$g(A \cup B) = g(A) + g(B) + \lambda g(A)g(B), \quad \lambda \in (-1, \infty). \tag{4.56}$$

Various fuzzy measures g can be derived using t-conorms. Since t-conorms are monotonic by definition, for any pair of disjoint sets A and B, crisp subsets of E, we can define a fuzzy measure g by

$$g(A \cup B) = s(g(A), g(B)). \tag{4.57}$$

This class of fuzzy measures are called **s-decomposable** fuzzy measures [164].

From associativity of the t-conorms, it follows that, for any A, a crisp subset of E,

$$g(A) = \mathcal{S}_{E_j \in A} g(\{E_j\}), \tag{4.58}$$

where $\mathcal{S}$ is the L-place extension of the t-conorm in (4.57). Similar formulas can be derived using t-norms [356].

To calculate g by (4.58) it is enough to know the values of the measure for the individual elements of E. These values, $g(\{E_1\}), \ldots, g(\{E_L\})$ are called

---

[5] g is a common notation for a fuzzy measure. Although g was also used for the discriminant functions in Chapter 2, the two concepts should not be confused.

fuzzy densities, and are denoted by $g^1, \ldots, g^L$. We often have some estimates of $g^1, \ldots, g^L$, $g^j \in [0,1]$. We can find a λ-fuzzy measure (4.56) which is consistent with these densities. The value of λ is obtained as the unique root greater than -1 of the polynomial

$$\lambda + 1 = \prod_{j=1}^{L}(1 + \lambda g^j), \quad \lambda \neq 0. \tag{4.59}$$

Let $A = \{E_{i_1}, \ldots, E_{i_m}\}$ be a (crisp) subset of E, $\{i_1, \ldots, i_m\} \subset \{1, \ldots, L\}$. We form a sequence of nested sets $A_1, \ldots, A_m$, starting from $A_1 = \{E_{i_1}\}$, and adding subsequently the elements E_{i_2} to E_{i_m} , one at a time (then $A_m = A$). The measure $g(A)$ is calculated through the recursive formula

$$\begin{aligned} & g(A_1) = g^{i_1}, \\ \text{For } k = 2, \ldots, m, \quad & g(A_k) = g^{i_k} + g(A_{k-1}) + \lambda\, g^{i_k}\, g(A_{k-1}), \\ & g(A) = g(A_m). \end{aligned} \tag{4.60}$$

Example 4.3.5. Assume the fuzzy densities expressing the unbiasedness of the 5 members of the jury in example 4.3.4 are $[0.7,\ 0.5,\ 0.4,\ 0.7,\ 0.8]^T$. Solving equation (4.59) with these densities, we get $\lambda = -0.9943$. ■

Having defined the fuzzy measure g, we can now define **fuzzy integral**. Let H be a fuzzy set on E. We are looking for *one* representative value of μ_H, showing how *all* elements of E comply with the characteristic H and taking into account the *importance* of each element. To simplify the notations, we shall use $a_i = \mu_H(E_i)$.

Two basic types of fuzzy integrals have been proposed. The **Sugeno fuzzy integral** with respect to a fuzzy measure g is obtained by

$$\mathcal{A}_g^{FI}(a_1, \ldots, a_L) = \sup_{\alpha}\{t(\alpha, g(H_\alpha))\}, \tag{4.61}$$

where H_α is the α-cut of H, and t is a t-norm. In the original formula by Sugeno, the t-norm was the *minimum*. Keller et al. [164] point out that to use (4.61), the t-norm should be mutually distributive with respect to the *maximum*. Examples of such t-norms are minimum, product, bounded difference, and drastic product (see Table 4.2).

Since E is finite, H has at most L different α-cuts, ranging from $H_0 = E$ to $H_{height(H)}$ containing only the element(s) for which μ_H reaches its maximum. Let us arrange the elements of E so that $a_{i_1} \geq a_{i_2} \geq \ldots \geq a_{i_L}$. Let the sequence of nested subsets, as explained earlier, be denoted by $A_1 = \{E_{i_1}\}, A_2 = \{E_{i_1}, E_{i_2}\}, \ldots, A_L = E$. Thus, each $A_j \subseteq E, a_{i_j} > a_{i_{j+1}}$ is the a_{i_j}-cut of H. Then (4.61) becomes

$$\mathcal{A}_g^{FI}(a_1, \ldots, a_L) = \max_{j=1}^{L}\left\{t(a_{i_j}, g(A_j))\right\}, \tag{4.62}$$

which is computationally simpler than (4.61) because $g(A_j)$ can be found through (4.60).

The second type fuzzy integral is the **Choquet fuzzy integral**, calculated by

$$\mathcal{A}_g^{FI}(a_1, \ldots, a_L) = a_{i_L} + \sum_{j=2}^{L} \left(a_{i_{j-1}} - a_{i_j}\right) g(A_{j-1}). \qquad (4.63)$$

Example 4.3.6. Using the jury votes for u (example 4.3.4) and the fuzzy densities (example 4.3.5), we shall calculate the aggregated value $\mathcal{A}_g^{FI}$ according to the Sugeno fuzzy integral (4.61). First, by arranging the marks, and the set with densities correspondingly, we obtain:

$$\begin{array}{l}[0.6, 0.7, 0.2, 0.6, 0.6]^T \rightarrow [0.7, 0.6, 0.6, 0.6, 0.2]^T \text{ (marks)} \\ [0.7, 0.5, 0.4, 0.7, 0.8]^T \rightarrow [0.5, 0.7, 0.7, 0.8, 0.4]^T \text{ (densities)}\text{'}\end{array}$$

i.e., the new arrangement of the jury members is $\{E_2, E_1, E_4, E_5, E_3\}$. Following the recursive procedure (4.60),

$$\begin{array}{ll} g(A_1) & = 0.5 \\ g(A_2) = 0.7 + 0.5 - 0.9943\,(0.7)\,0.5 & = 0.8520 \\ g(A_3) = 0.7 + 0.8520 - 0.9943\,(0.7)\,0.8520 & = 0.9590 \\ g(A_4) = 0.8 + 0.9590 - 0.9943\,(0.8)\,0.9590 & = 0.9962 \\ g(A_5) = 0.4 + 0.9962 - 0.9943\,(0.4)\,0.9962 & = 1.0 \end{array}$$

Juxtaposing g and the sorted marks, and taking the minimum as the t-norm in (4.62), the following vector is obtained

$$\min \left\{ \begin{array}{l} [0.7,\ 0.6,\quad 0.6,\quad 0.6,\quad 0.2]^T \\ [0.5,\ 0.8250,\ 0.9590,\ 0.9962,\ 1.0]^T \end{array} \right. \rightarrow [0.5, 0.6, 0.6, 0.6, 0.2]^T.$$

The maximum is 0.6, which is the aggregated value $\mathcal{A}_g^{FI}$ by the Sugeno fuzzy integral.

For the Choquet integral, using formula (4.63), we obtain

$$\begin{array}{l} \mathcal{A}_g^{FI} = 0.2 + \\ (0.7 - 0.6)\,0.5 + (0)\,0.8520 + (0)\,0.9590 + (0.6 - 0.2)\,0.9962 \\ = 0.6485 \end{array}$$

■

Grabisch [112] studies the connection between various aggregation operations and fuzzy integrals.

4.3.6 Using consensus in fuzzy aggregation

Consider again the holiday planning example 4.3.1. If a place has high degrees of membership on both criteria (A and B), this should be a good candidate, so the resultant degree of ***desirability*** of the place can be even higher than the maximum of the two. Thus, the value of Golden Sands can go up from 0.6 to something else, e.g., 0.7. In such situations the aggregation operator should have a disjunctive behavior. Conversely, for places where both degrees are low, the overall value should be even smaller than the smaller of the two – there is no attraction on either of the criteria. Therefore, a proper aggregation operator should have a conjunctive behavior. For places which have very different degrees of membership in A and B, we have no reason to either strengthen or weaken the aggregated value. This logic demands a new operation, which, using the terminology introduced in [44], is of **variable behavior**. We shall use in the aggregation formula an external parameter measuring the degree of **consensus** between the values.

Let $a_1, \ldots, a_L$, $a_i \in [0,1]$ be the set of values to be aggregated. Various consensus-based aggregation operations are developed in [192, 190, 204]. We assess consensus (general agreement on an opinion) gradually within the interval [0,1], with 0, meaning total dissensus, and 1, meaning unanimity. Five **measures of consensus** $\gamma_1, \ldots, \gamma_5$ are proposed in [187]

- Highest coincidence

$$\gamma_1 = 1 - \min_{i,j,i\neq j} \{|a_i - a_j|\}. \tag{4.64}$$

- Highest discrepancy

$$\gamma_2 = 1 - \max_{i,j,i\neq j} \{|a_i - a_j|\}. \tag{4.65}$$

- Integral mean coincidence. Let

$$\bar{a} = \frac{1}{L}\sum_{i=1}^{L} a_i.$$

The integral mean coincidence is defined as

$$\gamma_3 = 1 - \frac{1}{L}\sum_{i=1}^{L} |a_i - \bar{a}|. \tag{4.66}$$

- Integral pairwise coincidence

$$\gamma_4 = 1 - \frac{2}{L(L-1)}\sum_{i=1}^{L-1}\sum_{j=i+1}^{L} \{|a_i - a_j|\}. \tag{4.67}$$

- Integral highest discrepancy

$$\gamma_5 = 1 - \max_{i=1}^{L}\{|a_i - \bar{a}|\}. \tag{4.68}$$

Example 4.3.7. In the competition jury example, we have the following values to be aggregated: $\{0.6, 0.7, 0.2, 0.6, 0.6\}$. For this set:

$\gamma_1 = 1 - |0.6 - 0.6| = 1;$
$\gamma_2 = 1 - |0.7 - 0.2| = 0.5;$
$\gamma_3 = 1 - 1/5(0.06 + 0.16 + 0.34 + 0.06 + 0.06) = 0.864 \quad (\bar{a} = 0.54);$
$\gamma_4 = 1 - 2/20(0.1 + 0.4 + 0.5 + 0.1 + 0.5 + 0.4 + 0.4) = 1 - 0.240 = 0.760;$
$\gamma_5 = 1 - 0.34 = 0.66.$

■

For $L = 2$, $\gamma_1 = \gamma_2 = \gamma_3 = \gamma_4 = \gamma_5 = 1 - |a_1 - a_2|$. These measures can be used for aggregating expert opinions expressed as numbers in the interval [0,1], cf. [27].

Let $\mathcal{A}(a_1, \ldots, a_L)$ be the aggregated value by some aggregation connective $\mathcal{A}$. The **consensus operation** $\mathcal{K}$ should depend on $a_1, \ldots, a_L$ through the aggregated value, and on the consensus γ between the individual values, i.e.,

$$\mathcal{K}(a_1, \ldots, a_L) = \mathcal{K}(\mathcal{A}(a_1, \ldots, a_L), \gamma(a_1, \ldots, a_L)). \tag{4.69}$$

To simplify notation we shall write $\mathcal{K} = \mathcal{K}(\mathcal{A}, \gamma)$. We define $\mathcal{K}$ by a set of axioms [204]

- K1: **Commutativity** on $a_1, , \ldots, a_L$. This is satisfied iff both $\mathcal{A}$ and γ are commutative.
- K2: **Selective monotonicity** on the degree of consensus γ.

$$\mathcal{K}(\mathcal{A}, \gamma) \quad \text{is} \quad \begin{cases} \text{monotonically nondecreasing on } \gamma, & \text{if } \mathcal{A} > 0.5 \\ \text{monotonically nonincreasing on } \gamma, & \text{if } \mathcal{A} \leq 0.5 \end{cases} \tag{4.70}$$

One of the following two alternative properties can be added to this set:

- K3a: **Unanimity**.

$$\mathcal{K}(\mathcal{A}, 1) = \mathcal{A}. \tag{4.71}$$

- K3b: **Strengthened unanimity**.

$$\mathcal{K}(\mathcal{A}, 1) \quad \text{is} \quad \begin{cases} > \mathcal{A}, & \text{if } \mathcal{A} \in (0.5, 1) \\ < \mathcal{A}, & \text{if } \mathcal{A} \in (0, 0.5) \\ = \mathcal{A} = 0.5, & \text{if } \mathcal{A} = 0.5. \end{cases} \tag{4.72}$$

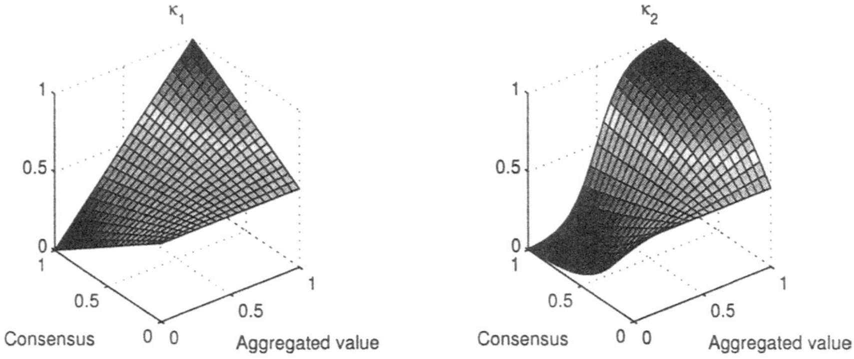

Fig. 4.4. Graph of the two consensus operations $\mathcal{K}_1$ and $\mathcal{K}_2$ (with $\alpha = 10$)

A consensus operation that satisfies K1, K2 and K3a is [204]

$$\mathcal{K}_1(\mathcal{A}, \gamma) = 0.5(1-\gamma) + \mathcal{A}\,\gamma, \tag{4.73}$$

and a consensus operation that satisfies K1, K2 and K3b is, e.g.,

$$\mathcal{K}_2(\mathcal{A}, \gamma) = \frac{1}{1+\exp\{-\alpha\gamma(\mathcal{A}-0.5)\}}, \tag{4.74}$$

where $\alpha > 0$ is a scaling constant.

Figure 4.4 shows the two surfaces corresponding to the consensus operations with respect to $\mathcal{A}$ and γ. If the consensus γ is 0, the aggregated value is always 0.5, regardless of $\mathcal{A}$. This reflects the idea that if there is no consensus between the values (expert opinions), the *fuzziest* value 0.5 is retrieved. Interestingly, even if the aggregated value $\mathcal{A}$ is zero, the consensus operator can return a value between 0 and 0.5. A non-zero value of $\mathcal{K}$ when $\mathcal{A} = 0$ will indicate that there have been disagreements in the pool $a_1, \ldots, a_L$. In the case of complete agreement ($\gamma = 1$), $\mathcal{K}$ either retrieves $\mathcal{A}$ ($\mathcal{K}_1$, on the lefthand-side graph) or exhibits the strengthened unanimity K3b ($\mathcal{K}_2$, on the righthand-side graph). There are two constant lines on both graphs – for $\mathcal{A} = 0.5, \mathcal{K}_1 = \mathcal{K}_2 = 0.5$, regardless of γ; and for $\gamma = 0$, $\mathcal{K}_1 = \mathcal{K}_2 = 0.5$, regardless of $\mathcal{A}$.

Example 4.3.8. Why do we need the consensus operations? Think of the consensus operators as a support for labeling an object $\mathbf{x}$ in a class $\omega \in \Omega$. Let $\mathcal{A}$ be the support for ω, calculated through one of the aggregation connectives. We assume that there are two fixed thresholds T_1 and T_2 in [0,1]. If the value of support for ω is higher than T_1, we assign label ω to $\mathbf{x}$. If the value of support is smaller than T_2, we conclude that $\mathbf{x}$ is definitely not in ω, and if the value is in-between, our classifier refuses to decide (see Chapter 2, the minimum risk classifier). If consensus is not taken into account, we assign the class label according to the value of $\mathcal{A}$.

Let $\mathcal{A}$ be the minimum aggregation, and let $T_1 = 0.6, T_2 = 0.3$ be the values of the thresholds. Assume that a_1 and a_2 express the support given by two classifiers for the hypothesis that $\mathbf{x}$ comes from $\omega \in \Omega$. The values of $\mathcal{A}$ (minimum) are plotted in Figure 4.5 (on the left). The right-hand-side plot in Figure 4.5 is the surface of $\mathcal{K}_1$ for $\gamma = \gamma_1 = \ldots = \gamma_5 = 1 - |a_1 - a_2|$.

Figure 4.6 shows the classification regions on the unit square $[0,1]^2$ (the feature space of a_1 and a_2). The light grey regions are obtained when using only $\mathcal{A}$ for classification, and the dark grey regions are obtained with $\mathcal{K}_1$. We see that $\mathcal{K}_1$ offers a more "conservative" decision than $\mathcal{A}$, and this decision depends on the agreement between a_1 and a_2. If we used $\mathcal{K}_2$, the regions for acceptance and rejection would be more "generous", shaped again as the darker regions in Figure 4.6. ∎

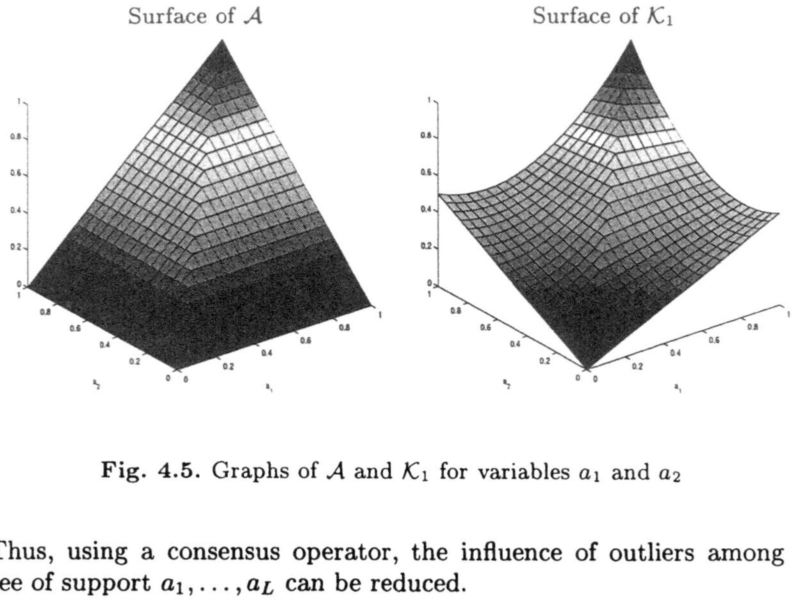

Fig. 4.5. Graphs of $\mathcal{A}$ and $\mathcal{K}_1$ for variables a_1 and a_2

Thus, using a consensus operator, the influence of outliers among the degree of support $a_1, \ldots, a_L$ can be reduced.

4.3.7 Equivalence, inclusion, similarity, and consistency

Fuzzy sets A and B on U are **equivalent** if and only if $\mu_A(u) = \mu_B(u)$, for all $u \in U$.

A is **included** in B, denoted $A \subseteq B$, if and only if $\mu_A(u) \leq \mu_B(u)$, for all $u \in U$.

A is **strictly included** in B if and only if $\mu_A(u) < \mu_B(u)$, for all $u \in U$.

Similarity is a central concept in pattern recognition. In designing fuzzy classifiers we will need to compare fuzzy sets, e.g., corresponding to an object

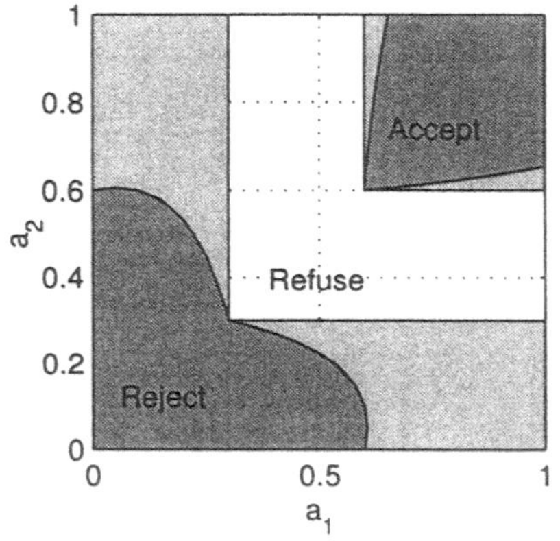

Fig. 4.6. Accept-Refuse-Reject regions in the feature space $[0,1]^2$ spanned by a_1 and a_2, and thresholds $T_1 = 0.6, T_2 = 0.3$. Light grey regions correspond to $\mathcal{A}$, and dark gray regions, to $\mathcal{K}_1$.

to be classified and the prototype of the class. A set of useful **measures of similarity** are listed in [81]

$$S_1(A,B) \equiv \frac{\| A \cap B \|}{\| A \cup B \|}, \tag{4.75}$$

where $\| \cdot \|$ is the relative cardinality (4.16). S_1 can be extended for more than two fuzzy sets, e.g.,

$$S_1(A_1,\ldots,A_L) \equiv \frac{\| \bigcap_{i=1}^{L} A_i \|}{\| \bigcup_{i=1}^{L} A_i \|}. \tag{4.76}$$

$$S_2(A,B) \equiv 1 - \| A \nabla B \|, \tag{4.77}$$

where $A \nabla B$ is the symmetric difference

$$\mu_{A\nabla B}(u) = |\mu_A(u) - \mu_B(u)|, \quad u \in U. \tag{4.78}$$

$$S_3(A,B) \equiv 1 - \| A \Delta B \|, \tag{4.79}$$

where

$$\mu_{A\Delta B}(u) = \max\{\mu_{A\cap\bar{B}}(u), \mu_{\bar{A}\cap B}(u)\}, \quad u \in U. \tag{4.80}$$

$$S_4(A,B) \equiv 1 - \sup_{u\in U} \{\mu_{A\nabla B}(u)\}. \tag{4.81}$$

Below are five **indices of inclusion** of A in B [81]

$$I_1(A,B) \equiv \frac{\| A \cap B \|}{\| A \|}. \tag{4.82}$$

$$I_2(A,B) \equiv 1 - \| A | - | B \| . \tag{4.83}$$

where $|-|$ is the bounded difference

$$\mu_{A|-|B}(u) = \max\{0, \mu_A(u) - \mu_B(u)\}, \quad u \in U. \tag{4.84}$$

$$I_3(A,B) \equiv \| \bar{A} \cup B \|, \tag{4.85}$$

$$I_4(A,B) \equiv \inf_{u \in U} \left\{ \mu_{\overline{A|-|B}}(u) \right\}, \tag{4.86}$$

$$I_5(A,B) \equiv \inf_{u \in U} \{ \mu_{\bar{A} \cup B}(u) \}. \tag{4.87}$$

Finally, a **consistency index** is defined by

$$C(A,B) \equiv \sup_{u \in U} \{ \mu_{A \cap B}(u) \}. \tag{4.88}$$

For intersection and union we use the minimum and maximum, respectively, and for complement, $\mu_{\bar{A}}(u) = 1 - \mu_A(u)$, $u \in U$.

4.3.8 Fuzzy relations

Let U and V be two universal (crisp) sets. A **fuzzy relation** $\mathcal{R}$ on $U \times V$ is any fuzzy set with membership function[6]

$$\mu_{\mathcal{R}} : U \times V \to [0,1], \tag{4.89}$$

i.e., $\mu_{\mathcal{R}}(u,v) \in [0,1]$, $u \in U$, $v \in V$. Fuzzy relations can be *composed* . Let $\mathcal{R}$ be a fuzzy relation on $U \times V$ and let $\mathcal{Q}$ be a fuzzy relation on $V \times W$, where U, V, and W are crisp sets. A **composition** of the fuzzy relations $\mathcal{R}$ and $\mathcal{Q}$, denoted $\mathcal{R} \circ \mathcal{Q}$, is a fuzzy relation $\mathcal{T}$ defined on $U \times W$ according to the composition rule $\circ$. The composition consists of two components, e.g., max-min, max-product, or generally any pair of aggregation operations. The calculation of degrees of membership of $\mu_{\mathcal{T}}(u,w)$, $u \in U$, $w \in W$ is performed as matrix multiplication but the summation is replaced by the first component of the composition and the product is replaced by the second component.

Example 4.3.9. Let $U = \{a,b,c\}$, $V = \{1,2\}$ and $W = \{x,y,z\}$. Define $\mathcal{R}$ on $U \times V$ and $\mathcal{Q}$ on $V \times W$ as shown below. The entries in the matrices are the degrees of membership of the pair of elements specifying the matrix cell. The max-min composition of the two relations $\mathcal{T} = \mathcal{R} \circ \mathcal{Q}$ is also shown. ■

6 Any relation (fuzzy or crisp) defined on the Cartesian product of *two* sets is called a *binary* relation.

$\mathcal{R}$

	1	2
a	0.3	0.0
b	0.7	0.4
c	0.1	0.8

$\mathcal{Q}$

	x	y	z
1	0.6	0.0	0.6
2	0.4	0.4	1.0

$\mathcal{T} = \mathcal{R} \circ \mathcal{Q}$

	x	y	z
a	0.3	0.0	0.3
b	0.6	0.4	0.6
c	0.4	0.4	0.8

Fuzzy relations will be needed to explain fuzzy relational classifiers in Chapter 7.

4.4 Determining membership functions

4.4.1 Modeling issues, approaches and difficulties

The key issue in all fuzzy sets applications is how to determine (design, estimate, tune) membership functions. There are two general strategies for estimation: **by experts** or **automatically**. Combinations of these are also sought by initializing the functions by experts, and then tuning the values automatically, using the available data.

There are two ways to specify membership functions:

1. Assigning individual **membership grades**. For example, let U be a group of students and A be a characteristic, such as "good student". The task is then to assign an individual degree of membership of each element of U in A. We call this **soft labeling** of U. Soft labeling is usually subjective. Is is the only option when the universal set U has no quantitative characteristic associated with A, e.g., if A is "inventiveness" or "motivation".
2. Designing a **membership function**. Consider the example above. Now we are interested in finding a formal expression, so that we can assign a degree of membership in A to any $u \in U$, e.g., as a function of student's marks on the exams. In this case, soft labeling is done automatically. We can use this second approach only for A's which have a numerical dimension associated with them. Thus, the fuzzy set *middle-aged* can be defined by a triangular membership function over an appropriate interval on $\Re^+$.

Regarding the interpretation of the membership functions, there are again two issues:

- **Linguistic modeling**. The membership function models some linguistic category, e.g., *rough, precise, cheerful.* The elements of U are assessed with respect to a certain *linguistic* label.
- **Nonlinguistic modeling**. Sometimes the membership function is needed inside the pattern recognition model, and it is not necessary that it bears any linguistic interpretation. A typical example is assigning the degree of membership by the similarity of the object to a prototype.

The following two examples outline the difficulties in designing membership functions.

Example 4.4.1. On the scale of human height in cm, the value $\mu_{tall}(183)$ shows to what extent we are likely to call a person of height 183 cm *tall.* $\mu_{tall}(183)$ is

- **Context-specific.** This means that $\mu_{tall}(u)$ will be different if u is a measurement of a male or a female height. Assume that the person is a male. Then $\mu_{tall}(183)$ will be different again if this person is an Australian Aborigine or Dutch.
- **Subject-specific.** Different people will design different membership functions for the same notion, according to their own opinion.
- **Problem-specific.** The membership function depends on the task that we want to solve. Thus, the value $\mu_{tall}(183)$ will be high if we use it for selecting youngsters for a team of gymnasts, and low, if we are selecting basketball players.

■

Example 4.4.2. Due to its notorious subjectivity and uncertainty, medicine has been a fertile domain for fuzzy set applications [7, 140, 205]. Take for example systolic blood pressure and its linguistic adjective *low*, represented as a fuzzy set on the set of all possible values of blood pressure (e.g., in $mmHg$). The membership function μ_{low} is ***context-specific***, which in this case means "patient-specific". We can have $\mu_{low}(135) = 0.95$ for a patient who is normally hypertonic, and $\mu_{low}(135) = 0.20$ for a hypotonic patient. μ_{low} also has to be ***subject-specific*** because, despite the known boundaries for normal and abnormal values, different clinicians may have different views of the precise quantification of "low". The membership function is also ***problem-specific***, with regard to what we are planning to do with this particular reading: put the patient on a therapy, or select candidate-astronauts. ■

The above examples look at a single degree of membership. When the entire function μ is concerned, there are more requirements to be met. For example, in order to be consistent, the function corresponding to a linguistic label should peak at the most typical value(s) (see the design of a fuzzy number in Example 4.2.1).

Example 4.4.3. Not all consistent membership functions need to be unimodal in order to conform with the underlying linguistic terms. An example given in [68] shows a bimodal function expressing the set "High driving risk" defined over the age of the driver. Both very young drivers (presumably unexperienced), and old drivers (with deteriorating vision and ability to concentrate) are at higher driving risk, and therefore have high membership. Thus, the membership function peaks once at the lefthand-side and once at the righthand-side of the age axis. ■

The problem becomes even more complicated when we have to design a sequence of connected linguistic labels, e.g., {*small*, *medium*, *large*}. (Such a sequence is called a "frame of cognition" in [64]). The fuzzy sets corresponding to the labels should be properly placed on the universal set, so that the sequence keeps its meaning, e.g., *small* should be on the left of *medium*, and should not be a subset of it, they should overlap to a certain (reasonable in the domain context) degree. This is not always easy to observe, especially when membership functions are obtained automatically from data. Figure 4.7 shows an example of acceptable and unacceptable sequences of related labels.

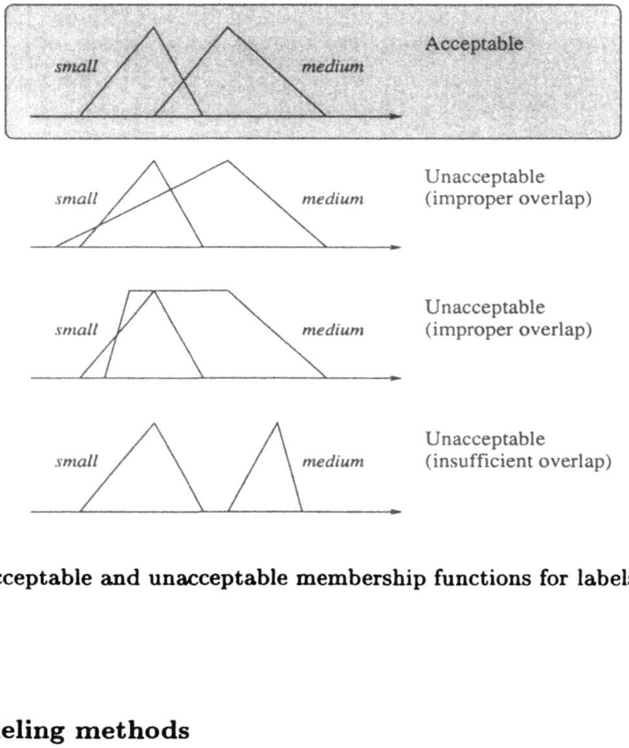

Fig. 4.7. Acceptable and unacceptable membership functions for labels *small* and *medium*.

4.4.2 Modeling methods

Table 4.9 summarizes five ways of **expert** estimation of degrees of membership [42, 41, 64]. Not all of these methods can be used interchangeably. For example, the interval estimate cannot be applied to a universal set which is not ordered, e.g., a set of ice-cream deserts. Also, there is no point in interviewing the public about who of two persons is taller and by how much (the pairwise estimate) because the answer is straightforwardly measurable. In some of the methods, the interviewed person looks on one element of U at a time (viz. polling and direct estimate), pairwise estimation uses two elements at a time, reverse estimation requires all of U, and interval estimation requires none. All methods which involve human assessment use linguistic categories, (see the example questions in Table 4.9), and therefore belong to the linguistic group. Except for interval estimation, all methods eventually specify a *degree of membership*, either for a single object or for a set of objects. Interval estimation can produce a *membership function*.

Table 4.9. Expert estimation of degrees of membership

Method	Example question	Comment
Polling (Horizontal method)	Do you think Engelbert Humperdink is famous? (Y/N)	Needs a group of people. The estimate is obtained as $\mu_{famous}(\text{Engelbert}) = $ $= N_{Yes}/N_{All}$.
Direct estimation	To what degree are the BBC news programmes unbiased?	Can be done by a group (and averaged) or by a single individual.
Reverse estimation (Vertical method)	Identify from a given set of actors, all actors whom you would label *talented* with degree 0.3 or more.	This method reconstructs the fuzzy set by its α-cuts.
Interval estimation	Specify an interval on the axis of car speed, corresponding to your idea of *low* speed.	Needs a group. The intervals are combined similarly to the polling method.
Pairwise comparison	Which of two ice-cream deserts is more delicious, and by how much?	The result is a matrix of pairwise preferences from which the degrees of membership are derived. Can be done by a group (and averaged) or by a single individual.

Defining membership functions for linguistic labels assumes that we have a *numerical universal set*, and we can use $u \in U$ to calculate the degree of membership. In some cases we can build up this universal set using a list of binary questions related to a specific variable [105, 292].

Example 4.4.4. Saitta and Torasso [292] study a set of compound risk factors influencing coronary disease. One such factor, not directly measurable, is *satisfaction*. The authors construct a score-type variable using a questionnaire with 41 questions, such as "Were you disappointed at leaving your studies?" and "Are you pleased with your home?", with several possible answers, varying from *not at all* (worth -1) to *very much* (worth 1). Similar questionnaires are used for other factors like ***adequacy, interpersonal relations, social mobility, stress, etc.*** Different weights are assigned to the value of each answer, reflecting the relevance of the question towards the factor. On the axis of the total weighted score for that factor (e.g., satisfaction) we define a sigmoidal membership function with values close to 0 on the lefthand-side (meaning *dissatisfaction*) and values close to 1 on the righthand-side (meaning *total satisfaction*). This approach can be very useful in medical practice, because the features (there are usually hundreds of them, most of which of categorical type (see Figure 2.1)) are grouped in meaningful constellations, characterizing a reasonably small number of factors. ■

In some environments, experiments to collect data entail risk, potential danger, high cost, or are just impossible to carry out. Examples of such areas are medicine, economics, power plant control. Then getting the membership functions by interviewing experts might be the only solution to our problem. Yet, eliciting membership functions from experts is a difficult and thankless job. Even world leading experts, most willing to cooperate, highly mathematically minded, might get confused when asked to redress their knowledge and intuition in a mathematical form. A membership function has to be designed on a separate feature axis, usually out of context, which is not a routine task for the expert. It could be difficult to specify the shape of the membership function, the overlap with the other linguistic terms, etc. However, membership function design is crucial for the fuzzy system performance.

Automatic (non-expert) estimation of membership functions has been used as a viable alternative to expert estimation [165, 235, 260, 267]. Methods for tuning membership functions are discussed in Chapter 6 in the context of training fuzzy if-then classifiers. Two cases can be detailed

- Approximating *one* membership function, given a data set for it.
- Approximating *several related* membership functions based on a labeled data set.

Medasani et al. [235] summarized methods for automatic estimation of membership functions, discussing methods based on heuristics, probability to possibility transformations, histograms, nearest neighbor techniques, feed-forward neural networks, clustering, and mixture decomposition. Here are four simple methods for membership function estimation

1. *End-points approximation* [260]. This is the simplest, and possibly the least accurate estimate. Let $U = \{u_1, \ldots, u_n\} \subset \Re$ be a data set on which we estimate the membership function for a fuzzy set A. Let

$$u_{max} = \max_i\{u_i\}, \quad u_{min} = \min_i\{u_i\}, \quad \bar{u} = \frac{1}{n}\sum_{i=1}^{n} u_i.$$

We define the parameters of a π-shaped membership function as

$$a_r = \bar{u} + \max\{|\bar{u} - u_{min}|, |\bar{u} - u_{max}|\}, \quad a_l = 2\bar{u} - a_r, \tag{4.90}$$

and the function itself

$$\mu_A(u) = \begin{cases} 0, & u \le a_l \text{ or } u \ge a_r, \\ 2\left(\frac{u-a_l}{\bar{u}-a_l}\right)^2, & a_l < u \le \frac{a_l+\bar{u}}{2}, \\ 1 - 2\left(\frac{u-\bar{u}}{\bar{u}-a_l}\right)^2, & \frac{a_l+\bar{u}}{2} < u \le \frac{a_r+\bar{u}}{2}, \\ 2\left(\frac{u-a_r}{\bar{u}-a_r}\right)^2, & \frac{a_r+\bar{u}}{2} < u < a_r. \end{cases} \tag{4.91}$$

The π-membership functions corresponding to the three classes from the Cone-torus data projected on the x-axis are shown in Figure 4.8.

2. *Clustering* [34, 235]. Here we assume that we have c clusters of data with centroids $\bar{u}_1, \ldots, \bar{u}_c$ (in the multidimensional case, both the data points and the centroids are vectors). For any point $u \in \Re$ we calculate the degree of membership in cluster A_i as

$$\mu_{A_i}(u) = \frac{\frac{1}{(u-\bar{u}_i)^2}}{\frac{1}{(u-\bar{u}_1)^2} + \ldots + \frac{1}{(u-\bar{u}_c)^2}}. \tag{4.92}$$

Thus, the closer u is to the cluster centroid, the higher the degree of membership in this cluster. Notice that the clusters *share* the membership, i.e.,

$$\sum_{i=1}^{c} \mu_{A_i}(u) = 1, \quad \forall u \in U. \tag{4.93}$$

Having a labeled data set, we calculate the cluster centroids $\bar{u}_i, i = 1, \ldots, c$, as the means of the data from the respective classes (clusters). The three membership functions for the Cone-torus data (x axis only) are shown in Figure 4.8.

3. *Histograms.* We can scale and use the histogram of the data straightforwardly as a membership function. The histogram membership functions of the three classes from Cone-torus data (x axis only) are shown in Figure 4.8.

4. ***k-nearest neighbor*** estimate. Assume we have a labeled data set $\mathbf{Z}$. The problem is to estimate c degrees of membership $\mu_{A_1}(u), \ldots, \mu_{A_c}(u)$, for any $u \in \Re$. First we find the k points in $\mathbf{Z}$ closest to u. Then we calculate

$$\mu_{A_i}(u) = \frac{k_i}{k}, \tag{4.94}$$

where k_i is the number of elements amongst the k closest neighbors of u which are labeled in class ω_i. Again, the membership is shared between the classes, and (4.93) holds. The k-nn membership functions for the

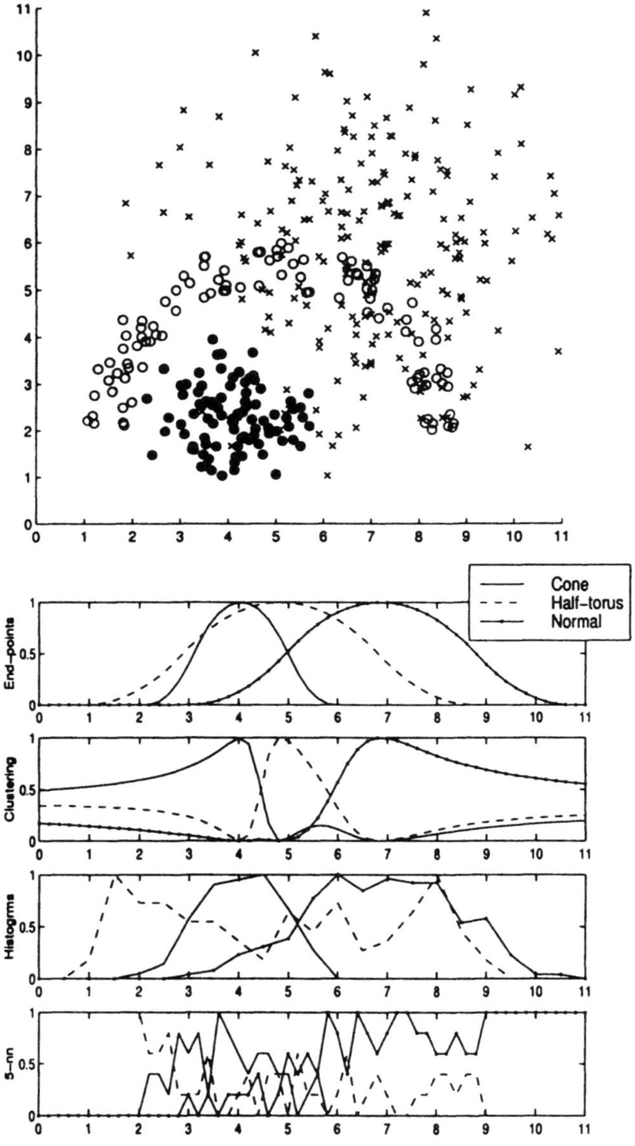

Fig. 4.8. Results from four automatic methods for membership function estimation applied on the Cone-torus data, projected on the x-axis.

three classes from Cone-torus data (x axis only) are shown in Figure 4.8 (bottom diagram). This method was proposed by Jóźwik [159] for calculating *soft* labels of a (crisply) labeled data set by an iterative procedure (see also Chapter 7). Keller et al. [165, 167] propose a scheme for soft relabeling of a (crisply) labeled data set **Z**. The scheme guarantees that all objects retain their true class labels if the soft labels are "hardened" by the maximum membership rule (2.3). According to this scheme

$$\mu_{A_i}(u) = \begin{cases} 0.51 + 0.49\frac{k_i}{k}, & \text{if } \omega_i \text{ is the true class label of } u, \\ 0.49\frac{k_i}{k}, & \text{otherwise.} \end{cases} \quad (4.95)$$

In summary, it seems that estimation of membership functions is an art rather than a technology. Different methods can produce very different membership functions: smooth or spiky, shared or independent, consistent with linguistic labels, or context free. To what extent do we have to dig in this quarry? This depends on what we intend to do with the membership functions. In some problems interviewing a group of people or a single expert is the only way of devising the membership functions. The inconsistencies should then be smoothed out by tuning the fuzzy system built upon these initial estimates. In problems where data is available, it may not pay off to set up an interview. In this case it is better to start from ad-hoc membership functions and tune them to match the data. This approach has been widely adopted in fuzzy neural networks. Together with tuning (or even instead of tuning!) we can *select* the membership functions from a large initial pool. Genetic algorithms come in handy for such problems [149] (Chapter 6). And finally, we can construct the functions automatically from data. The problem of how to encode and handle membership functions computationally is discussed in great detail and with many examples by Cox [68].

5. Fuzzy if-then classifiers

5.1 Fuzzy if-then systems

A **fuzzy if-then system** has n inputs ($\mathbf{x} = [x_1, \ldots, x_n]^T \in \Re^n$) and c outputs ($\mathbf{y} = [y_1, \ldots, y_c]^T \in \Re^c$). Here are three popular acronyms for fuzzy (and also non-fuzzy) systems

- **SISO**. Single input – single output systems ($n = c = 1$).
- **MISO**. Multiple input – single output systems ($n > 1, c = 1$).
- **MIMO**. Multiple input – multiple output systems ($n > 1, c > 1$).

It is convenient to view the c outputs as the discriminant functions $g_1(\mathbf{x}), \ldots, g_c(\mathbf{x})$, as in the canonical model of a classifier (see Figure 2.2). When a vector $\mathbf{x}$ is submitted to the input, the system operates as shown in Figure 5.1. The input is "fuzzified" and submitted to the fuzzy inference block. A *rule base* is used to calculate a fuzzy decision which is subsequently "defuzzified" to get the final output. The processes of fuzzification, inference and and defuzzification are described in the next subsections.

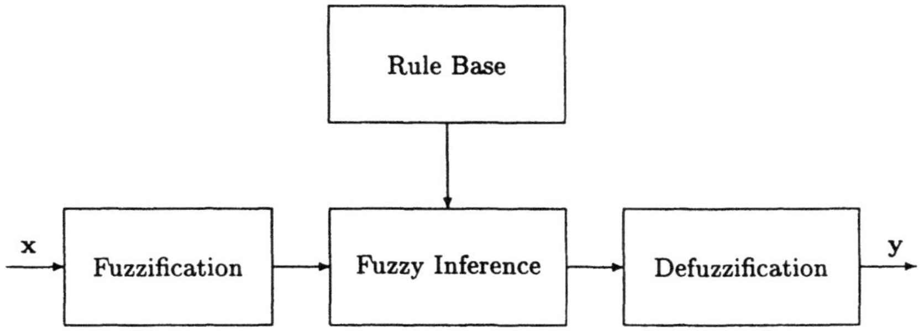

Fig. 5.1. Operation of a fuzzy if-then system

In this Chapter we explain only *how fuzzy if-then classifiers operate.* The problem of *how to design a fuzzy if-then classifier* is deferred to Chapter 6.

5.1.1 Fuzzification

The first step is the **fuzzification**. To quote a friend, "fuzzification" is the second ugliest word in the fuzzy set jargon, the ugliest being "defuzzification". Nonetheless, "fuzzification" is a widely accepted notion, and is a compact expression for "finding the degree of match of the input to a set of linguistic terms". Let $A = \{A_{i,1}, \ldots, A_{i,K_i}\}$ be a set of **linguistic terms** for the i-th component x_i of the input $\mathbf{x}$. For example, the feature *age* can be described by {*very young, young, middle-aged, old*}, the feature *profit*, by {*small, reasonable, very good, appreciable, huge*}, etc. Each of the linguistic terms is represented as a fuzzy set on the set of possible values of x_i. Fuzzification replaces x_i by a set of K_i numbers $\mu_{A_{i,1}}(x_i), \ldots, \mu_{A_{i,K_i}}(x_i)$ in [0,1], showing how well the value x_i matches each of the K_i linguistic terms.

Example 5.1.1. Figure 5.2 plots the membership functions of a set of linguistic terms for "student performance" (feature x_i), defined on the axis of the exam results. Let $\mathbf{x}$ be a student, whose exam result is $x_i = 46$ % (the thick line). Through fuzzification, the value 46 is replaced by the following four-tuple [0.0 (Unsatisfactory), 0.4 (Pass), 0.6 (Good), 0.0 (Excellent)]. These values can be then submitted to the second block of the diagram in Figure 5.1: Fuzzy Inference. ■

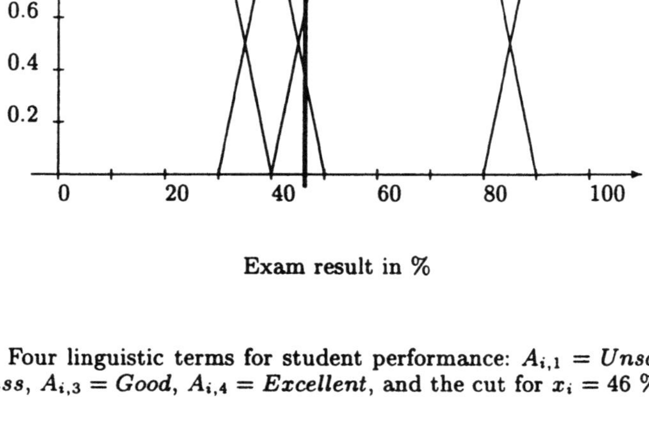

Fig. 5.2. Four linguistic terms for student performance: $A_{i,1} = Unsatisfactory$, $A_{i,2} = Pass$, $A_{i,3} = Good$, $A_{i,4} = Excellent$, and the cut for $x_i = 46$ %.

5.1.2 Fuzzy inference

The **inference engine** uses a set of M **fuzzy if-then rules (a fuzzy rule base)**, built up in advance. The rule base can be specified by a domain expert or extracted from input-output data pairs. Each rule has an **antecedent part** (if-part, or premise), and a **consequent part** (then part, or consequent). The antecedent part is a Boolean expression of simple **clauses** on the individual features $x_1, \ldots, x_n$. A simple clause is, for example, "x_i is *small*".

There are two basic types of fuzzy if-then systems that have crystallized out of the large amount of "fuzzyware" developed since 1965 [356].

- The **Mamdani-Assilian (MA)** model (the *logical* model). In MA systems, both the input and the output are represented by linguistic terms. The antecedent and the consequent of an if-then rule are typically Boolean expressions of simple clauses.
- The second type of fuzzy systems is the **Takagi-Sugeno-Kang (TSK)** model [315] (the *functional* model). The antecedent part of each rule is again a Boolean expression of simple clauses but the consequent is a function of the input $\mathbf{x}$ (most often a polynomial).

Mamdani-Assilian (MA) fuzzy systems. The antecedent part in this model is a conjunction of simple clauses. For the time being we shall assume that all features are used in each rule (this means that the antecedent part is a conjunction of n clauses).

The k-th rule in the database of an MIMO fuzzy if-then system has the following general form

$$R_k : \text{IF } x_1 \text{ is } A_{1,i(1,k)} \text{ AND } x_2 \text{ is } A_{2,i(2,k)} \text{ AND } \ldots \text{ AND } x_n \text{ is } A_{n,i(n,k)}$$

$$\text{THEN } y_1 \text{ is } B_{o(1,k)} \text{ AND } \ldots \text{ AND } y_c \text{ is } B_{o(c,k)},$$

$$\text{for } k = 1, \ldots, M.$$

In this expression, the subscript $i(j,k)$ specifies which linguistic term on the j-th feature is used in the k-th rule. The notations $i(.,.)$ and $o(.,.)$ are used for *index* functions for the *i*nput and *o*utput parts of the rules, respectively.

Example 5.1.2. Let x_3 have the three linguistic labels {*inappropriate, moderately good, excellent*} ($K_3 = 3$), expressed by the fuzzy sets $A_{3,1}$, $A_{3,2}$, and $A_{3,3}$. Assume that the if-part of the 5th rule contains

$$R_5 : \text{IF } \ldots \text{ AND } x_3 \text{ is } \textit{excellent} \text{ AND } \ldots$$

In this case $A_{3,i(3,5)}$ should be $A_{3,3}$, hence $i(3,5) = 3$. ■

The notation $A_{j,i(j,k)}$ might look overornate but it is necessary for keeping the notations consistent. Figure 5.3, included here for an easy reference, displays the notation.

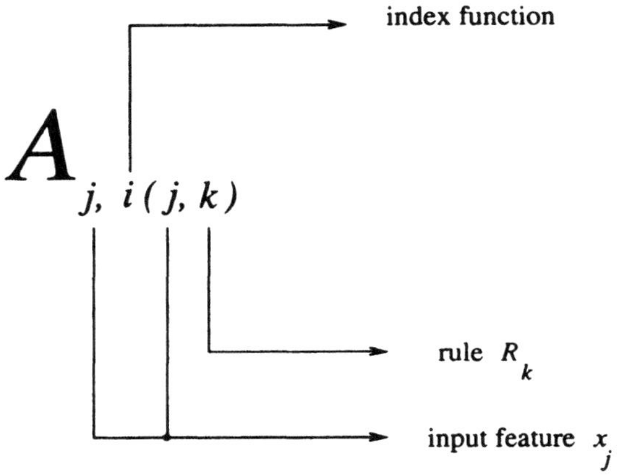

Fig. 5.3. Notation for the jth antecedent clause of the kth rule in a fuzzy rule-based system

We consider a MISO system, whose output is a numeric variable $y \in \Re$, equipped with a set of linguistic terms $B_1, \ldots, B_s$. The kth rule for such system is

$$R_k : \text{IF } x_1 \text{ is } A_{1,i(1,k)} \text{ AND } x_2 \text{ is } A_{2,i(2,k)} \text{ AND } \ldots \text{ AND } x_n \text{ is } A_{n,i(n,k)}$$

$$\text{THEN } y \text{ is } B_{o(k)},$$

where $B_{o(k)}$ is the consequent fuzzy set in rule R_k.

Let $\mu_{j,i(j,k)}(x_j)$ be the degree of membership of x_j in the fuzzy set $A_{j,i(j,k)}$. The value $\mu_{j,i(j,k)}$ can be interpreted as the degree of consistency of the value x_j with the linguistic term $A_{j,i(j,k)}$. The **firing strength** $\tau_k(\mathbf{x})$ of rule $R_k, k = 1 \ldots, M$, for an input $\mathbf{x}$ is calculated by a conjunction type of aggregation operation $\mathcal{A}_t$[1]. on $\mu_{1,i(1,k)}(x_1), \ldots, \mu_{n,i(n,k)}(x_n)$, i.e.,

$$\tau_k(\mathbf{x}) = \mathcal{A}_t\left(\mu_{1,i(1,k)}(x_1), \ldots, \mu_{n,i(n,k)}(x_n)\right). \tag{5.1}$$

Thus, the firing strength reflects to what extent $\mathbf{x}$ satisfies *all* antecedent clauses. The minimum and the product are the two most widely used operations. As pointed out by Cordón et al. [66], other operations can also be considered instead of conjunctions. A possible choice, for example, is a class of mean operations yielding values between the minimum and the average.

The firing strength is then used to "cut" the fuzzy set corresponding to the linguistic value in the consequence of R_k. Thus we use the minimum operation as the implication (THEN)[2]. Thus, an output fuzzy set C_k on y is obtained for each R_k, $k = 1, \ldots, M$ as

---

[1] The subscript t comes from t-norm. t-norms are the AND type operations and extensions thereof are used for aggregation (cf. Chapter 4)

[2] The suitability of minimum as the implication is discussed in Chapter 4.

$$\mu_{C_k}(y) = \min\{\tau_k(\mathbf{x}), \mu_{B_{o(k)}}(y)\} \tag{5.2}$$

Substituting the aggregation formula for τ in (5.2),

$$\mu_{C_k}(y) = \min\{\mu_{1,i(1,k)}(x_1), \ldots, \mu_{n,i(n,k)}(x_n), \mu_{B_{o(k)}}(y)\}. \tag{5.3}$$

The resultant fuzzy sets $C_1, \ldots, C_M$ are then aggregated through a disjunction type operation $\mathcal{A}_s$,

$$\mu_C(y) = \mathcal{A}_s\left(\mu_{C_1}(y), \ldots, \mu_{C_M}(y)\right), \tag{5.4}$$

or

$$C = \bigcup_{k=1}^{M} C_k. \tag{5.5}$$

Typically, the maximum is used as $\mathcal{A}_s$ but other formulas are also applicable. Some generalized mean operators lying between the average and the maximum can prove useful [66].

The functioning of a SISO MA fuzzy system with three if-then rules is shown in Figure 5.4. Each rule is depicted as a horizontal layer and the input x is indicated by the thick line. The fuzzification process is marked by an upward arrow. The rule base consists of

R_1	if	x	is	small	then	y	is	medium.
R_2	if	x	is	medium	then	y	is	bad.
R_3	if	x	is	high	then	y	is	good.

There is only one clause in each rule, so the firing strength is the degree of membership $\mu_k(x)$, where $k = 1$ for rule R_1, $k = 2$ for R_2, and $k = 3$ for R_3. The fuzzy sets C_1, C_2 and C are shaded in grey color ($C_3 = \emptyset$ because the firing strength $\mu_3(x) = 0$). They are found by "cutting" the consequent fuzzy sets with the firing strength value (horizontal grey arrows). C is the *fuzzy output* of the MA fuzzy system obtained as the union of C_1 and C_2. The representative value y, which is also the output of the system, is found by *defuzzification* (discussed later) of the resultant fuzzy set C.

Takagi-Sugeno-Kang (TSK) fuzzy systems. A MIMO Takagi-Sugeno-Kang fuzzy system has the following type of rules

R_k : IF x_1 is $A_{1,i(1,k)}$ AND x_2 is $A_{2,i(2,k)}$ AND ... AND x_n is $A_{n,i(n,k)}$

THEN $\mathbf{y} = \mathbf{f}_k(\mathbf{x})$,

$k = 1, \ldots, M$.

where $\mathbf{f} : \Re^n \to \Re^c$ is a vector function of the input $\mathbf{x}$ with c components. MISO TSK systems are most often used with a polynomial as $f : \Re^n \to \Re$.

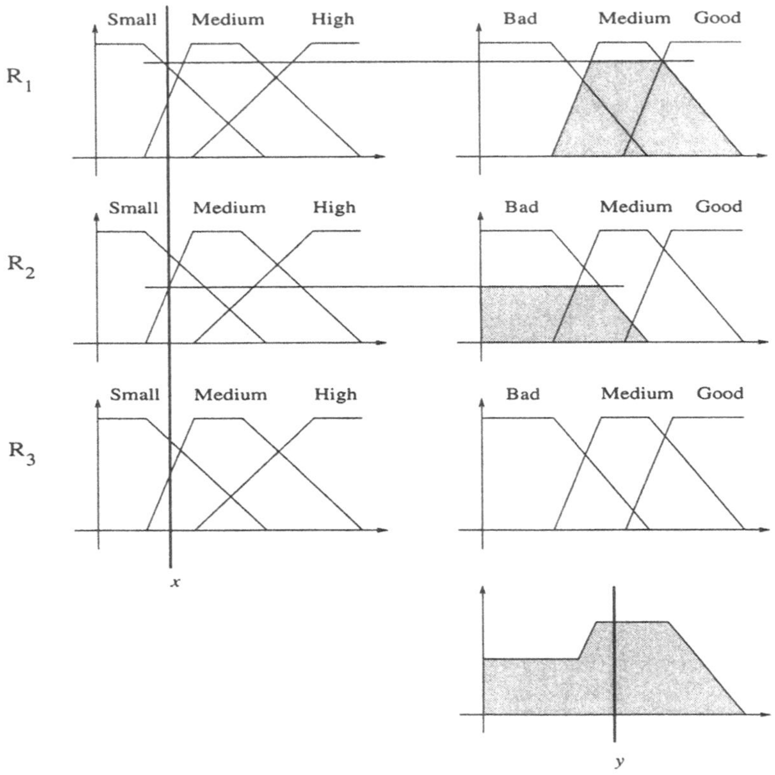

Fig. 5.4. Functioning of a SISO MA fuzzy system with 3 rules. The output y is obtained for the input x.

A TSK system of order d is a TSK system in which the polynomial f is of order d. First-order TSK systems for which

$$y_k = f_k(\mathbf{x}) = a_{k,0} + \sum_{j=1}^{n} a_{k,j} x_j, \quad k = 1, \ldots, M \tag{5.6}$$

are the most popular choice in fuzzy control [356].

The firing strength is calculated as in the MA model. The overall output of a TSK fuzzy system is

$$y(\mathbf{x}) = \frac{\sum_{k=1}^{M} \tau_k(\mathbf{x}) f_k(\mathbf{x})}{\sum_{k=1}^{M} \tau_k(\mathbf{x})}. \tag{5.7}$$

Example 5.1.3. Let the output of a MISO TSK model with 3 rules be

R_1: ... THEN $y_1 = 3x_1 - x2$.
R_2: ... THEN $y_2 = 2x_1$.
R_3: ... THEN $y_3 = x_1 + x2 - 4$.

Assume that for $\mathbf{x} = [x_1, x_2]^T = [2, 1]^T$, the firing strengths have been

$$\tau_1(\mathbf{x}) = 0.1; \ \ \tau_2(\mathbf{x}) = 0.9 \ \ \tau_3(\mathbf{x}) = 0.2.$$

Let us calculate the output y for the input $\mathbf{x}$. First, find, using the consequents, $y_1 = 5$, $y_2 = 4$ and $y_3 = -1$. Apply 5.7 to find y

$$y(\mathbf{x}) = \frac{0.1 \times 5 + 0.9 \times 4 + 0.2 \times (-1)}{0.1 + 0.9 + 0.2} = 3.25 \tag{5.8}$$

■

TSK model is related to the MA model. For each input $\mathbf{x}$ we can regard the consequence of rule R_k as a singleton fuzzy set $B_{o(k)}$ on $\Re$, such that

$$\mu_{B_{o(k)}}(y) = \begin{cases} 1, & \text{if } y = y_k = f_k(\mathbf{x}), \\ 0, & \text{otherwise} \end{cases}, \quad y \in \Re. \tag{5.9}$$

The implication, implemented as minimum, gives a subnormal singleton fuzzy set C_k

$$\mu_{C_k}(y) = \begin{cases} \tau_k(\mathbf{x}), & \text{if } y = y_k = f_k(\mathbf{x}), \\ 0, & \text{otherwise} \end{cases}, \quad y \in \Re. \tag{5.10}$$

The rule outputs y_k are then pooled in a set $Y = \{y_1, \ldots, y_M\}$. Thus we obtain the resultant fuzzy set C over the crisp set Y with degrees of membership $\mu_C(y_k) = \tau_k(\mathbf{x}), k = 1 \ldots, M$. The set Y may contain coinciding values of y_k. However, we treat them as *separate* members of the universal set Y of cardinality M. The representative value of Y is derived by defuzzification (next subsection), and coincides with the TSK output (5.7).

5.1.3 Defuzzification

"**Defuzzification**" is a procedure to calculate the single representative value of a fuzzy set. Ideally, this element should best characterize the fuzzy set. This operation is needed to find the system's output y from the resultant fuzzy set C [214, 356].

The two most popular defuzzification methods are the *Center of Gravity* (COG), and the *Mean of Maxima* (MOM).

If the output fuzzy sets are defined over a finite universal set $Y = \{y_1, \ldots, y_S\}$, the resultant fuzzy set C is also defined on Y, and the COG defuzzified value is

$$y = \frac{\sum_{i=1}^{S} \mu_C(y_i) y_i}{\sum_{i=1}^{S} \mu_C(y_i)}. \tag{5.11}$$

Here we assumed that although C is defined over the finite domain $\{y_1, \ldots, y_M\} \subset \Re$, the defuzzified y can take any value (presumably between $\min_i\{y_i\}$ and $\max_i\{y_i\}$).

If the output fuzzy sets $C_1, \ldots, C_M$, are defined on some interval $[a, b] \subset \Re$, the resultant fuzzy set C is also defined on $[a, b]$. We note that $\mu_C(y)$ has a finite non-zero integral over $[a, b]$. Then the COG defuzzification gives

$$y = \frac{\int_a^b \mu_C(\zeta)\,\zeta\,d\zeta}{\int_a^b \mu_C(\zeta)\,d\zeta}. \tag{5.12}$$

Example 5.1.4. Figure 5.5 shows a resultant fuzzy set C defined over the interval $[0, 10]$.

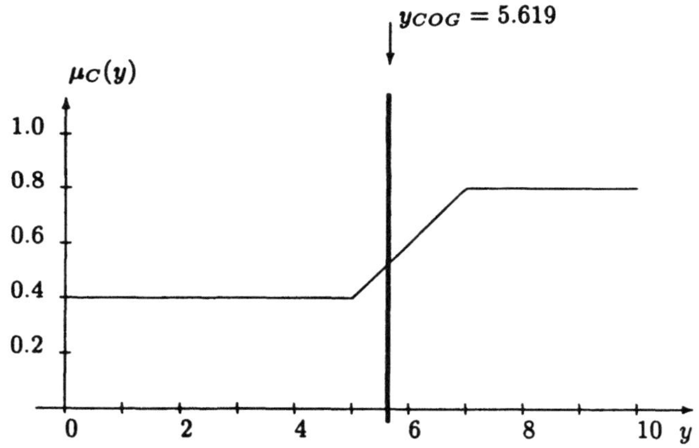

Fig. 5.5. A resultant fuzzy set and its defuzzified value

We use (5.12) to calculate the output y_{COG}

$$\begin{aligned} y_{COG} &= \frac{\int_0^{10} \mu_C(\zeta)\,\zeta\,d\zeta}{\int_0^{10} \mu_C(\zeta)\,d\zeta} \\ &= \frac{\int_0^5 0.4\zeta\,d\zeta + \int_5^7 (0.1\zeta - 0.1)\zeta\,d\zeta + \int_7^{10} 0.8\zeta\,d\zeta}{5 \times 0.4 + 2 \times \frac{0.8+0.4}{2} + 3 \times 0.8} \\ &\approx \frac{31.4667}{5.6} \approx 5.619. \end{aligned} \tag{5.13}$$

The output y_{COG} is shown with a thick line in Figure 5.5. Assume that the membership function of the resultant set C was discrete, defined over $\{0, 1, \ldots, 10\}$. Then we use (5.11) to get y_{COG}

$$y_{COG} = \frac{0.4 \times (0+1+2+3+4+5) + 0.5 \times 6 + 0.8 \times (7+8+9+10)}{0.4 \times 6 + 0.5 \times 1 + 0.8 \times 4}$$
$$= \frac{36.2}{6.1} \approx 5.934. \tag{5.14}$$

■

The defuzzified value for y in Figure 5.4 is marked with a thick line.

Let Y_{max} be the set of all elements of y_i for which μ_C attains its maximal value. The MOM defuzzification yields

$$y = \frac{1}{|Y_{max}|} \sum_{y_i \in Y_{max}} y_i. \tag{5.15}$$

Example 5.1.5. Applying the MOM defuzzification formula for the resultant set from example 5.1.4, we obtain

$$y_{MOM} = \frac{\int_7^{10} 0.8\zeta \, d\zeta}{0.8 \times 3} = 8.5, \tag{5.16}$$

for the continuous-valued $\mu_C(y)$, and

$$y_{MOM} = \frac{7+8+9+10}{4} = 8.5, \tag{5.17}$$

for the discrete version. ■

Figure 5.6 shows a brief summary of all characteristics of a MIMO fuzzy system. They all need to be specified for the system to produce an output $\mathbf{y} \in \Re^c$ for a given input $\mathbf{x} \in \Re^n$. The notable amount of parameters and operations that need to be chosen suggests that fuzzy systems are very flexible. It is this property that makes a comparison between fuzzy models infeasible. If, say, model A has been shown to fare better than model B on some data set, then probably there is a different choice of parameters for model B so that it outperforms model A on the same data. While many statistical classifiers (cf. Chapter 3), e.g., linear discriminant classifier and nearest neighbor rule, are fixed in their structure and parameter training, and are therefore good benchmarks, there are no benchmark settings for fuzzy systems.

5.2 Function approximation with fuzzy if-then systems

5.2.1 MA approximation

In this section we present an illustration of approximating y on x using the values shown in Table 5.1.

COMPLETE CHARACTERIZATION OF A MIMO FUZZY SYSTEM

1. Type of the model (MA or TSK).
2. Rule-base ($R_1, \ldots, R_M$) in linguistic form for the antecedents and the respective (MA or TSK) consequents.
3. Membership functions of all linguistic terms for each input feature: $A_{i,1}, \ldots, A_{i,K_i}, i = 1, \ldots, n$.
4. a) For MA fuzzy systems: Membership functions of all linguistic terms for each output variable: $B_{j,1}, \ldots, B_{j,K_j}, j = 1, \ldots, c$.
   b) For TSK fuzzy systems: Coefficients of the polynomials in the consequents of the rules $R_1, \ldots, R_M$.
5. The conjunction type aggregation $\mathcal{A}_t$ to calculate the firing strength (5.1).
6. The disjunction type aggregation $\mathcal{A}_s$ to aggregate the resultant fuzzy sets for MA systems (Chapter 4, p. 96).
7. The association operation for MA systems (implication, Table 4.5).
8. The defuzzification operation for MA systems.

Fig. 5.6. Characterization of a MIMO fuzzy system

Table 5.1. Input-output (x, y) pairs for function approximation

x	0	1	2	3	4	5	6	7	8	9	10	11	12
y	0	2	4	5	6	6.25	6.5	6.75	7	7	7	7	7

Figure 5.7 shows a piece-wise linear approximation of $y(x)$. From the graph we formulate the following set of $M = 3$ rules

R_1	if	x	is	small	then	y	is	small.
R_2	if	x	is	medium	then	y	is	medium.
R_3	if	x	is	high	then	y	is	medium.

The membership functions on x and y, corresponding to {*low, medium, high*} are also guessed by eye and plotted. The functions are as follows

$$\mu_{small}(x) = \begin{cases} 1, & \text{if } x \leq 2, \\ \frac{4-x}{2}, & \text{if } 2 < x \leq 4, \\ 0, & \text{if } x > 4, \end{cases} \tag{5.18}$$

$$\mu_{medium}(x) = \begin{cases} 1, & \text{if } 4 < x \leq 7, \\ \frac{x-2}{2}, & \text{if } 2 < x \leq 4, \\ \frac{9-x}{2}, & \text{if } 7 < x \leq 9, \\ 0, & \text{if } x > 9 \text{ or } x \leq 2, \end{cases} \tag{5.19}$$

$$\mu_{high}(x) = \begin{cases} 1, & \text{if } x > 9, \\ \frac{x-7}{2}, & \text{if } 7 < x \leq 9, \\ 0, & \text{if } x \leq 7. \end{cases} \tag{5.20}$$

The membership functions for the output y are formalized in a similar manner. The dashed lines split the xy space into regions, one for each pair of input-output linguistic terms. The values of x between 0 and 3 correspond to *small*, between 3 and 8, to *medium*, and between 8 and 12, to *high*. Similarly, we can define three regions for y. Then the lower lefthand-side corner of the graph will be labeled $\{small,\ small\}$. This region corresponds to rule R_1.

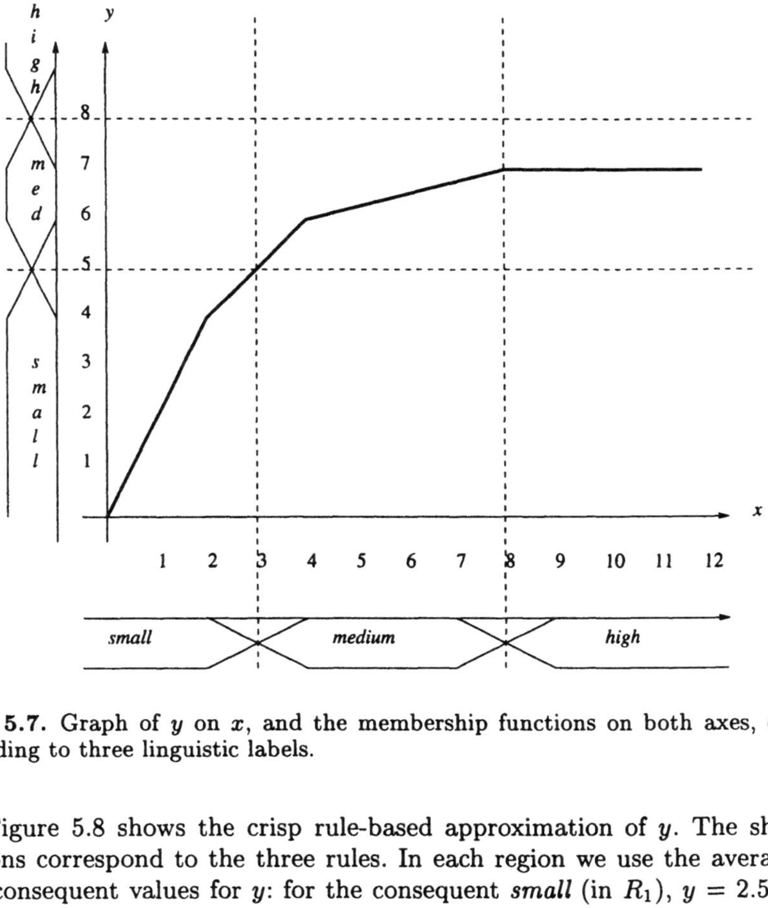

Fig. 5.7. Graph of y on x, and the membership functions on both axes, corresponding to three linguistic labels.

Figure 5.8 shows the crisp rule-based approximation of y. The shaded regions correspond to the three rules. In each region we use the average of the consequent values for y: for the consequent *small* (in R_1), $y = 2.5$, and for *medium* (in R_2 and R_3), $y = 6.5$. The approximation is depicted with a dashed line.

Figure 5.9 plots the MA approximation of y using minimum to calculate the firing strength and the implication; maximum, to derive the final fuzzy set; and COG defuzzification.

Example 5.2.1. The system output y for $x = 3.18$ is calculated below. For illustration purposes we discretize y and constitute a universal set $U = \{0, 1, \ldots, 8\}$. For $x = 3.18$, $\tau_1(x) = \mu_{small}(3.18) = \frac{4-3.18}{2} = 0.41$. Similarly, $\tau_2(x) = \mu_{medium}(3.18) = \frac{3.18-2}{2} = 0.59$ and $\tau_3(x) = \mu_{high}(3.18) = 0$, i.e.,

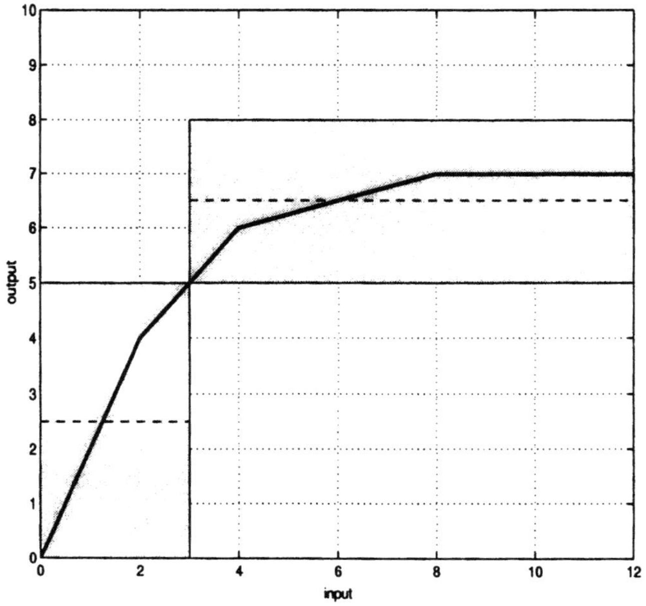

Fig. 5.8. Crisp rule-based approximation of y

$$\tau_1(x) = 0.41, \quad \tau_2(x) = 0.59, \quad \tau_3(x) = 0.$$

Both rules R_1 and R_2 are fired to different extent. For rule R_1 we "cut" the fuzzy set *small* (defined on U) with the value of $\tau_1(x) = 0.41$. The following set is obtained

y	0	1	2	3	4	5	6	7	8
$\mu_{C_1}(y)$	0.41	0.41	0.41	0.41	0.41	0.41	0	0	0

,

and for the consequent of R_2

y	0	1	2	3	4	5	6	7	8
$\mu_{C_2}(y)$	0	0	0	0	0	0.5	0.59	0.59	0.5

.

By aggregating the two through maximum we obtain the resultant fuzzy set C

y	0	1	2	3	4	5	6	7	8
$\mu_C(y)$	0.41	0.41	0.41	0.41	0.41	0.5	0.59	0.59	0.5

.

The COG defuzzification gives the output of the system

$$y = \frac{0.41(0+1+2+3+4) + 0.5(5+8) + 0.59(6+7)}{0.41+0.41+0.41+0.41+0.41+0.5+0.59+0.59+0.5} = 4.3191.$$

■

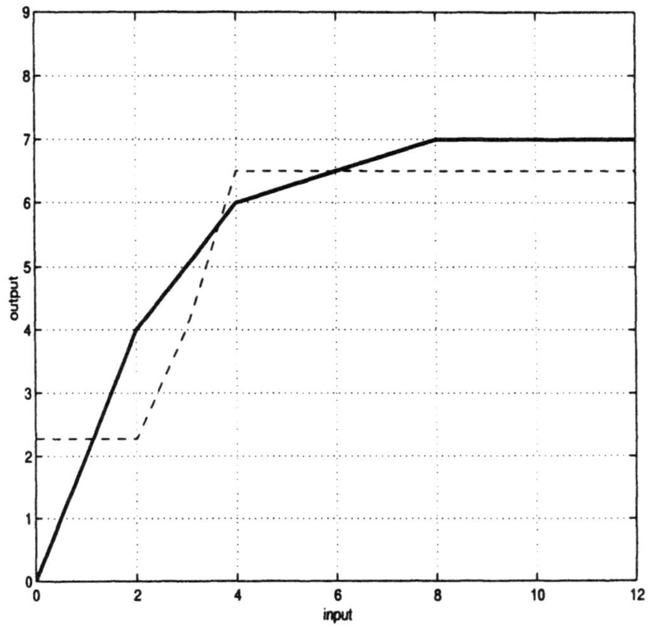

Fig. 5.9. MA approximation of y.

5.2.2 TSK approximation

Using the same linguistic labels on x and their membership functions shown in Figure 5.7 we build a Takagi-Sugeno-Kang fuzzy system with the following three rules

R_1	if	x	is	small	then	$y = 5x/3$
R_2	if	x	is	medium	then	$y = 0.4x + 3.8$
R_3	if	x	is	high	then	$y = 7$

For the consequent parts of the rules we calculated the segments between the end-points of the intervals for *small, medium* and *high* for x. For example, the line $y = 0.4x + 3.8$ (for rule R_2) goes through points (3,5) and (8,7) because the interval *medium* for x is [3,8] and the respective values of y are 5 and 7. The result is shown in Figure 5.10, where the approximation is depicted with a dashed line.

Example 5.2.2. Below we show the calculations of TSK output for $x = 3.18$. As with the MA model,

$$\tau_1(x) = 0.41, \quad \tau_2(x) = 0.51, \quad \tau_3(x) = 0.$$

The consequent parts give

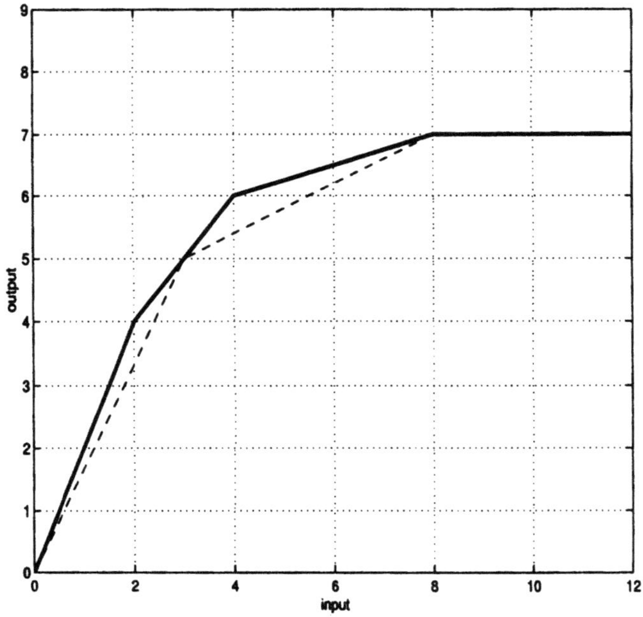

Fig. 5.10. TSK approximation of y.

$$y_1 = 5(3.18)/3 = 5.3, \quad y_2 = 0.4(3.18) + 3.8 = 5.072, \quad y_3 = 7.$$

Applying COG defuzzification, the output of the system is

$$y = \frac{\tau_1 y_1 + \tau_2 y_2 + \tau_3 y_3}{\tau_1 + \tau_2 + \tau_3} = 5.1655.$$

■

Table 5.2 shows the numerical results from the three approximations for discrete values of x from 0 to 12. The **Mean Squared Error** (MSE) of approximation is shown in the bottom row. Let $\hat{y}$ denote the approximation of y. For n input-output pairs $(x_1, y_1), \ldots, (x_n, y_n)$, the MSE is calculated by

$$MSE = \frac{1}{n} \sum_{i=1}^{n} (y_i - \hat{y}(x_i))^2. \tag{5.21}$$

The TSK fuzzy system appeared to be the best of the three approximations (smallest MSE). This, however, is not a guarantee that TSK systems are generally better than MA systems. Both types have numerous degrees of freedom that can be used to tune the system, e.g., the membership functions of the antecedent and consequent, the types of conjunction, implication and disjunction, the consequence parameters and function type of the TSK model, the rule base, the defuzzification method, etc.

Table 5.2. Approximation results

x	y	$\hat{y}_{crisp}$	$\hat{y}_{MA}$	$\hat{y}_{TSK}$
0	0	2.5	2.2727	0
1	2	2.5	2.2727	1.6667
2	4	2.5	2.2727	3.3333
3	5	2.5	4	5
4	6	6.5	6.5	5.4
5	6.25	6.5	6.5	5.8
6	6.5	6.5	6.5	6.2
7	6.75	6.5	6.5	6.6
8	7	6.5	6.5	7
9	7	6.5	6.5	7
10	7	6.5	6.5	7
11	7	6.5	6.5	7
12	7	6.5	6.5	7
MSE		1.2788	0.8345	0.0947

5.3 Fuzzy if-then classifiers

5.3.1 Function approximation and classification

Can we use function approximation for classifier design? In fuzzy control, often the output variables are independent, and a MIMO model can be decomposed as a collection of MISO models, which are significantly easier to handle [356]. In pattern recognition, the classes (corresponding to the outputs) are not independent – conversely, they are dependent, usually mutually exclusive. Consider a classification problem with $\mathbf{x} \in \Re^n$ and set of classes $\Omega = \{\omega_1, \ldots, \omega_c\}$. Ideally, the consequent part of a rule distinguishing class ω_k from the rest of Ω should be as the following rule

R_k : IF x_1 is $A_{1,i(1,k)}$ AND x_2 is $A_{2,i(2,k)}$ AND ... AND x_n is $A_{n,i(n,k)}$

THEN $g_k(\mathbf{x}) \geq g_i(\mathbf{x}), \forall i = 1 \ldots, c$,

$k = 1, \ldots, M$,

where $g_i(\mathbf{x})$ is the ith output of the fuzzy system. There are various ways to represent this type of consequence. An equivalent representation of the THEN part is

... THEN *class* is ω_k.

The above is a single-output system but the output takes nominal categorical values in Ω and is not directly implementable in terms of function approximation. Neither are the inequalities in the previous model computationally tractable in the function approximation framework. One solution is to approximate c binary discriminant functions as in the following rule

R_k : IF x_1 is $A_{1,i(1,k)}$ AND x_2 is $A_{2,i(2,k)}$ AND ... AND x_n is $A_{n,i(n,k)}$

THEN $g_{k,1}(\mathbf{x}) = 0$ AND ... AND $g_{k,o(k)}(\mathbf{x}) = 1$ AND ... AND $g_{k,c}(\mathbf{x}) =$ 0,

$k = 1, \ldots, M.$

In terms of function approximation, the problem will be to approximate the binary valued discriminant functions $g_1(\mathbf{x}), \ldots g_c(\mathbf{x})$.

5.3.2 The general fuzzy if-then classifier model

We call the **general model** a system with the following type of rules

R_k : IF x_1 is $A_{1,i(1,k)}$ AND ... AND x_n is $A_{n,i(n,k)}$

THEN $g_{k,1} = z_{k,1}$ AND ... AND $g_{k,c} = z_{k,c}$.

The values $z_{k,j} \in \Re$ are interpreted as "support" for class ω_j given by rule R_k if the premise part is completely satisfied. If necessary, we can scale the support values in the interval [0,1] by some bijective transformation $f : \Re \rightarrow [0, 1]$, so that the discriminant functions be regarded as membership functions. Cordón et al. [66] distinguish between three type of fuzzy classification systems depending on the consequent :

1. Fuzzy rules with a class label in the consequent, e.g.,

   R_k : ... THEN class is $\omega_{o(k)}$.

   where $o(k)$ is the ***output*** indicator function giving the index of the class associated with rule R_k. In our general type model this translates to a c-dimensional binary output vector with 1 at $o(k)$ and 0, elsewhere.
2. Fuzzy rules with a class and a certainty degree in the consequent, e.g.,

   R_k : ... THEN class is $\omega_{o(k)}$ with $z_{k,o(k)}$.

   This corresponds to $g_{k,1} = 0$ AND ... AND $g_{k,o(k)} = z_{k,o(k)}, \ldots,$ AND $g_{k,c} = 0$. In fuzzy set terminology, the output is a possibly subnormal singleton over Ω.
3. Fuzzy rules with certainty degrees for all classes in the consequent, i.e., the general model, where $z_{k,i}$ are certainty degrees, typically in the interval [0,1].

To complete the list, we have to add a fuzzy classifier model where the outputs are linguistic labels expressing the certainty, e.g., $g_ik, i = low$.

4. Fuzzy rules with linguistic labels for the c outputs

   THEN $g_{k,1} = B_{o(1,k)}$ AND ... AND $g_{k,c} = B_{o(c,k)}$,

   where $B_{o(i,k)}$ are linguistic labels defined over a set of certainty values, e.g., the interval [0,1].

The first three groups comply with the TSK system model whereas the fourth one requires MA type of reasoning.

5.3.3 Rules as prototypes

A vague, short, general, and useful postulate in fuzzy pattern recognition is that *any if-then rule defines a prototype and any prototype defines an if-then rule.* Taking this correspondence into account many fuzzy designs can be placed in the prototype classifier paradigm. Consider the point-prototype $\mathbf{v} = [v_1, \ldots, v_n]^T \in \Re^n$ from a set of prototypes $\mathbf{V} \subset \Re^n$. Assume that $\mathbf{v}$ is a typical example of some class $\omega_i \in \Omega$. We can design the following intuitive if-then rule:

$$R : \text{IF } \mathbf{x} \text{ is } \textit{like } \mathbf{v} \text{ THEN class is } \omega_i.$$

If $\mathbf{x}$ is not exactly $\mathbf{v}$ but is close to it, then the rule is not substantially violated. *Closeness* in $\Re^n$ is typically expressed by some metric, e.g., using the Euclidean distance. Points $\mathbf{x}$ far away from $\mathbf{v}$ should not "activate" the rule, i.e., the firing strength should be close to zero.

Breaking R into clauses, we get

$$R : \text{IF } x_1 \text{ is } \textit{like } v_1 \text{ AND } \ldots \text{AND } x_n \text{ is } \textit{like } v_n$$

$$\text{THEN class is } \omega_i.$$

The separate clauses are aggregated by a conjunction type operation $\mathcal{A}_t$. It is reasonable to choose as the membership functions for *like* v_i some kernel-type function on $|x_i - v_i|$ peaking at 0 (see Chapter 3, Section 3.2.4 on page 50). Depending on the individual closeness and the conjunction operation, the closeness may or may not be inversely proportional to the Euclidean distance. This however is not specified in the above postulate whose meaning is only informal.

The inverse correspondence, i.e., from a rule to a prototype, can also be easily constructed. Consider a rule with the same consequent as above and ith antecedent clause "x_i is *medium*". Most often, the fuzzy set corresponding to the linguistic term "medium" defined over a set of values for x_i peaks at a typical value and has a kernel shape. Let the most typical value be called x_i^{medium}. Then the ith clause can be translated into "x_i is *like* x_i^{medium}", which relates the rule with the prototype. Of course, the correspondence just delineated is rather simplistic but we will not be wrong assuming that the majority of the fuzzy systems exploit exactly this type of intuition.

5.3.4 Types of class labels

Having established the relationship between prototypes and rules, we can think of the consequent of a rule as the class label of the respective prototype.

In the representation above the class label is **crisp**, i.e., the rule (prototype) votes for (belongs to) ω_k only. The representation of the output $\mathbf{y}$ as a c-dimensional vector enables us to use **soft class labels** in the rule consequent. Jim Bezdek defines three sets of label vectors [38]

$$N_{pc} = \{\mathbf{y} \in \Re^c \mid y_i \in [0,1]\ \forall i, \text{ and } \exists i, y_i > 0\} = [0,1]^c - \{\mathbf{0}\}; \qquad (5.22)$$

$$N_{fc} = \left\{ \mathbf{y} \in N_{pc} \;\middle|\; \sum_{i=1}^{c} y_i = 1 \right\}; \qquad (5.23)$$

$$N_{hc} = \{\mathbf{y} \in N_{fc} \mid y_i \in \{0,1\},\ \forall i\} = \{\mathbf{e}_1, \ldots, \mathbf{e}_c\}. \qquad (5.24)$$

The $\mathbf{0}$ denotes the zero vector in $\Re^c$. N_{pc} contains **possibilistic** c-dimensional labels, N_{fc} contains **fuzzy** (and **probabilistic**) labels, and N_{hc} contains **crisp** labels.[3] The set $\{\mathbf{e}_1, \ldots, \mathbf{e}_c\}$ is a canonical basis in $\Re^c$.

Example 5.3.1. We reproduce the peach/plum/nectarine example illustrating the use and usefulness of soft labels [38]. Let $\mathbf{Z}$ consist of $\{\mathbf{z}_1 = \textit{peach}$, $\mathbf{z}_2 = \textit{plum}$, $\mathbf{z}_3 = \textit{nectarine}\}$, and let $c = 2 : \omega_1 = \textit{Peaches}$ and $\omega_2 = \textit{Plums}$. The table below shows three partitions of $\mathbf{Z}$ with crisp, fuzzy and possibilistic labels.

	Crisp labels N_{h2}			Fuzzy labels N_{f2}			Possibilistic labels N_{p2}		
Object	$\mathbf{z}_1$	$\mathbf{z}_2$	$\mathbf{z}_3$	$\mathbf{z}_1$	$\mathbf{z}_2$	$\mathbf{z}_3$	$\mathbf{z}_1$	$\mathbf{z}_2$	$\mathbf{z}_3$
Peaches	1	0	0	1	0.2	0.4	1	0.2	0.5
Plums	0	1	1	0	0.8	0.6	0	0.8	0.6

The nectarine $\mathbf{z}_3$ gets (erroneously) full membership in class *Plums* if we consider crisp labels. Thus, soft labeling is an adequate model for situations where assigning class membership is not straightforward. ■

5.3.5 MA fuzzy if-then classifiers

The MA fuzzy if-then classifier is an MA fuzzy if-then system with input $\mathbf{x} \in \Re^n$ and output $g_1(\mathbf{x}), \ldots, g_c(\mathbf{x})$. The rules associate the linguistic values for $\mathbf{x}$ with the class labels. The operation of the MA classifier is exactly the same as the operation of an MA system for function approximation.

[3] Note that this label classification excludes linguistic labels defined over a set of certainty values.

We illustrate the MA classifier with the 15-point synthetic set (Figure 5.11). We can set up (by eye!) the following set of $M = 4$ rules (L standing for *low*, and H for *high*)

R_1 if x_1 is L AND x_2 is H	then	g_1 is H; g_2 is L
R_2 if x_1 is H AND x_2 is L	then	g_1 is L; g_2 is H
R_3 if x_1 is H AND x_2 is H	then	g_1 is L; g_2 is H
R_4 if x_1 is L AND x_2 is L	then	g_1 is L; g_2 is H

We define the two linguistic labels for g_1 and g_2 on the interval [0,1]. The membership function definition is a powerful tuning option for the classifier. Shown in Figure 5.11 are the data set and the membership functions for for x_1 and x_2. The membership functions for g_1 and g_2 were set as in Figure 5.12. To fully specify the MA fuzzy system we choose

- Conjunction = minimum.
- Disjunction = maximum.
- Implication (association) = minimum.
- Defuzzification method = COG.

This MA classifier committed one mistake with the 15-point data set: $\mathbf{z}_8$ (circled in Figure 5.11) was labeled in class ω_1 instead of class ω_2.

5.3.6 TSK fuzzy if-then classifiers

First we illustrate the functioning of a TSK fuzzy if-then classifier on the 15-point set. The same set of rules and membership functions as with the MA classifier is used. The difference is in the consequent part. Instead of being a fuzzy set on [0,1], here the consequent is a polynomial of order 0, corresponding to a target class label. For crisp labeling (N_{h2}) the set of rules becomes

R_1 if x_1 is L AND x_2 is H	then	$g_1 = 1; g_2 = 0$
R_2 if x_1 is H AND x_2 is L	then	$g_1 = 0; g_2 = 1$
R_3 if x_1 is H AND x_2 is H	then	$g_1 = 0; g_2 = 1$
R_4 if x_1 is L AND x_2 is L	then	$g_1 = 0; g_2 = 1$

Applying formula (5.7) and then the maximum membership rule to each of the 15 points, the TSK model misclassified only $\mathbf{z}_8$. We notice that in both MA and TSK systems, logically, rule R_1 and its negation suffice to get the same classification result. Instead of the initial four, the following set of two rules can be used

R_1 if x_1 is L AND x_2 is H	then	g_1 is H; g_2 is L
R_2 if NOT(x_1 is L AND x_2 is H)	then	g_1 is L; g_2 is H

for the MA classifier, and

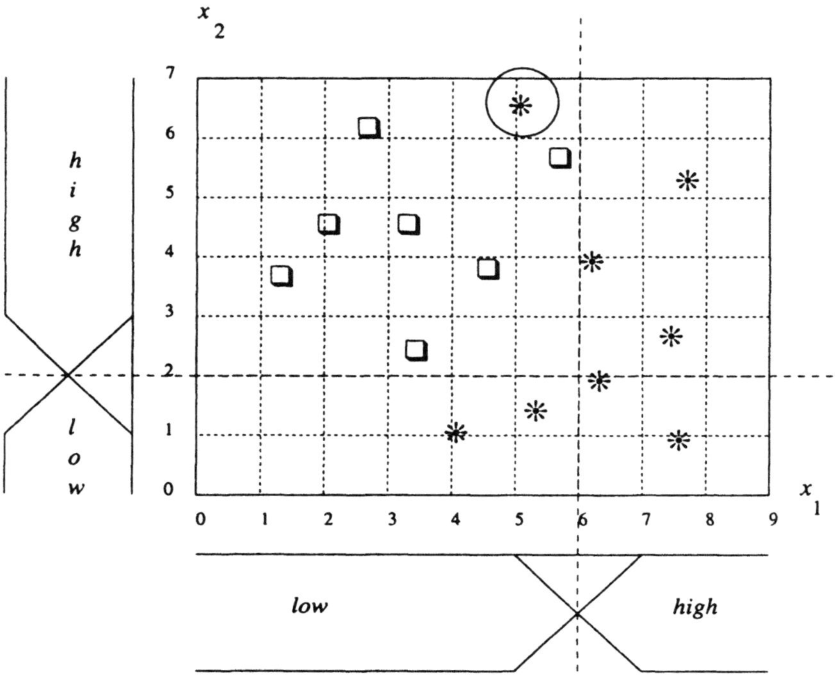

Fig. 5.11. Scatterplot of the 15 labeled points and the membership functions for the two inputs and the class labels for the MA fuzzy if-then classifier.

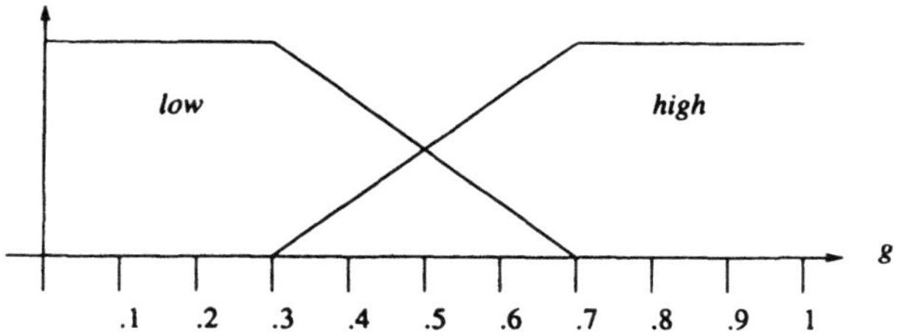

Fig. 5.12. Membership functions for the two inputs and the class labels for the MA fuzzy if-then classifier.

R_1 if x_1 is L AND x_2 is H	then	$g_1 = 1;\ g_2 = 0$
R_2 if NOT(x_1 is L AND x_2 is H)	then	$g_1 = 0;\ g_2 = 1$

for the TSK classifier. With the standard complement for "NOT" applied to the firing strength ($\tau_2 = 1 - \tau_1$) both models misclassified only $\mathbf{z}_8$.

TSK classifier is in fact a richer design than equation (5.7) suggests. The generic TSK model is characterized by

1. The rule-base consisting of M if-then rules of the general type.

2. The conjunction (AND connective): $\mathcal{A}_t$. The *firing strength* of rule R_k is as (5.1)

$$\tau_k(\mathbf{x}) = \mathcal{A}_t\left\{\mu_{1,i(1,k)}(x_1), \ldots, \mu_{n,i(n,k)}(x_n)\right\}. \tag{5.25}$$

3. The calculation of the output.

Five popular fuzzy classifier variants are detailed below. The first four are subsequently used in this chapter for proving various fuzzy it-then classifier properties. The fifth model is used in Chapter 6 for explaining practical designs.

The **TSK1 classifier** is derived from the generic TSK model by specifying

- $z_{k,i} \in \{0,1\}$, $k = 1,\ldots,M$, $i = 1,\ldots,c$, $\sum_{i=1}^{c} z_{k,i} = 1$; (crisp labels)
- $\mathcal{A}_t$ is minimum;
- The ith output of TSK1 is

$$g_i^{TSK1}(\mathbf{x}) = \max_{k=1}^{M}\{z_{k,i}\cdot\tau_k(\mathbf{x})\} = \max_{k=1}^{M}\left\{z_{k,i}\cdot\min_{j=1}^{n}\{\mu_{j,i(j,k)}(x_j)\}\right\}. \tag{5.26}$$

The **TSK2 classifier** is derived from the generic TSK model by specifying

- $z_{k,i} \in \Re$, $k = 1,\ldots,M$, $i = 1,\ldots,c$;
- $\mathcal{A}_t$ is product;
- The ith output of TSK2 is

$$g_i^{TSK2}(\mathbf{x}) = \frac{\sum_{k=1}^{M} z_{k,i}\cdot\tau_k(\mathbf{x})}{\sum_{k=1}^{M}\tau_k(\mathbf{x})} = \frac{\sum_{k=1}^{M} z_{k,i}\prod_{j=1}^{n}\mu_{j,i(j,k)}(x_j)}{\sum_{k=1}^{M}\prod_{j=1}^{n}\mu_{j,i(j,k)}(x_j)} \tag{5.27}$$

The **TSK3 classifier** is obtained from the generic TSK model by specifying

- $z_{k,i} \in \{0,1\}$, $k = 1,\ldots,M$, $i = 1,\ldots,c$, $\sum_{i=1}^{c} z_{k,i} = 1$; (crisp labels)
- $\mathcal{A}_t$ is product;
- The ith output of TSK3 is

$$g_i^{TSK3}(\mathbf{x}) = \max_{k=1}^{M}\{z_{k,i}\cdot\tau_k(\mathbf{x})\} = \max_{k=1}^{M}\left\{z_{k,i}\cdot\prod_{j=1}^{n}\{\mu_{j,i(j,k)}(x_j)\}\right\}. \tag{5.28}$$

The only difference between TSK1 and TSK3 is $\mathcal{A}_t$. Both of them are common fuzzy classifier designs. They assign $\mathbf{x}$ to the most supported class, irrespectively of how many rules vote for each class.

The **TSK4 classifier** differs from TSK2 only by the type of its consequent constants.

The **TSK4 classifier** is obtained from the generic TSK model by specifying

- $z_{k,i} \in \{0,1\}$, $k = 1,\dots,M$, $i = 1,\dots,c$, $\sum_{i=1}^{c} z_{k,i} = 1$; (crisp labels)
- $\mathcal{A}_t$ is product;
- The ith output of TSK4 is

$$g_i^{TSK4}(\mathbf{x}) = \frac{\sum_{k=1}^{M} z_{k,i} \cdot \tau_k(\mathbf{x})}{\sum_{k=1}^{M} \tau_k(\mathbf{x})} = \frac{\sum_{k,z_{k,i}=1} \prod_{j=1}^{n} \mu_{j,i(j,k)}(x_j)}{\sum_{k=1}^{M} \prod_{j=1}^{n} \mu_{j,i(j,k)}(x_j)} \qquad (5.29)$$

The **TSK5 classifier** is derived from the generic TSK model by specifying

- $z_{k,i} \in [0,1]$, $k = 1,\dots,M$, $i = 1,\dots,c$, (soft labels)
- $\mathcal{A}_t$ is product;
- The output is

$$g_i^{TSK5}(\mathbf{x}) = \max_{k=1}^{M} \{z_{k,i} \cdot \tau_k(\mathbf{x})\} = \max_{k=1}^{M} \left\{ z_{k,i} \cdot \prod_{j=1}^{n} \{\mu_{j,i(j,k)}(x_j)\} \right\}. \qquad (5.30)$$

5.4 Universal approximation and equivalences of fuzzy if-then classifiers

While there is an abundance of theoretical results about fuzzy rule-based control systems, few publications explore rigorously the architectural or theoretical aspects of fuzzy if-then classifiers [66, 173, 193]. How good are fuzzy classifiers? In this section we bring together some theoretical results which address this question.

5.4.1 Exact match of the classification boundary

Fuzzy if-then classifiers can produce regions of non-rectangular shape. Consider first an example.

Example 5.4.1. If the true classification region is close to circular, a look-up classifier with rectangular cells will need many rules while only two fuzzy rules with Gaussian membership functions and product as the conjunction operation will match the region. Shown in Figure 5.13 are two classes generated with equal prior probabilities from a normal and a uniform distribution

$$p(\mathbf{x}|\omega_1) \sim N\left([4,2]^T, I\right); \quad \text{and} \quad p(\mathbf{x}|\omega_2) = \begin{cases} \frac{1}{64}, & \text{if } \mathbf{x} \in [-1,7]^2, \\ 0, & \text{otherwise.} \end{cases} \tag{5.31}$$

By I we denote the identity matrix.

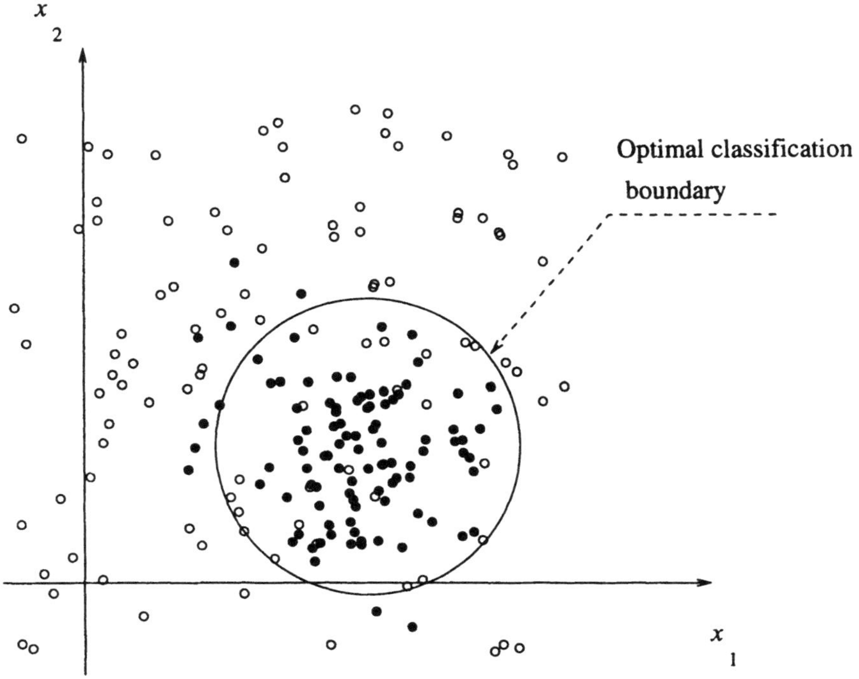

Fig. 5.13. Scatterplot of two classes with a normal and a uniform distribution in $\Re^2$. The circle depicts the Bayes-optimal classification boundary separating the two classes in $[-1,7]^2$.

The Bayes-optimal classification boundary between the classes in $[-1,7]^2$ (ω_1, shown with solid dots and ω_2, with open dots) is a circle centered at the vector of the expectation of ω_1, $[4,2]^T$, and with radius 2.15. The TSK4 classifier that produces this classification boundary uses the following rule base

R : IF x_1 is about 4 AND x_2 is about 2 THEN $g_1(\mathbf{x}) = 1; g_2(\mathbf{x}) = 0.0$.
R : IF x_1 is any AND x_2 is any THEN $g_1(\mathbf{x}) = 0; g_2(\mathbf{x}) = 0.0982$.

The membership function of "any" is 1 for any value of x_j. Consider one prototype located at $[4,2]^T$. The membership functions for A_1 = "about 4" and A_2 = "about 2" are

$$\mu_1(x_1) = \exp\left\{-\frac{(x_1-4)^2}{2}\right\}; \quad \mu_2(x_2) = \exp\left\{-\frac{(x_2-2)^2}{2}\right\}. \tag{5.32}$$

The firing strengths (for product as $\mathcal{A}_t$) are

$$\tau_1(\mathbf{x}) = \exp\left\{-\frac{(\mathbf{x}-[4,2]^T)^T(\mathbf{x}-[4,2]^T)}{2}\right\}, \quad \tau_2(\mathbf{x}) = 1. \tag{5.33}$$

Applying (5.27), the overall output for input $\mathbf{x}$ is

$$g_1(\mathbf{x}) = \frac{1}{C}\exp\left\{-\frac{(\mathbf{x}-[4,2]^T)^T(\mathbf{x}-[4,2]^T)}{2}\right\}; \tag{5.34}$$

$$g_2(\mathbf{x}) = \frac{1}{C}0.0982, \tag{5.35}$$

C being the common denominator. According to the maximum membership rule, $\mathbf{x}$ will be assigned to class ω_1 for $g_1(\mathbf{x}) \geq g_2(\mathbf{x})$, and to ω_2, otherwise. To find the classification boundary we use $g_1(\mathbf{x}) = g_2(\mathbf{x})$, which leads to an equation of a circle with radius $R = 2.15$, as shown in Figure 5.13. ∎

Fuzzy classifiers can match exactly a large class of classification boundaries in $\Re^2$. Here we follow the constructive proof of the Lemma by Klawonn and Klement [173]. They consider a monotonic function $f(x)$ as the classification boundary between two classes in $\Re^2$. We extend the result for an arbitrary f defined on some interval $[a_1, b_1] \subset \Re$.

Lemma 5.4.1. *Let $a_1, b_1, a_2, b_2 \in \Re$ and $a_1 < b_1, a_2 < b_2$. Consider a 2-class problem in $\Re^2$. Let $g(x,y) = 0$ be the equation of the classification boundary between the two classes ω_1 and ω_2. Assume that the function $g(x,y)$ can be represented as $g(x,y) = y - f(x)$ (and the boundary by function $y = f(x)$, respectively) where $f(x)$ is defined on $[a_1, b_1]$ with a range $f([a_1, b_1]) \subseteq [a_2, b_2] \subset \Re$. There exists a TSK1 classifier with 2 rules (a rule and its negation) which produces $g(x,y) = 0$ as the classification boundary.*

To illustrate the proof, Figure 5.14 depicts 300 points from two classes in $\Re^2$ and the classification boundary.

Proof. The classification boundary $g(x,y) = 0$ splits the region of interest $[a_1, b_1] \times [a_2, b_2] \subset \Re^2$ into two classification regions: C_1 (for ω_1) and C_2 (for ω_2). Without loss of generality, assume that the points in C_1 get the positive value of g. We have to show that there exists a TSK1 classifier such that for any point (x,y) in $\Re^2$, such that $g(x,y) > 0$, the classifier yields class ω_1, and for any point (x,y) such that $g(x,y) < 0$, the TSK1 classifier yields class ω_2. For the points on the border ($g(x,y) = 0$), any class label is acceptable.

Consider a TSK1 classifier with the following membership functions (shown on the bottom two plots in Figure 5.14 for the above example)

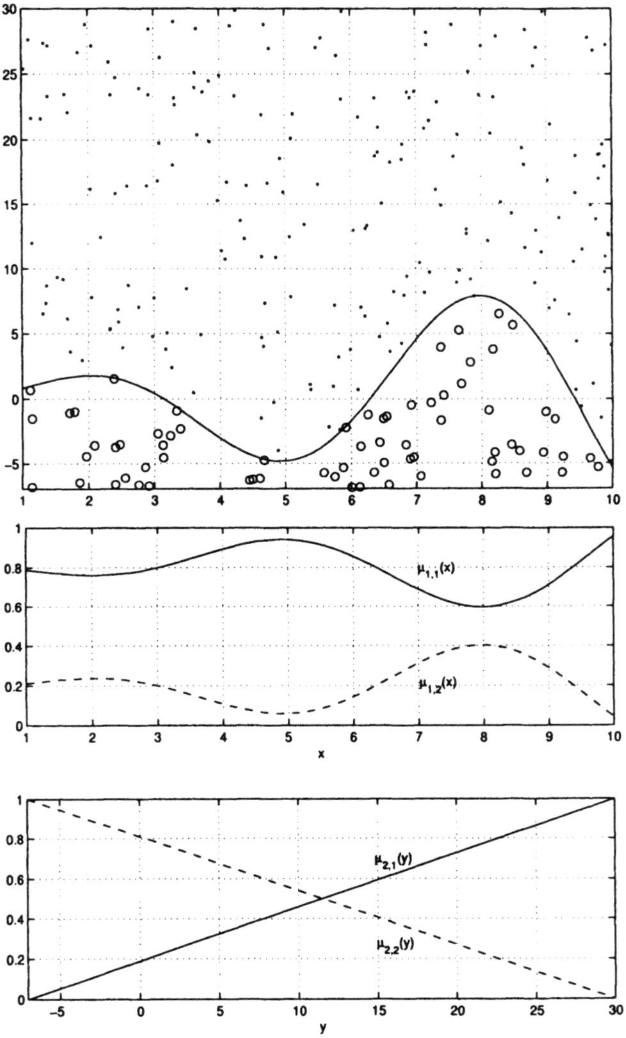

Fig. 5.14. Classification boundary approximated by a TSK1 classifier. Points from ω_1 are depicted with dots, and these from class ω_2 with circles.

$$\mu_{1,1}(x) = \frac{b_2 - f(x)}{b_2 - a_2}, \quad \text{and} \quad \mu_{1,2}(x) = 1 - \mu_{1,1}(x),$$

and

$$\mu_{2,1}(y) = \frac{y - a_2}{b_2 - a_2}, \quad \text{and} \quad \mu_{2,2}(y) = 1 - \mu_{2,1}(y);$$

and rule-base

R_1 : IF x is $A_{1,1}$ AND y is $A_{2,1}$ THEN $g_1 = 1$ AND $g_2 = 0$;

R_2 : IF x is $A_{1,2}$ AND y is $A_{2,2}$ THEN $g_1 = 0$ AND $g_2 = 1$;

Let $\mathbf{x} = [x, y]^T$ be an input vector. Assume that $g(x, y) > 0$, i.e., the input is from class ω_1. Using equation (5.26), the two discriminant functions are

$$g_1(x,y) = \min\{\mu_{1,1}(x), \mu_{2,1}(y)\} = \min\left\{\frac{b_2 - f(x)}{b_2 - a_2}, \frac{y - a_2}{b_2 - a_2}\right\} \tag{5.36}$$

and

$$g_2(x,y) = \min\{\mu_{1,2}(x), \mu_{2,2}(y)\} = \min\left\{1 - \frac{b_2 - f(x)}{b_2 - a_2}, 1 - \frac{y - a_2}{b_2 - a_2}\right\}$$

$$= 1 - \max\left\{\frac{b_2 - f(x)}{b_2 - a_2}, \frac{y - a_2}{b_2 - a_2}\right\}. \tag{5.37}$$

According to the maximum membership rule, the TSK1 classifier assigns ω_1 iff $g_1(x, y) \geq g_2(x, y)$, and class ω_2, otherwise. To derive the class label, form

$$g_1(x,y) - g_2(x,y) = \min\{\mu_{1,1}(x), \mu_{2,1}(y)\} - 1 + \max\{\mu_{1,1}(x), \mu_{2,1}(y)\}$$

$$= \mu_{1,1}(x) + \mu_{2,1}(y) - 1 \tag{5.38}$$

which leads to

$$g_1(x,y) - g_2(x,y) = \frac{y - f(x)}{(b_2 - a_2)}. \tag{5.39}$$

By definition $b_2 > a_2$, and also for the input $[x, y]^T$, from $g(x, y) > 0$, follows that $y > f(x)$, therefore $g_1(x, y) > g_2(x, y)$ and TSK1 labels $[x, y]^T$ in class ω_1.

For points in class ω_2 the numerator of (5.39) takes negative values, which completes the proof. ■

Example 5.4.2. The example in Figure 5.15 shows a classification boundary in $\Re^2$ for the 15-point data. Notice that x and y can be chosen in any way. In this example we choose $x_1 = f(x_2)$ as the discriminant function and construct the TSK1 classifier respectively. The membership functions are shown on the left and below the plot. Notice also that the fuzzy sets for x (in this case x_2) are not necessarily normal. The limits of these membership functions depend on the interval $[a_2, b_2]$ which we select for y (in this case x_1). ■

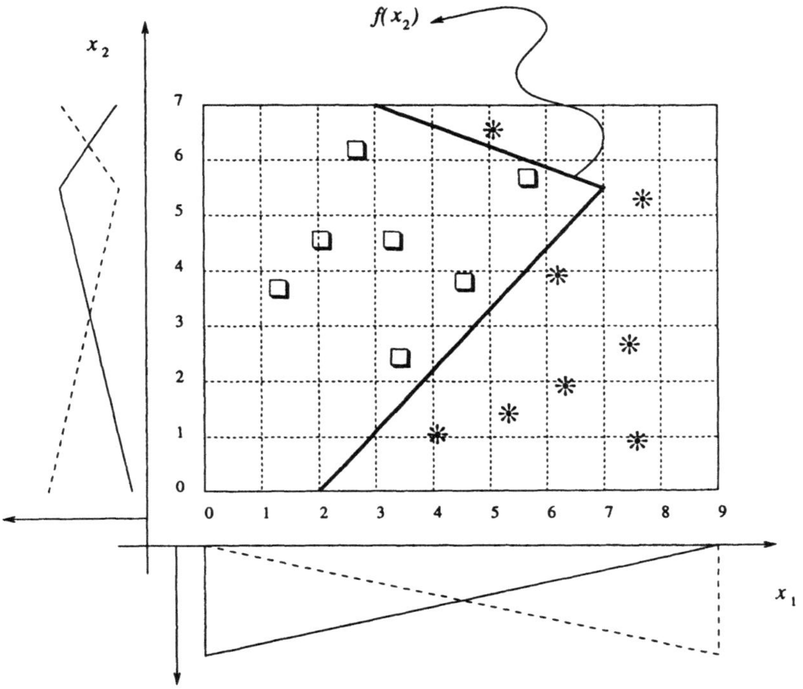

Fig. 5.15. Classification boundary approximated by the TSK1 classifier

The problem with this Lemma (and most of the theoretic results in general) is that they do not give you the tool to build the classifier. The proof is constructive but notice that we *must know* the true discriminant function $g(x, y)$ to build the corresponding TSK1 classifier. If we knew $g(x, y)$ we might wish to use it directly. The value of this Lemma is that it shows that TSK1 is versatile enough to fit any such classification boundary.

Unfortunately, this type of constructive proof cannot be extended beyond the two dimensional case. Let $g(x_1, x_2, x_3) = 0$ be a classification boundary in $\Re^3$. Klawonn and Klement [173] show that even if g is a plane, the exact match by TSK1 (as in $\Re^2$) with a finite number of rules is impossible. They suggest other types of conjunction operations $\mathcal{A}_t$, e.g., product, and then a class of functions g can be matched with a finite number of rules.

Lastly, constructing membership functions as in the Lemma will hardly lead to linguistic interpretability.

5.4.2 Universal approximation

Many proofs exist showing that a type of fuzzy if-then systems can approximate to an arbitrary precision any continuous function g on a compact domain U, i.e., $g : U \to \Re$, $U \subset \Re^n$ **(universal approximation property)** [231, 286, 337, 339, 365, 364, 369]. An extensive account of such methods is presented in [180]. We consider a TSK2 model with Gaussians as the membership functions of the antecedents, centered at points $x_{j,k}$ with widths $\sigma_{j,k}$, $j = 1, \ldots, n$, $k = 1, \ldots, M$, i.e.,

$$\mu_{j,k}(x_j) = a_{j,k} \exp\left\{ -\frac{1}{2} \left(\frac{x_j - x_{j,k}}{\sigma_{j,k}} \right)^2 \right\}, \quad 0 < a_{j,k} \leq 1. \tag{5.40}$$

Notice that each rule has its own "center" or *prototype*. Thus, each feature has M different fuzzy sets defined on its axis (one for each rule). The antecedent part of the rules for an input $\mathbf{x}$ can be interpreted as

IF $\mathbf{x}$ is *like* $\mathbf{x}_k$

with clauses

IF ... x_j is *like* $x_{j,k}$ ...

The fuzzy sets $A_{j,i(j,k)}$ (see the notation in Figure 5.3) are rule-specific and can be denoted simply by $A_{i,k}$. Without losing generality consider a MISO TSK2 classifier approximating one discriminant function $g(\mathbf{x})$. Since there is only one output, denote by z_k the constant in the consequent of rule R_k, $k = 1, \ldots, M$. The output of the TSK2 classifier is

$$g(\mathbf{x}) = \frac{\sum_{k=1}^{M} z_k \prod_{j=1}^{n} \mu_{j,k}(x_j)}{\sum_{k=1}^{M} \prod_{j=1}^{n} \mu_{j,k}(x_j)}. \tag{5.41}$$

Let $\mathcal{G}$ be the class of all functions of the type (5.41), and let d_∞ be a metric on $\mathcal{G}$

$$d_\infty(g^1, g^2) = \sup_{\mathbf{x} \in \Re^n} \{|g^1(\mathbf{x}) - g^2(\mathbf{x})|\}, \quad g^1, g^2 \in \mathcal{G}. \tag{5.42}$$

The proof of the universal approximation of TSK2 classifier is based on the Stone-Weierstrass Theorem.

Stone-Weierstrass Theorem [306]. *Let Z be a set of real continuous functions on a compact set U and let $C[U]$ be the set of all real continuous functions defined on the compact set U. If*

1. *Z is an algebra, i.e., Z is closed under addition, multiplication and scalar multiplication;*
2. *Z separates points on U, i.e., for every $\mathbf{x}, \mathbf{y} \in U$, $\mathbf{x} \neq \mathbf{y}$, there exists a function $f \in Z$ such that $f(\mathbf{x}) \neq f(\mathbf{y})$;*
3. *Z vanishes at no point on U, i.e., for each $\mathbf{x} \in U$ there exists $f \in Z$ such that $f(\mathbf{x}) \neq 0$;*

then the uniform closure of Z consists of all real functions on U, i.e., (Z, d_∞) is dense in $(C[U], d_\infty)$.

Below we reproduce the rest of the proof by Wang [337]. By design $\mathcal{G}$ is nonempty. Also, $\mathcal{G}$ is well defined because the requirement about the coefficients $a_{j,k}$ of the membership functions $(0 < a_{j,k} \le 1)$ and the fact that Gaussians are used as the membership functions ensures that the denominator of (5.27) is always non-zero.

Lemma 5.4.2. *$\mathcal{G}$ is an algebra.*

Proof. Let $g^1, g^2 \in \mathcal{G}$ such that

$$g^1(\mathbf{x}) = \frac{\sum_{k=1}^{M^1} z_k^1 \prod_{j=1}^n \mu_{j,k}^1(x_j)}{\sum_{k=1}^{M^1} \prod_{j=1}^n \mu_{j,k}^1(x_j)}. \tag{5.43}$$

$$g^2(\mathbf{x}) = \frac{\sum_{k=1}^{M^2} z_k^2 \prod_{j=1}^n \mu_{j,k}^2(x_j)}{\sum_{k=1}^{M^2} \prod_{j=1}^n \mu_{j,k}^2(x_j)}. \tag{5.44}$$

1. Show that $\mathcal{G}$ is closed under summation.

$$g^1(\mathbf{x}) + g^2(\mathbf{x}) = \frac{\sum_{k_1=1}^{M^1} \sum_{k_2=1}^{M^2} (z_{k_1}^1 + z_{k_2}^2) \prod_{j=1}^n \mu_{j,k_1}^1(x_j) \prod_{j=1}^n \mu_{j,k_2}^2(x_j)}{\sum_{k_1=1}^{M^1} \sum_{k+2=1}^{M^2} \prod_{j=1}^n \mu_{j,k_1}^1(x_j) \prod_{j=1}^n \mu_{j,k_2}^2(x_j)}. \tag{5.45}$$

The membership functions μ_{j,k_1}^1 and μ_{j,k_2}^2 are Gaussians, therefore their product is also a Gaussian of the class of (5.40). To verify this, consider the following expression corresponding to the one obtained at the power of the resulting exponent

$$\alpha(x-a)^2 + \beta(x-b)^2 = (\alpha+\beta)(x-c)^2 + \frac{\alpha\beta(a-b)^2}{\alpha+\beta}, \tag{5.46}$$

where c is

$$c = \frac{\alpha a + \beta b}{\alpha + \beta}. \tag{5.47}$$

The result is a Gaussian centered at c with width $\sigma = 1/\sqrt{2(\alpha+\beta)}$. Thus the sum of g^1 and g^2 is a function g^3 in $\mathcal{G}$ with $M_1 \times M_2$ terms in the summations.

2. Show that $\mathcal{G}$ is closed under multiplication.

$$g^1(\mathbf{x}) g^2(\mathbf{x}) = \frac{\sum_{k_1=1}^{M^1} \sum_{k_2=1}^{M^2} (z_{k_1}^1 z_{k_2}^2) \prod_{j=1}^n \mu_{j,k_1}^1(x_j) \prod_{j=1}^n \mu_{j,k_2}^2(x_j)}{\sum_{k_1=1}^{M^1} \sum_{k_2=1}^{M^2} \prod_{j=1}^n \mu_{j,k_1}^1(x_j) \prod_{j=1}^n \mu_{j,k_2}^2(x_j)}, \tag{5.48}$$

and this function is also in $\mathcal{G}$ (see the above item).

3. Show that $\mathcal{G}$ is closed under scalar multiplication. For any $c \in \Re$,

$$cg(\mathbf{x}) = \frac{\sum_{k=1}^{M}(cz_k)\prod_{j=1}^{n}\mu_{j,k}(x_j)}{\sum_{k=1}^{M}\prod_{j=1}^{n}\mu_{j,k}(x_j)}, \tag{5.49}$$

hence, $cg \in \mathcal{G}$.

■

Lemma 5.4.3. *$\mathcal{G}$ separates points of U.*

Proof. Let $\mathbf{a}, \mathbf{b} \in \Re^n, \mathbf{a} \neq \mathbf{b}$. Construct a TSK2 classifier with membership functions centered at the components of $\mathbf{a}$ and $\mathbf{b}$, and widths 1 for all antecedent membership functions. Consider the following two rules:

R_1: If $\mathbf{x}$ is *like* $\mathbf{a}$ then $g = 1$;
R_2: If $\mathbf{x}$ is *like* $\mathbf{b}$ then $g = 0$.

The output of the TSK2 classifier is then

$$g(\mathbf{x}) = \frac{\exp\left\{-\frac{1}{2}(\mathbf{x}-\mathbf{a})^T(\mathbf{x}-\mathbf{a})\right\}}{\exp\left\{-\frac{1}{2}(\mathbf{x}-\mathbf{a})^T(\mathbf{x}-\mathbf{a})\right\} + \exp\left\{-\frac{1}{2}(\mathbf{x}-\mathbf{b})^T(\mathbf{x}-\mathbf{b})\right\}}. \tag{5.50}$$

Then

$$g(\mathbf{a}) = \frac{1}{1+\exp\left\{-\frac{1}{2}(\mathbf{a}-\mathbf{b})^T(\mathbf{a}-\mathbf{b})\right\}}; \quad g(\mathbf{b}) = \frac{1}{2} \tag{5.51}$$

Unless $\mathbf{a} = \mathbf{b}$, the two values $g(\mathbf{a})$ and $g(\mathbf{b})$ are different. Therefore, for any two different points in $\Re^n$ there exists a TSK2 design which separates them.
■

Lemma 5.4.4. *$\mathcal{G}$ does not vanish on any point of U.*

Proof. We can design a TSK2 classifier with $M = 1$ rule and nonzero consequent constant. Since all antecedent Gaussians have positive values for any $\mathbf{x} \in \Re^n$, (5.27) yields a nonzero output $g(\mathbf{x})$. ■

The universal approximation is stated as the following theorem.

Theorem 5.4.1. *For any continuous function $f(\mathbf{x})$ on a compact set $U \subset \Re^n$ and any $\epsilon > 0$ there exists a TSK2 classifier (with finite number of rules) with output $g(\mathbf{x})$ such that*

$$\sup_{\mathbf{x}\in U} |f(\mathbf{x}) - g(\mathbf{x})| < \epsilon. \tag{5.52}$$

Proof. The outputs of all TSK2 classifiers form the set $\mathcal{G}$ which has been shown to satisfy the conditions of the Stone-Weierstrass Theorem. The Theorem 5.4.1 is a direct consequence of it. ■

As with the result about the exact match in the previous subsection, this theorem does not tell you how to build the classifier – it is only about the existence of a classifier.

5.4.3 Equivalence between fuzzy TSK classifiers and 1-nn and Parzen classifier

Some popular classifier designs can be shown to be examples of TSK models. The following two theorems [193] establish this relationship.

Theorem 5.4.2. *Let* $\mathbf{V} = \{\mathbf{v}_1, \ldots, \mathbf{v}_M\}$ *be a set of prototypes,* $\mathbf{v}_k \in \Re^n, k = 1, \ldots, M$. *A fuzzy TSK3 classifier with* n *inputs,* c *outputs,* M *rules and membership functions*

$$\mu_{j,k}(x_j) = \exp\left\{-\frac{1}{2}(x_j - v_{j,k})^2\right\}, \quad j = 1, \ldots, n, \ k = 1, \ldots, M \tag{5.53}$$

is equivalent to the nearest neighbor classifier (1-nn).

Proof. The ith output of the TSK3 classifier for class ω_i is

$$g_i(\mathbf{x}) = \max_{k=1}^{M}\left\{z_{k,i} \cdot \prod_{j=1}^{n} \mu_{j,k}(x_j)\right\}$$

$$= \max_{k=1}^{M}\left\{z_{k,i} \cdot \exp\left\{\frac{1}{2}(\mathbf{x} - \mathbf{v}_k)^T(\mathbf{x} - \mathbf{v}_k)\right\}\right\}. \tag{5.54}$$

Since the maximum is used, the value of g_i is determined by the closest neighbor to $\mathbf{x}$ from all prototypes $\mathbf{v}_k$ from class ω_i. This neighbor will produce the highest value of the exponent. The terms corresponding to prototypes from other classes are multiplied by $z_{k,i} = 0$.

Comparing the c discriminant functions, the winner in the maximum membership rule will be the function where the overall closest neighbor (prototype) has been found. This function assigns $\mathbf{x}$ to the class of its nearest neighbor. ∎

Theorem 5.4.3. *Let* $\mathbf{V} = \{\mathbf{c}_1, \ldots, \mathbf{c}_M\}$ *be a set of centers,* $\mathbf{c}_i \in \Re^n, i = 1, \ldots, M$. *A fuzzy TSK2 classifier with* n *inputs,* c *outputs,* M *rules and membership functions*

$$\mu_{j,k}(x_j) = \exp\left\{-\frac{1}{2h^2}(x_j - v_{j,k})^2\right\} \tag{5.55}$$

is equivalent to the Parzen classifier (see Chapter 3).

Proof. The firing strength of rule R_k is

$$\tau_k(\mathbf{x}) = \prod_{j=1}^{n} \mu_{j,k}(x_j) = \exp\left\{-\frac{1}{2h^2}(\mathbf{x} - \mathbf{c}_k)^T(\mathbf{x} - \mathbf{c}_k)\right\}. \tag{5.56}$$

$\tau_k(\mathbf{x})$ differs from the Gaussian kernel K_G (3.35) by a constant A which does not depend on the class label or the rule number k. Hence, substituting AK_G

for τ and absorbing the constant A and the common denominator by $C(\mathbf{x})$, the ith output of the TSK4 classifier (corresponding to class ω_i) is

$$g_i(\mathbf{x}) = C(\mathbf{x}) \sum_{k=1}^{M} z_{k,i} K_G \left(\frac{\mathbf{x} - \mathbf{c}_k}{h} \right). \tag{5.57}$$

Here $z_{k,i}$ acts as an indicator function. Also by dropping $C(\mathbf{x})$ which does not depend on i, we arrive at c discriminant functions equivalent to those of the Parzen classifier (3.40). ■

Notice that the universal approximation and the equivalences have been proven on fuzzy TSK classifiers that have a "prototype flavor". These are not the typical ***transparent*** architectures of fuzzy systems. In the models used for the proofs the linguistic meaning of the fuzzy sets in the antecedents is obscured (prototype-based). Instead of having 3 or 5 linguistic labels for each variable, which is a comprehensible amount, there are M labels per variable in this model.

5.4.4 A caveat: fuzzy classifiers are look-up tables with hyperbox cells

Consider again the example with the 15-point data set. The classification result obtained using the MA and TSK fuzzy classifier models can be easily recovered using a ***crisp*** look-up table as shown in Table 5.3. The classification regions specified in the table are shown with dashed lines in Figure 5.11. Each cell in the look-up table corresponds to a rule with a crisp consequent for one of the classes. So, why fuzzy?

Table 5.3. Crisp classifier (a look-up table) giving the same result (and the same classification regions) as the MA fuzzy if-then classifier in Figure 5.11 on page 136

	$x_1 \leq 6$	$x_1 > 6$
$x_2 \leq 3$	ω_2	ω_2
$x_2 > 3$	ω_1	ω_2

Sometimes we may "overdo" the fuzzy classifier by designing a large rule base. Such fuzzy classifiers are expected to be more accurate but at the same time they become less transparent and approach the basic look-up table classifier. Then the whole point in introducing fuzzy semantics and inference becomes unclear.

Example 5.4.3. Let us apply the results from Lemma 5.4.1 to the 15-point dataset as in example 5.4.2 on page 142. Since the membership functions in this example are not designed to have any linguistic meaning, we shall denote them by s for the solid-line membership functions and by d for the dashed-line membership functions. Then the two rules in the example are represented as

R_1 : IF x is s AND y is s THEN $g_1 = 1$ AND $g_2 = 0$; (class "squares")

R_2 : IF x is d AND y is d THEN $g_1 = 0$ AND $g_2 = 1$; (class "snowflakes")

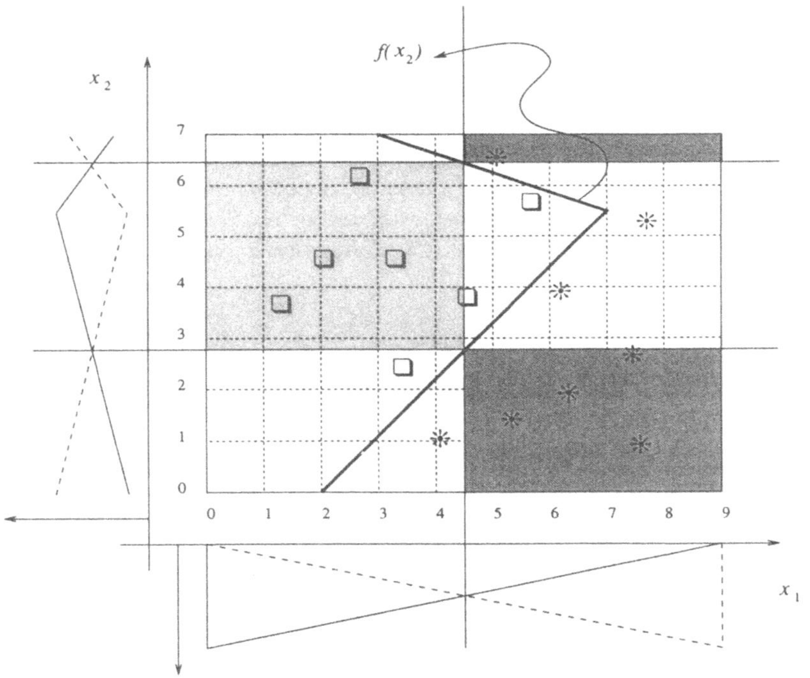

Fig. 5.16. Regions specified by the TSK1 classifier

The shaded regions in Figure 5.16 correspond to the two rules. The light-grey region R_{light} is specified by rule R_1. For any point in R_{light}, the firing strength of R_1 (minimum of the two (solid-line) degrees of membership) is greater than the firing strength of R_2. Therefore all points in R_1 are labeled as "squares". Similarly, the two dark-grey regions correspond to R_2 and the class label for each point in either of them will be "snowflakes". For the rest of the feature space, the firing strengths of the two rules *compete* to label $\mathbf{x}$.

Let us add a third rule R_3 to the above two:

R_3 : IF x is d AND y is s THEN $g_1 = 0$ AND $g_2 = 1$; (class "snowflakes")

By doing this this we fix the region between the two dark ones with class label "snowflakes". Every point in this newly labeled region will be assigned to this class, regardless of the shape of the membership functions. The right-hand side of the feature space is therefore "over-specified" and is a look-up table whose cells are defined by the intercepts of the membership functions. If *all* possible rules constitute the rule base, the result is exactly a look-up table. ■

The following Lemma and Theorem give the formal proof of the observation in the above example.

Lemma 5.4.5. *Consider a TSK1 fuzzy classifier. Let Δ_j be the coincidence set of all membership functions defined on x_j, i.e., the set of all x_j's such that one or more membership functions have the same value. We assume that for any j=1,...,n, Δ_j is a finite union of closed bounded intervals, possibly degenerate*[4]. *Let $B_{j,l}$ be the subset of $\Re$ where $\mu_{j,l}(x_j)$ is the maximal one amongst the K_j membership functions defined on x_j, i.e.,*

$$B_{j,l} = \{x_j, \mid \mu_{j,l}(x,j) > \{\mu_{j,t}(x_j)\},\ t = 1, \ldots, K_j,\ t \neq l\}. \tag{5.58}$$

For simplicity we assume for now that $B_{j,l}$ are compact subsets of $\Re$. Figure 5.17 illustrates the regions $B_{j,l}$.

Let $\mathbf{B}_k \subset \Re^n$ be a hyperbox formulated as

$$\mathbf{B}_k = B_{1,i(1,k)} \times \ldots \times B_{n,i(n,k)}. \tag{5.59}$$

and let R_k be the corresponding rule in the rule-base, i.e.,

R_k : IF x_1 is $A_{1,i(1,k)}$ AND ... AND x_n is $A_{n,i(n,k)}$
THEN ...

Then the firing strength $\tau_k(\mathbf{x})$ satisfies

$$\tau_k(\mathbf{x}) > \tau_s(\mathbf{x}),\ \ \forall \mathbf{x} \in \mathbf{B}_k,\ s = 1, \ldots, M,\ s \neq k. \tag{5.60}$$

Proof. Let $\mathbf{x}$ be a point in $\mathbf{B}_k$. Then $\forall j = 1, \ldots, n$

$$\mu_{j,i(j,k)}(x_j) > \mu_{j,i(j,s)}(x_j),\ \ \forall s = 1 \ldots, M, s \neq k. \tag{5.61}$$

Then

$$\tau_k(\mathbf{x}) = \min_{j=1}^{n} \{\mu_{j,i(j,k)}(x_j)\} > \min_{j=1}^{n} \{\mu_{j,i(j,s)}(x_j)\} = \tau_s(\mathbf{x}),$$
$$\forall s = 1, \ldots, M,\ s \neq k. \tag{5.62}$$

■

[4] Degenerate intervals account for the point intersection of membership functions

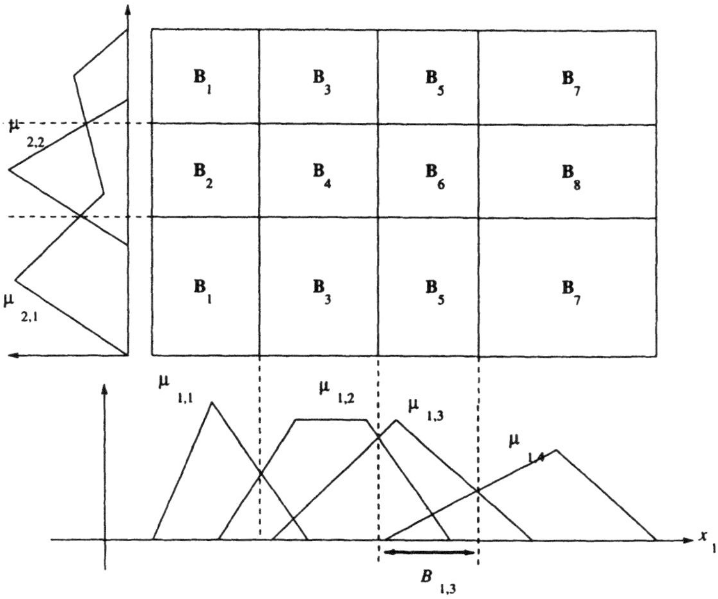

Fig. 5.17. The look-up classifier regions equivalent to the TSK1 regions with all rules in the rule-base.

This lemma shows that each rule R_k defines a hyperbox in $\Re^n$, for any point of which the firing strength τ_k dominates the firing strengths of all other rules.

Theorem 5.4.4. *Consider a TSK1 classifier. Assume that the coincidence sets Δ_j, $j = 1, \ldots, n$ are finite unions of closed bounded intervals, possibly degenerate. If the TSK1 classifier contains all possible rules in its rule-base (all combinations of linguistic labels of the inputs) then it is a look-up classifier with rectangular cells, regardless of the shape of the membership functions used.*

Proof. Let $\mathcal{B}$ denote the feature space covered by the fuzzy sets defined over the feature axes, i.e., for any $\mathbf{x} \in \mathcal{B}$ and for any $j = 1, \ldots, n$, there exists at least one membership function $\mu_{j,t(j)}$ such that $\mu_{j,t(j)} > 0$. (If for some j all degrees of membership are zero, then $\mathbf{x}$ cannot be described by any linguistic term on x_j, and is therefore outside the region of interest). That is,

$$\mathcal{B} = supp\left(\cup_{i=1}^{K_1} A_{1,i}(x_1) \times \ldots \times \cup_{i=1}^{K_n} A_{n,i}(x_n)\right) \subset \Re^n. \quad (5.63)$$

Let $\mu_{j,s(j)}(x_j), j = 1, \ldots, n$, be the n largest nonzero degrees of membership for some $\mathbf{x} \in \mathcal{B}$. Then the rule with the highest firing strength for $\mathbf{x}$ has the antecedent

IF x_1 is $A_{1,s(1)}$ AND ... AND x_n is $A_{n,s(n)}$.

Whatever the class label, this rule is in the rule base by definition, and therefore each $\mathbf{x}$ in $\mathcal{B}$ belongs in a box. Alternatively, $\mathbf{x}$ belongs in the boundary region $\Delta_B \subset \Re^n$ defined as

$$\mathbf{x} \in \Delta_B \iff \text{one or more components of } \mathbf{x} \text{ are in } \Delta_j. \tag{5.64}$$

Then

$$\mathcal{B} = \cup_{k=1}^{M} \mathbf{B}_k \cup \Delta_B. \tag{5.65}$$

The region $\mathcal{B}$ in Figure 5.17 is composed of all boxes $\mathbf{B}_k$ and their borders.

Let $\mathbf{x} \in \mathbf{B}_k$. From the lemma, the firing strength $\tau_k(\mathbf{x})$ dominates the firing strengths of all other rules in $\mathbf{B}_k$ and the TSK1 outputs are

$$g_{o(k)}(\mathbf{x}) = \tau_k(\mathbf{x}), \tag{5.66}$$

and

$$g_i(\mathbf{x}) = \max_{s=1}^{M}\{g_{s,i} \cdot \tau_s(\mathbf{x})\} < \tau_k(\mathbf{x}), \;\; i = 1, \ldots, c, \; i \neq o(k). \tag{5.67}$$

Using the maximum membership rule, the class label assigned to $\mathbf{x}$ (an arbitrary point in box $\mathbf{B}_k$) is $\omega_{o(k)}$. For the points in Δ_B, the classification decision can be made for any of the bordering boxes.

Taking together equation (5.65) and the result that each hyperbox produces one and only one class label, the TSK1 classifier is shown to be a look-up table regardless of the shape of the membership functions used. ■

Three points need further comments.

1. There can be more than one hyperbox per rule. This number is determined by how many times the membership functions of the antecedent clauses dominate the remaining membership functions on the respective features axes. In the example in Figure 5.17, $\mu_{2,1}(x_2)$ is higher than $\mu_{2,2}(x_2)$ on two compact sets thereby defining two boxes when used with any of the fuzzy sets on x_1.
2. Points inside the hyperboxes are covered by *one and only one* rule, and points on the borders, by more than one rule.
3. Equation (5.62) holds for any aggregation based on extended t-norms due to the monotonicity property of t-norms. Therefore, Theorem 5.4.4 also holds for the TSK3 classifier where $\mathcal{A}_t$ is product instead of minimum.

Corollary 5.4.1. *TSK1 and TSK3 fuzzy if-then classifiers are universal approximators.*

The idea of the proof: An integral-based norm d_{int} can be defined, such that by using look-up tables (fuzzy classifiers in this capacity) we can approximate any integrable function and thereby any classification boundary $f(\mathbf{x}) = 0, \mathbf{x} \in \Re^n$ on a compact $U \subset \Re^n$ with an arbitrary precision.

The requirement about coincidence sets $\Delta_j, j = 1, \ldots, n$, merely ensures that the fuzzy classifier has a finite number of regions. Most of the classifiers do, e.g., with triangular, trapezoidal or Gaussian membership function. Can there be fuzzy classifiers with an infinite number of regions? Yes, as the following counter-example suggests. We borrow the example from [333] to show that there exists a setup where the TSK1 classifier has infinitely many regions, and therefore does not meet the definition of a look-up table. Let $x \in [0, \frac{1}{3}]$ be the feature interval, and let the following two membership functions be defined on it

$$\mu_1(x) = 3x \tag{5.68}$$

and

$$\mu_2(x) = 3x - x^2 \cdot \sin\left(\frac{1}{x}\right). \tag{5.69}$$

Assume $\mu_2(0) = 0$. When $\mu_2(x)$ approaches zero from the right, it oscillates around $\mu_1(x)$ with a frequency approaching infinity at zero. Therefore, since the two membership functions intersect infinitely many times, there should be infinitely many alternating regions where one of the functions dominates the other. Figure 5.18 shows μ_1 and μ_2. For clarity we plotted also

$$\mu_3(x) = 3x - x \cdot \sin\left(\frac{1}{x}\right) \tag{5.70}$$

with a higher amplitude of the oscillations.

To avoid situations like this, the authors of [333] require that the membership functions have a local one-sided Taylor expansion everywhere. This means that for each point on the feature axis x_j we can expand the membership function to the left and to the right in a vicinity of the point. The authors point out that most widely used membership functions satisfy this. In the condition parts of Lemma 5.4.5 and Theorem 5.4.4 we do not specify what functions we use but use Δ_j's to restrict their relationship. This is less specific than the assumption in [333] (so functions such as μ_2 are still allowed provided there is no μ_1 on the same feature axis). However, our assumption is more difficult to verify.

Another important consequence of Theorem 3 is that no improvement on the accuracy can be gained by altering the membership functions from one type to another as long as they intersect for the same value of the argument. It is not even helpful to switch to the product instead of minimum as $\mathcal{A}_t$. Thus a fuzzy classifier with Gaussian membership functions and product as $\mathcal{A}_t$, and a classifier with triangular membership functions and minimum as $\mathcal{A}_t$ can give exactly the same result on $\mathcal{B} \subset \Re^n$. Both fuzzy classifiers are identical to a look-up classifier on $\mathcal{B}$.

If we use a look-up table, however, $\mathcal{B}$ is restricted to the area of the table only. Fuzzy classifiers can smooth the edges between regions covered by the rules and the rest of the feature space. If the membership functions do not vanish anywhere on the feature space (e.g., Gaussians), there will be no blank

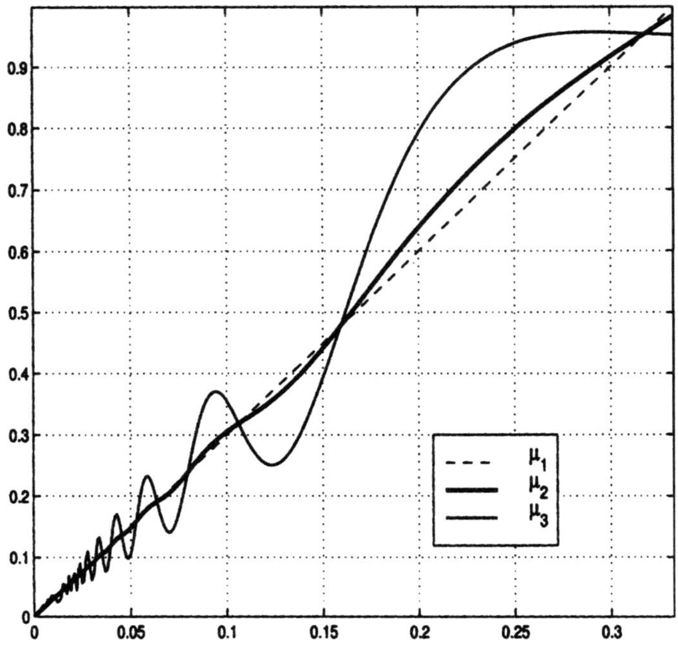

Fig. 5.18. A counter example showing membership functions (μ_1 and μ_2 or μ_1 and μ_3) generating infinitely many boxes.

spots on the feature space, i.e., the fuzzy system will be able to infer a class label for any point in $\Re^n$.

5.4.5 Why fuzzy if-then classifiers then?

Table 5.4 summarizes the results brought about in this section with classifier models TSK1 to TSK4. The universal approximation, the equivalence with prototype-based statistical designs and the look-up table isomorphism pull fuzzy classifiers out of their initial philosophical context. At the beginning, interpretability was perceived to be the most essential bonus of fuzzy classifiers. Now little attention is paid to that, and accuracy renders the main concern.

How can we achieve good interpretability? One way is by using a small number of rules. The lemma by Klawonn and Klement [173] shows that a small number of rules can suffice. However, interpretability of the membership functions associated with these rules is not straightforward. A small number of rules usually implies specific and irregular shapes of the membership functions (needed to achieve a good accuracy) which cannot be associated with linguistic labels on the feature axes. On the other hand, if we adopt simple

Table 5.4. Fuzzy if-then classifier models TSK1 to TSK4 and the main results

Classifier model	TSK1	TSK2	TSK3	TSK4
$\mathcal{A}_t$	minimum	product	product	product
Consequent	binary	real-valued	binary	binary
Universal approximation	look-up table Corollary 5.4.1	Stone-Weierstrass [337]	look-up table Corollary 5.4.1	Bayes-optimality
Prototype-related	No	Yes	Yes	Yes
Main result	Lemma 5.4.1 (Approximation of classification boundaries in $\Re^2$) Theorem 5.4.4 (Reduces to a look-up table)	Universal approximation	Theorem 5.4.2 (Equivalent to 1-nn) Theorem 5.4.4 (Reduces to a look-up table)	Theorem 5.4.3 (Equivalent to Parzen classifier)

models such as triangular or trapezoidal functions, we might need many of them per axis and the inevitable exponential explosion of rules to achieve high accuracy. Again, interpretability will be lost.

6. Training of fuzzy if-then classifiers

6.1 Expert opinion or data analysis?

Assume we have a difficult pattern recognition problem which can easily be handled by a human but not by a machine. Assume also that the human recognition process is difficult to articulate or express in any functional or algorithmic form. Examples of such tasks are face recognition and speaker verification. In some problems we have some knowledge about the classes. An example is handwriting recognition where the theoretical shapes, connections, loops, etc. for each symbol are known, so the "ideal" prototype for each class is described by a set of rules. Nevertheless, handwriting recognition by a machine (and sometimes by a human) is still a challenge. Two natural approaches to designing a classifier are

- Ask an expert how they solve the problem and try to encapsulate the knowledge in a fuzzy rule-base classifier.
- Collect input-output data (i.e., a labeled data set) and extract the classifier parameters from the data.

While the first model is interpretable in the domain context (the classifier is said to be **transparent**), the model based on data may or may not be interpretable. Fuzzy models are deemed to be able to integrate both human and data sources. There is no unique way for this integration, so the literature offers a whole spectrum of possible fuzzy classifier designs ranging from entirely expert-driven to entirely data-driven ones. Data-driven classifiers are usually more accurate and less interpretable than expert-driven classifiers. Practice has given rise to the interpretability-accuracy dilemma [33, 301]. It seems that the two characteristics have a compensatory behavior. Highly accurate classifiers usually need a sophisticated classification mechanism (e.g., many if-then rules, intricate formulas, etc.) which cannot easily be translated into the domain jargon.

Most of the recent effort in fuzzy classifier design is focused on deriving the classifier from data. There are studies that look at a combination of expert and data sources in classifier design (e.g., [144, 338]) but we will confine this chapter to data-driven fuzzy designs only.

A hallmark of fuzzy classifiers is their heuristic nature which makes it difficult to group them in a coherent taxonomy. Several lines have been explored so far.

1. *Tuning the consequents.* Fix the rule-base, specify the respective membership functions of the antecedents and tune the consequent parts of the rules only.
2. *Selecting rules.* Keep the antecedent membership functions fixed and choose the best subset of rules. Genetic algorithms are a common choice for this task.
3. *Tuning the whole rule-base.* The system is built and tuned by simultaneous adaptation of the rule-base and the antecedent membership functions.

The next sections present examples from the above groups. A few representatives have been picked from the big family of fuzzy if-then designs (not necessarily the best ones!). The criteria for selecting the representatives have been

- easy to understand and code,
- take reasonable training and execution time,
- have been a basis for modifications by other authors.

6.2 Tuning the consequents

6.2.1 Crisp consequents

Consider a fuzzy if-then classifier with M rules of the general type

R_k : IF x_1 is $A_{1,i(1,k)}$ AND ... AND x_n is $A_{n,i(n,k)}$

THEN $g_{k,1} = z_{k,1}$ AND ... AND $g_{k,c} = z_{k,c}$.

Let $z_{k,i} \in \{0,1\}, i = 1,\ldots,c,\ k = 1,\ldots,M$, and $\sum_{i=1}^{c} z_{k,i} = 1$ (crisp labels). Let the AND be implemented as an n-place extension of a t-norm (see Chapter 4, page 95). As shown in the previous chapter (lemma 5.4.5), each such rule defines a hyperbox or a set of hyperboxes in $\Re^n$, where the firing strength $\tau_k(\mathbf{x})$ dominates all firing strengths $\tau_s(\mathbf{x}), s = 1,\ldots,M, s \neq k$. To determine the crisp class label for the consequent of rule R_k, we use the available data set $\mathbf{Z} = \{\mathbf{z}_1, \ldots \mathbf{z}_N\}$.

Example 6.2.1. A two-dimensional 3-class example is shown in Figure 6.1: the classes are ω_1 (open circles), ω_2 (squares) and ω_3 (closed circle). Depicted are the relevant membership functions for rule R_k (solid lines), the corresponding 2-d box, and the points from $\mathbf{Z}$ that fall in the box. We can view each data point $\mathbf{z}_j$ in the box as a labeled prototype. Taking into account the correspondence between rules and prototypes outlined in 5.3.3, each $\mathbf{z}_j$ in the box generates a copy of R_k with its class label as the consequent. Therefore, there will be rules with the same antecedents and different consequents, called "contradicting rules". The contradiction can be resolved intuitively by assigning to R_k the label of the most represented class. Hence, the rule in Figure 6.1 is labeled to class ω_1 because the number of open circles is greater than the number of elements from either of the other two classes. ■

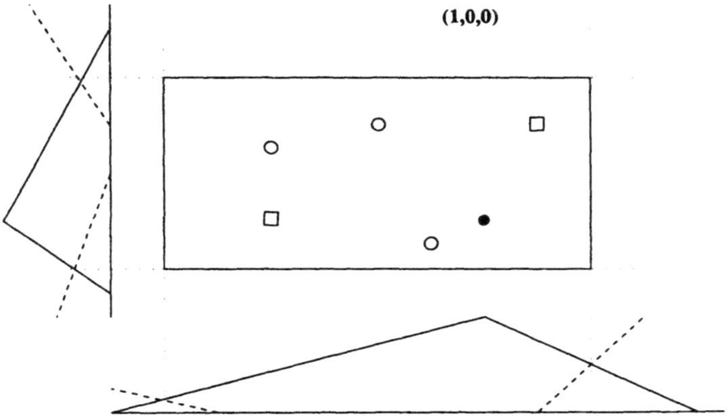

Fig. 6.1. Calculation of the consequent of the rule corresponding to the rectangle shown. A crisp label is obtained (shown in boldface) in favor of class ω_1 (open circles) because their number is greater than the number of elements from either of the other two classes.

Sztandera and Hayashi use this labeling in a TSK classifier for diagnosis of hepatitis [314].

6.2.2 Fuzzy consequents

Here we consider two different schemes for assigning a soft consequent to a rule: the *average support* method due to Ishibuchi and Nakashima [145] and the *maximum support* method which is a classification version of the Wang and Mendel rule generation method [338]. The names of the methods suggest how the consequents are obtained from the data points in the hyperbox.

Average support method. The soft class label in this model has the form

$$z_{k,i} = \begin{cases} \beta_{k,i} \in [0,1], & i = o(k) \\ 0, & \text{otherwise.} \end{cases} \tag{6.1}$$

Following the model from [145], each feature x_j, $j = 1, \ldots, n$, is scaled in the interval [0,1]. The antecedent membership functions are triangular, symmetric, and are placed as a grid on the hyperbox $[0,1]^n$. Figure 6.2 shows the membership functions for a 2-dimensional problem with two and five membership functions per axis. The shaded boxes correspond to the rules whose antecedent membership functions are shown with solid lines. Denote by t the number of membership functions per axis, called "resolution".

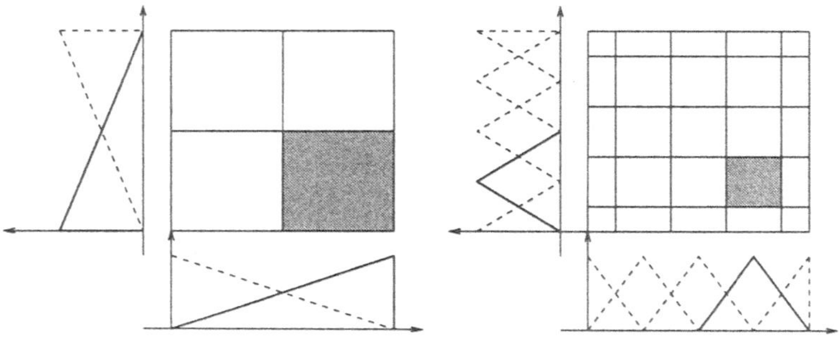

Fig. 6.2. Membership functions of the antecedents and the corresponding regions with the highest firing strength in $[0,1]^2$.

The sth membership function on x_j for resolution t ($t \geq 2$) is

$$\mu_{j,s}^t(x_j) = \max\left\{1 - \frac{|x_j - a^{t,s}|}{b^t}, 0\right\}, \tag{6.2}$$

where the constants $a^{t,s}$ and b^t determine the centers and the spreads of the membership functions

$$a^{t,s} = \frac{s-1}{t-1}, \quad b = \frac{1}{t-1} \quad s = 1, \ldots, t. \tag{6.3}$$

Let $\mu_{k,i(j,k)}(x_j)$ denote the antecedent membership function of the jth clause of rule R_k for some fixed t (the superscript will be omitted for convenience). To obtain the consequent value $\beta_{k,i}$, each possible rule is evaluated with respect to the existing labeled data set $\mathbf{Z} = \{\mathbf{z}_1, \ldots, \mathbf{z}_N\}$. The following procedure is applied.

1. Using the antecedent membership functions for rule R_k, calculate the c sums $\beta_1, \ldots, \beta_c$ of firing strengths, one for each class $\omega_i \in \Omega$

$$\beta_i = \sum_{\mathbf{z} \in \mathbf{Z}} Ind(\mathbf{z}, \omega_i) \tau_k(\mathbf{z}). \tag{6.4}$$

where Ind is the indicator function (3.2). The firing strength $\tau_k(\mathbf{z})$ is obtained according to the chosen AND operation.

2. Find the maximal one amongst the sums,

$$\beta_{i^*} = \max_{1\le i\le c} \beta_i \tag{6.5}$$

to determine the index of the winning class, $o(k) = i^*$. Set all other consequent constants to zero: $z_{k,i} = 0,\ i = 1,\dots,c,\ i \neq o(k)$. Calculate $z_{k,o(k)}$ as

$$z_{k,o(k)} = \frac{(c+1)\beta_{o(k)} - \sum_{i=1}^{c}\beta_i}{c\sum_{i=1}^{c}\beta_i}. \tag{6.6}$$

As soon as the rule-base is designed the classifier is ready to operate. The name average support comes from the fact that for any rule R_k, *all* data points from class ω_i give their support to calculate β_i, $i = 1,\dots,c$.

The TSK5 model with the average support scheme for tuning the consequents was applied to the Cone-torus and Normal mixtures data for resolution t varying from 2 to 15. The results are plotted in Figure 6.3. Training and testing errors are shown against the number of rules (t^2 for each data set because both data sets have 2 features) and are also presented in Tables A.5 and A.15 in the Appendix. The testing error is shown with a thick line.

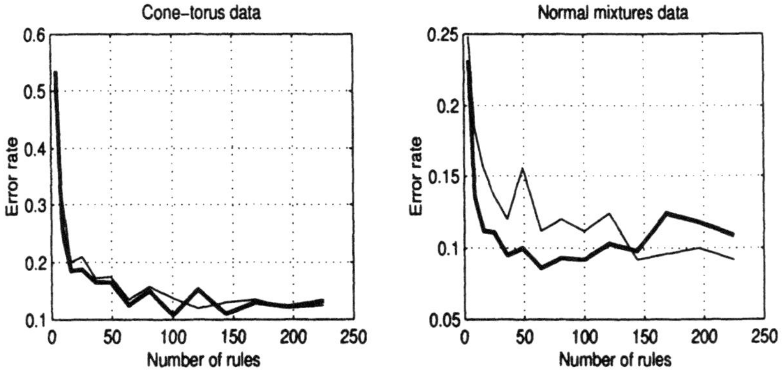

Fig. 6.3. Error rates of a TSK5 classifier with average support method for tuning the consequents. The error rates are plotted against the number of rules, the thick lines showing the testing error.

The best training error with the Cone-torus data was 12.0 %, obtained for $t = 11$ (121 rules) and the corresponding test error was 15.25 %. With the Normal mixtures data, the best training error was 9.20 % obtained for $t = 12$ (144 rules), and the corresponding testing error was 9.80 %. Experiments with the other two data sets (Phoneme and Satimage) were not feasible because of the *curse of dimensionality* involving significant execution time.

The calculation of β_i's is not related in any way with the type of the antecedent membership functions. Although the functions in the current model were defined through (6.2) and (6.3), any other specification will do as well.

Example 6.2.2. Take the hyperbox from Example 6.2.1. Depicted in Figure 6.4 is the box with the firing strength displayed next to each point. The average support method takes into account not only the points *inside* the box but all points with nonzero firing strength for the rule R_k. Three such outside points are shown in the figure with their firing strengths. We assume that the nine points in the are the only ones in $\mathbf{Z}$ with nonzero $\tau_k(\mathbf{z}_j)$.

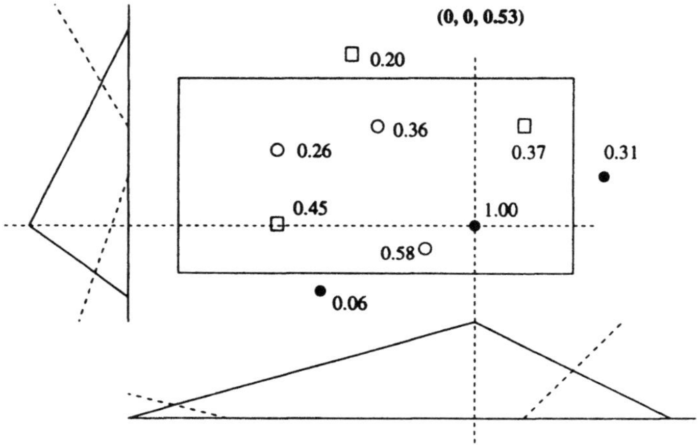

Fig. 6.4. Calculation of the c consequents of the rule by the average support method. The firing strength is displayed next to each sample and the obtained soft label of the rule is shown above the graph in boldface.

The calculation of the label goes through calculation of β_i's as the sums of the respective firing strengths

$$\begin{aligned} \beta_1 &= 0.58 + 0.26 + 0.36 = 1.20 \\ \beta_2 &= 0.45 + 0.37 + 0.20 = 1.02 \\ \beta_3 &= 0.06 + 0.31 + 1.00 = 1.37 \end{aligned} \tag{6.7}$$

The winning class is ω_3, i.e., $i^* = o(k) = 3$ and

$$\begin{aligned} z_{k,1} &= 0 \\ z_{k,2} &= 0 \\ z_{k,3} &= \frac{(3+1) \times 1.37 - (1.20 + 1.02 + 1.37)}{3 \times ((1.20 + 1.02 + 1.37)} \approx 0.53. \quad \blacksquare \end{aligned} \tag{6.8}$$

Maximum support method. Wang and Mendel [338] propose an algorithm for designing a fuzzy TSK MISO system from data. The algorithm allows incorporating if-then rules defined by the domain expert. The authors call the method "model free" which translates as "nonparametric classification" in the pattern recognition jargon of Chapter 2. The rule base is specified in advance, together with the antecedent membership functions.

The algorithm can be used for classifier design. The classifier model is TSK2 with $z_{k,i}$ constricted to the interval [0,1].

Wang and Mendel's algorithm trains the consequent values in the following way:

1. Use the membership functions on each axis to specify all possible (hyperbox) rules.
2. For each rule R_k,
   a) Find all elements of the data set $\mathbf{Z}$ that fall in the respective hyperbox, i.e., all points for which the firing strength of R_k is the highest amongst all rules.
   b) For each class ω_i, pick the highest firing strength $\tau_k(\mathbf{z}_j)$ among all $\mathbf{z}_j$ labeled in ω_k ***in the hyperbox***. Assign this value to the consequent $g_{k,i}$ of rule R_k. This procedure specifies a soft class label for rule R_k.

Example 6.2.3. Consider again the rule R_k represented by the 2-d box as in example 6.2.1. Figure 6.5 shows the data points in the box with their firing strengths. Wang and Mendel's method will assign a soft label with regard to *where* the samples are, and not *how many* of them are in the box. By this method, R_k gets the following consequent values

$$\begin{aligned} z_{k,1} &= \max\{0.26, 0.36, 0.58\} = 0.58 \\ z_{k,2} &= \max\{0.45, 0.37\} = 0.45 \\ z_{k,3} &= 1.00. \end{aligned}$$

■

Once the consequents are specified, the TSK2 classifier is ready to operate. For an input $\mathbf{x}$, the ith output is calculated by

$$g_i(\mathbf{x}) = \frac{1}{C(\mathbf{x})} \sum_{k=1}^{M} \tau_k(\mathbf{x}) z_{k,i}, \tag{6.9}$$

where $C(\mathbf{x})$ is the common denominator. The input $\mathbf{x}$ is assigned to the class with the highest support.

The maximum support model was tested on the Cone-torus and Normal-mixtures data with triangular membership functions of the antecedents, as in the average support method (see equation 6.2 and Figure 6.2). The two coordinate axes were split using t fuzzy sets, with the resolution t varying from 2 to 17, thereby defining 16 different classifiers with 4 to 289 rules. Figure 6.6 plots the training and testing classification errors for the two data

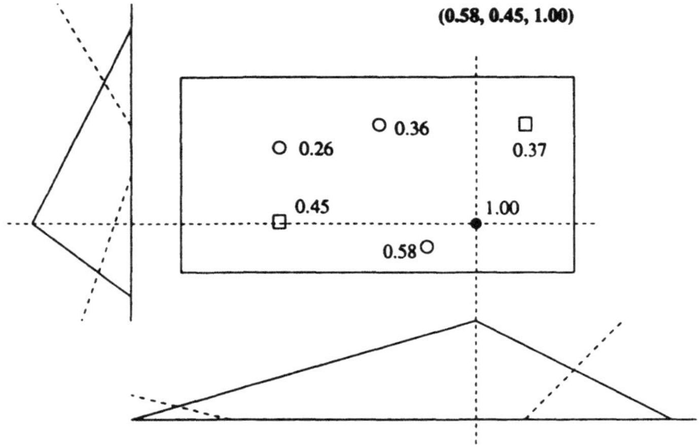

Fig. 6.5. Calculation of the c consequents of the rule corresponding to the shown rectangle by the maximum support method. The firing strength is displayed next to each sample and the obtained soft label of the rule is shown above the graph in boldface.

sets (testing error is the thick line). The lowest training error with the Cone-torus data, 10.50 %, was found at 169 rules and the corresponding test error was 9.75 %. For the Normal-mixtures data, the lowest training error was 8.40 % at 225 rules and the corresponding testing error was 10.10 %. The results are also displayed in Tables A.6 and A.16 in the Appendix.

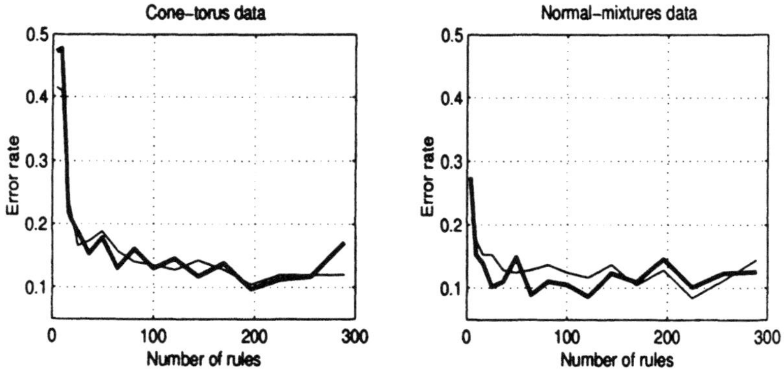

Fig. 6.6. Error rates of a TSK2 classifier with maximum support method for tuning the consequents with the Cone-torus and Normal-mixtures data. The error rates are plotted against the number of rules, the thick line showing the testing error.

6.2.3 Distributed fuzzy rules

The **distributed representation of fuzzy rules** was proposed by Ishibuchi [146]. This method employs a TSK5 classifier. $L-1$ rule-bases of different resolutions compete to label the input $\mathbf{x}$. Each rule-base consists of t^n rules (n is the dimensionality of the feature space), using t fuzzy sets on each feature axis. The rule-bases are constructed for values of t from 2 to L. They are then pooled and used as one single rule-base. The rules and their consequents are formed exactly as in the average support method. The difference is that now for a specified L we have $L-1$ different rule-bases with $t = 2, 3, \ldots, L$ fuzzy sets on each feature axis. So, the total number of rules M is calculated as

$$M = \sum_{t=2}^{L} t^n. \tag{6.10}$$

Thus, for $L = 4$ in a 4-dimensional feature space, $M = 2^4 + 3^4 + 4^4 = 353$.

Note that the spatial pattern of the classification regions is not as simple as that in Theorem 5.4.4 because the regions of the rules are superimposed, and also because the singletons in the consequents can be subnormal.

Example 6.2.4. The distributed fuzzy rule classifier has been applied to the Cone-torus and the Normal-mixtures data. The results with 5 values of L are shown in Tables 6.1 and 6.2. The training and the test error rates and the number of rules in the rule-base are given. ■

Table 6.1. Error rates of fuzzy distributed-rule classifier with the Cone-torus data.

	$L=3$	$L=4$	$L=5$	$L=6$	$L=7$
Number of rules	13	29	54	90	139
Training error [%]	35.75	22.00	19.75	16.50	16.50
Testing error [%]	37.50	18.50	18.50	17.00	17.00

The results show that the classifier does not offer any dramatical improvement over the basic statistical designs. Besides, the number of rules is rapidly growing with the dimensionality, and the application of this model to higher dimensions is infeasible. For example, with the Phoneme data, even for $L = 3$, the number of rules is already $2^5 + 3^5 = 275$, and the time to design and test the classifier is significant. Interpretability of the membership functions is possible, but the excessive number of rules hampers the overall transparency of the system. The distribution rule model is important as an example of the first generation fuzzy if-then classifiers that learn from data.

Table 6.2. Error rates of fuzzy distributed-rule classifier with the Normal-mixtures data.

	$L=3$	$L=4$	$L=5$	$L=6$	$L=7$
Number of rules	13	29	54	90	139
Training error [%]	22.80	12.80	12.80	12.40	12.80
Testing error [%]	20.90	10.20	9.50	8.80	9.10

An interesting follow-up of the partitioning method is the *selective partitioning method* [145, 147, 240]. The regions where the consequent singleton does not have a particularly high value (regions where the classes overlap) are further split with a finer resolution until the "level of contradiction" between the rules drops below a certain threshold. This classifier may overtrain, but this is true for any flexible classification model. The real disadvantage of this type of methods comes from the curse of dimensionality.

6.3 Tuning the antecedents

Methods in this group aim at generating a "good" rule base, i.e., specifying the antecedent clauses and the membership functions needed.

6.3.1 Means method

The *Means* method due to Ishibuchi and Nakashima [145] is perhaps the simplest in this group. Consider a TSK5 model. The c class means are calculated from the data set $\mathbf{Z}$ as in (3.6)

$$\mathbf{m}_i = \frac{1}{N_i} \sum_{\mathbf{z}_j} Ind(\mathbf{z}_j, \omega_i)\mathbf{z}_j \tag{6.11}$$

where N_i is the total number of elements of $\mathbf{Z}$ from class ω_i. The rule base consists of c rules, one per class. Each such rule has as its consequent a binary vector with 1 at the place of the class index and zeros elsewhere. The membership function for class ω_i on feature axis x_j is an exponent centered at the jth component of the mean $\mathbf{m}_i$

$$\mu_i(x_j) = \exp\left\{-\frac{1}{2}\left(\frac{x_j - m_{i,j}}{\sigma_{i,j}}\right)^2\right\}. \tag{6.12}$$

The width parameter $\sigma_{i,j}$ is the standard deviation of x_j for class ω_i (estimated on the elements of $\mathbf{Z}$ from class ω_i). Designed in this way, the TSK5 classifier is a simplified version of the quadratic discriminant classifier from Chapter 3. The difference is that instead of a full covariance matrix per class, here the covariance matrices are diagonal. Besides, the normalizing coefficient of the Gaussian is not included in the TSK classifier which can incur further errors compared to the QDC. Notice that the matrices are class-specific, therefore the classification boundaries will be quadratic curves as shown in Figure 3.4 (Chapter 2). Diagonal covariance matrices reflect the assumption that the classes are shaped as (possibly different) hyperellipses whose axes are parallel to the coordinate axes (independent features). The following results were obtained with this method on the four data sets: Cone-torus: training error 28.0 %, testing error 25.25 %; Normal mixtures data: training error 16.8 %, testing error 13.3 %; Phoneme data: training error 26.0 %, testing error 25.75 %; Satimage data, training error 75.0 %, testing error 72.62 %. The Means method is simple but not sufficiently accurate.

6.3.2 Histogram method

Ishibuchi and Nakashima [145] suggested that instead of using Gaussians, we can use histograms to design the membership functions of the antecedents. This idea corresponds to nonparametric classifier designs discussed in Chapter 3. The rule base consists of c rules, one per class, with binary consequents (as in the Means method). The ith membership function for the jth feature is obtained from the histogram of x_j taken over the elements of class ω_i from $\mathbf{Z}$. The number of bins per feature (the resolution t) is picked in advance. Triangular membership functions are associated with each bin, calculated through (6.2).

Let B_j^k denote the the kth bin on the jth feature axis with center a^k, as in equation (6.3). First we calculate

$$h_{i,j}^k = \frac{1}{N_i} \sum_{\mathbf{z}, z_j \in B_j^k} Ind(\mathbf{z}, \omega_i) \mu_{k,i(j,k)}(z_j). \tag{6.13}$$

The membership function of rule R_i for class ω_i on feature x_j is obtained in a tabular form, i.e. $\{(a^1, h_{i,j}^1), \dots, (a^t, h_{i,j}^t)\}$. The fuzzy set is then normalized so that the highest degree of membership is 1. For an input $\mathbf{x}$, c membership values for its jth component are calculated. Linear approximation between the tabulated values is used.

Example 6.3.1. Figure 6.7 illustrates the histogram method for tuning the antecedents. Class ω_1 from the Normal mixtures data was used to calculate the membership function $\mu_{1,1}$ on x_1 for class ω_1. The figure shows the initial partition of the x_1 axis into 6 bins by six membership functions with cutpoints 0.1, 0.3, 0.5, 0.7 and 0.9. The true (normalized) histogram using these bins is shown in the bottom plot and the membership function $\mu_{1,1}(x_1)$ is overlayed (the thick line).

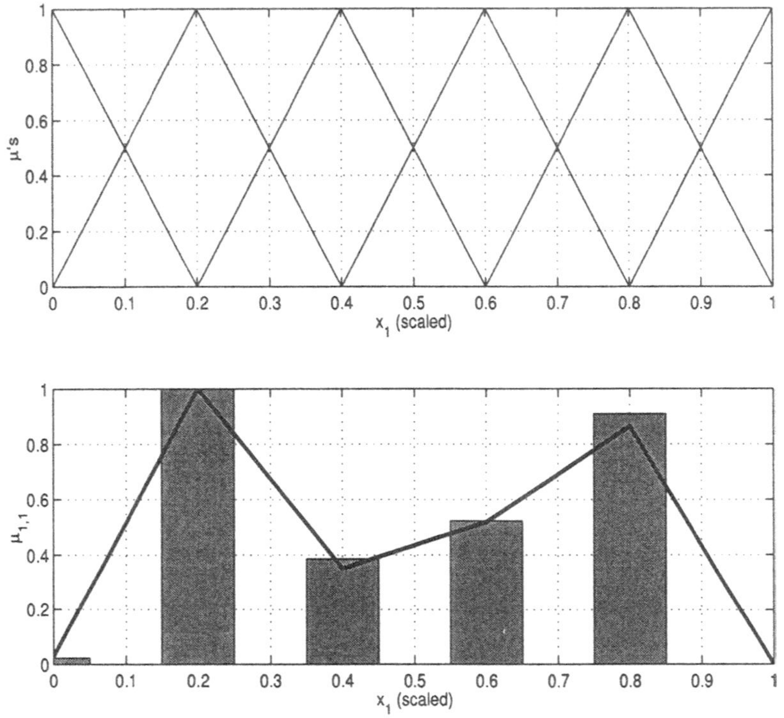

Fig. 6.7. Top plot: partitioning for feature x_1 with 6 fuzzy sets specified by (6.2). Bottom plot: the membership function for class ω_1 of the Normal mixtures data set obtained by the Histogram method (the thick line), and the true (normalized) histogram using the bins defined in the upper plot.

For an input $\mathbf{x}$ in the Normal mixtures example, the four degrees $\mu_{i,j}(x_j)$ are calculated, $i = 1, 2$, $j = 1, 2$, and the two discriminant functions, $g_1(\mathbf{x})$ and $g_2(\mathbf{x})$, are obtained by the TSK5 model. The two membership values, $\mu_{1,1}(x_1)$ and $\mu_{1,2}(x_2)$, are used in R_1, the rule responsible for class ω_1. R_2 determines the membership in class ω_2.

The histogram method was applied to the four data sets for number of bins varying from 3 to 48 with offset 5. Figure 6.8 plots the results from applying the histogram method to the four data sets. The exact figures are given in Tables A.7, A.17, A.25 and A.30 in the Appendix. Training and testing errors are shown against the number of rules and the testing error is shown with a thick line.

The best training error with Cone-torus data was 14.25 %, obtained for 43 bins per axis, and the corresponding test error was 19.00 %. With the Normal mixtures data, the best training error was 10.80 % obtained for 33

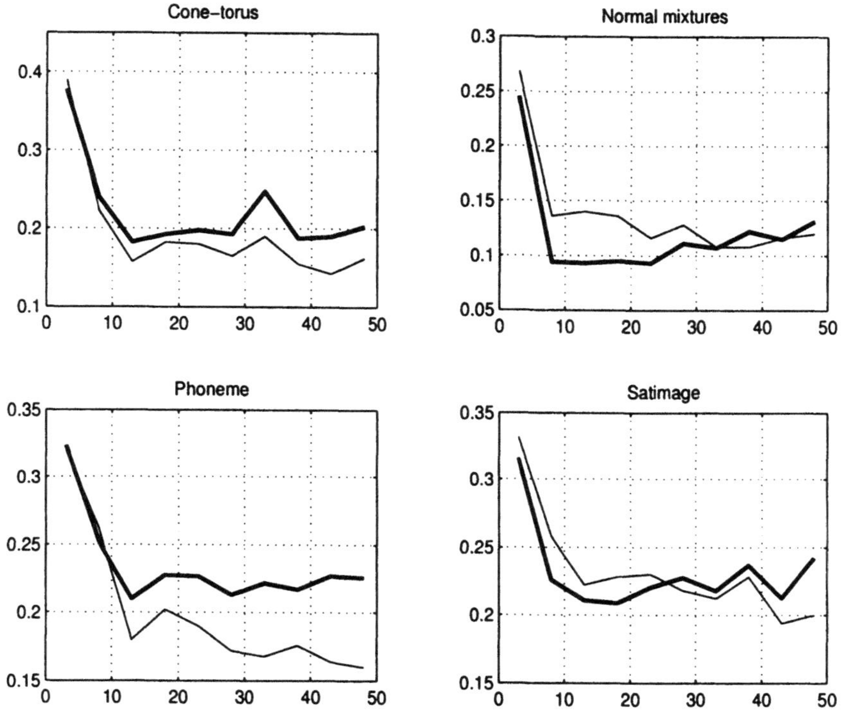

Fig. 6.8. Error rates for the TSK5 classifier using the Histogram method for training the antecedents, with the Cone-torus, Normal-mixtures, Phoneme and Satimage data. The error rates are plotted against the number of rules, the thick line showing the testing error.

bins per axis, and the corresponding testing error was 10.70 %. With the Phoneme data, the best training rate was 16.00 % with 48 bins per axis, with corresponding testing error of 22.55 %. The Satimage data gave 19.40 % training error with 43 bins per axis and 21.23 % corresponding test error. ∎

The Means and the Histogram methods are not expected to produce particularly good results. However, they are perhaps the two most natural and simple ideas for devising membership functions from a given data set.

6.3.3 Min-max (hyperbox) fuzzy designs

Simpson's Min-max fuzzy classifier. The Fuzzy Min-max classifier was proposed by Simpson [307] as a neural network model. It has a TSK1 representation with trapezoidal membership functions for the antecedents and average as the $\mathcal{A}_t$. The *Min-max* name of the scheme comes from the geometrical representation of the rules as hyperboxes with lower (min) and upper (max) corners in the n-dimensional feature space. The degree of membership (firing strength) inside the hyperbox is set to 1, and is declining along the "rim" of the hyperbox. The function has a peculiar shape with "stripe tails" across the feature space. Each rule is responsible for one class only. The number of hyperboxes (rules) is not specified in advance but is learned during training. Thus, each class may have many representatives in the rule base. The hyperboxes for different classes do not intersect, i.e., if $\mathbf{x}$ is inside hyperbox k (the "core" of rule R_k), the firing strength of R_k for $\mathbf{x}$, $\tau_k(\mathbf{x})$, is 1, and the (crisp) class label for $\mathbf{x}$ is the (crisp) consequent of R_k.

Let $V_k = [v_{k,1}, \ldots, v_{k,n}]^T$ be the "minimum" point of the kth hyperbox, and let $W_k = [w_{k,1}, \ldots, w_{k,n}]^T$ be its maximum point as illustrated in Figure 6.9. Clearly,

$$v_{k,j} \leq w_{k,j}, \; j = 1, \ldots, n; \; k = 1, \ldots, M. \tag{6.14}$$

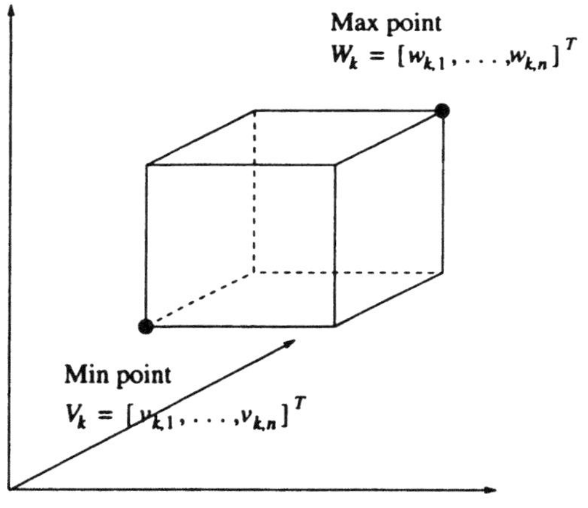

Fig. 6.9. Illustration of "minimal" and "maximal" points of a hyperbox in $\Re^3$.

All features are scaled into $[0, 1]$ so that the feature space is $[0, 1]^n \subset \Re^n$. The one-dimensional membership function of the Min-max fuzzy classifier is shown in Figure 6.10. The equation for the kth membership function (rule R_k) on axis x_j is

$$\mu_{k,j}(x_j) = \frac{1}{2}\left[\max\left(0, 1 - \max(0, \gamma \min(1, x_j - w_{k,j}))\right) + \right.$$

$$\left. \max\left(0, 1 - \max(0, \gamma \min(1, v_{k,j} - x_j))\right)\right]. \tag{6.15}$$

The user-specified parameter γ controls the spread of the "rim" of the hyperboxes. Smaller values for γ lead to covering wider regions.

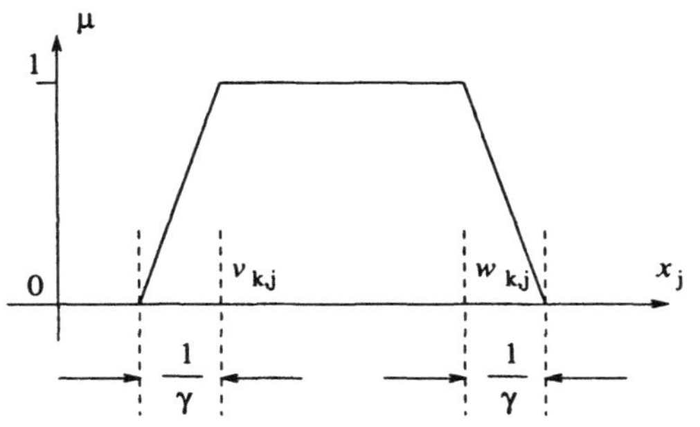

Fig. 6.10. Antecedent membership functions for the Min-max fuzzy classifier

The conjunction (AND) operation $\mathcal{A}_t$ in the Min-max model is the *average* (which is in fact in the mean family of fuzzy connectives introduced in Chapter 4). The firing strength of the kth rule is

$$\tau_k(\mathbf{x}) = \frac{1}{n}\sum_{j=1}^{n} \mu_{k,j}(x_j). \tag{6.16}$$

This choice for $\mathcal{A}_t$ leads to an interesting shape of the firing strength in $\Re^n$. Figure 6.11 shows an example of such a function for $V = [0.4, 0.2]^T$ and $W = [0.6, 0.4]^T$ (the same example as in [307]). On the lefthand-side is the firing strength over $\Re^2$ and on the righthand-side is the grey-scale projection of the function onto $\Re^2$. The higher the degree of firing (membership), the brighter the pixel. The plateau is the region inside the "core" hyperbox, i.e., where $\tau_k(\mathbf{x}) = 1$. The "stripe tails" of degree 0.5, parallel to the coordinate axes correspond to the points for which the membership on one of the features is one (inside the box) and for the other feature is zero (outside the box and the rim). The (averaged) firing strength for these points is 0.5 across the whole $\Re^2$. With more dimensions, these "stripes" will have smaller degrees of membership, i.e., $1/n$.

For designing the rule base of the Min-max classifier the training data set is scaled so that each feature takes values between 0 and 1. The algorithm is sketched in Figure 6.12 and explained in detail below.

- Expansion of hyperboxes to accommodate a new point from $\mathbf{Z}$.
- Overlap test to determine whether overlap exists between hyperboxes of different classes.
- Hyperbox contraction to eliminate the overlap.

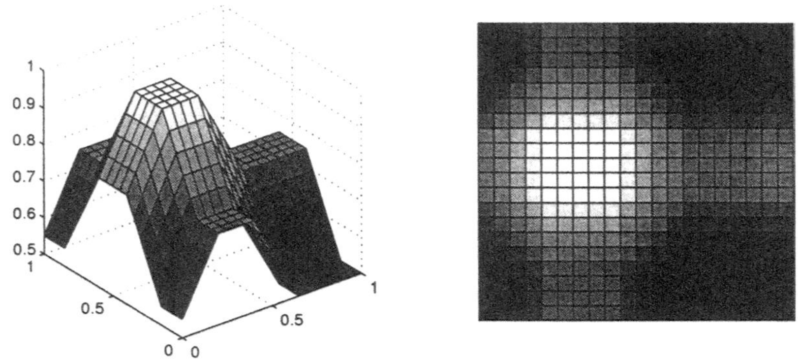

Fig. 6.11. The firing strength (membership function) for the Min-max fuzzy classifier ($V = [0.4, 0.2]^T$ and $W = [0.6, 0.4]^T$).

MIN-MAX FUZZY CLASSIFIER TRAINING

1. Pick $\theta \in [0, 1]$. Set the current rule base (set of boxes) **B** empty.
2. For j=1 to N % (a pass through **Z**)
   a) Check which box from **B** contains $\mathbf{z}_j$.
   b) If there is no such box or there is one, but the box label is not the same as $l(\mathbf{z}_j)$[a], THEN
      i. Identify the box in **B** labeled in $l(\mathbf{z}_j)$ with the closest "wall" to $\mathbf{z}_j$.
      ii. If such a box does not exist,
         A. THEN add to **B** a hyperbox centered at $\mathbf{z}_j$ with 0 volume and label $l(\mathbf{z}_j)$;
         B. ELSE check the expansion condition for the box (see the text). If the condition is met, expand the box to accommodate $\mathbf{z}_j$, else add to **B** a hyperbox centered at $\mathbf{z}_j$ with 0 volume and label $l(\mathbf{z}_j)$.
         C. End % (If)
      iii. Go through all boxes in **B** with a class label different from $l(\mathbf{z}_j)$. Check if there is an overlap with the newly created/expanded box
      iv. If overlap has been found, contract the respective hyperboxes to eliminate it (see the text).
      v. End % (If)
   c) End % (For j)
3. Return **B** with the labels.

[a] $l(\mathbf{z}_j)$ is the crisp label of $\mathbf{z}_j$.

Fig. 6.12. Training of the Min-max fuzzy classifier

1. Initialization. Pick the training parameter of the classifier: $\theta \in [0, 1]$ as the upper bound of the size of a hyperbox (θ is used later in the criterion for hyperbox expansion). Set the current rule base **B** empty. (Notice that the membership function parameter γ is not needed during this training stage.)

2. For each point in the data set, $\mathbf{z}_j \in \mathbf{Z}, j = 1, \ldots, N$, check whether the rule base needs modification. If the point falls in a hyperbox whose class label is the same as the label of $\mathbf{z}_j$, do not modify the rule base. Otherwise, implement the following steps
   a) Find the hyperbox $B_m \in \mathbf{B}$ with the closest "wall" to $\mathbf{z}_j$. Check the *expansion* constraint:

$$n\theta \geq \sum_{k=1}^{n} (\max(w_{m,k}, z_{j,k}) - \min(v_{m,k}, z_{j,k})) \,. \tag{6.17}$$

   If the constraint is satisfied by B_m, expansion takes place adjusting B_m so that $\mathbf{z}_j$ fits within it, i.e.,

$$v_{m,k} \leftarrow \min(v_{m,k}, z_{j,k}); \;\; w_{m,k} \leftarrow \max(w_{m,k}, z_{j,k}); \tag{6.18}$$
$$\forall k = 1, \ldots, n.$$

   If the expansion criterion is not met, a new hyperbox of volume 0 at $\mathbf{z}_j$ is created and added to the rule-base $\mathbf{B}$. Label the new box with the class label of $\mathbf{z}_j$.
   By expanding a box or creating a new one, intersection of hyperboxes with different class labels may occur. The following step detects and corrects this.
   b) Check for *overlap*. For each $B_t \in \mathbf{B}$ whose class label is different from the label of the adjusted/created hyperbox B_m and for each feature, check if one of the four cases in the upper plot of Figure 6.13 occurs. If there is a feature for which none of the four cases occurs, then the two hyperboxes do not intersect and the check proceeds with the next hyperbox in $\mathbf{B}$. If all features are marked with overlap, the smallest overlap δ determines which feature should be used to adjust the rule base. The type of adjustment depends on the type of the overlap on the particular feature. Shown in the lower four diagrams in Figure 6.13 are the way the boxes are *contracted* to meet the nonoverlapping requirement. This algorithm is based on the "minimal adjustment" principle, i.e., the smallest possible change of box size is made to meet the requirement.
3. In the original algorithm the training procedure sees the data set $\mathbf{Z}$ once. It looks reasonable to modify the procedure and feed the data T times, shuffling $\mathbf{Z}$ before each submission (epoch).

Figure 6.14 shows the result from running the Min-max training algorithm on the Normal mixtures data for 4 values of the maximal size limit θ as shown in the Figure. The boxes for the two classes are shaded in lighter and darker grey.

The Min-max classifier was applied to the four data sets: Cone-torus, Normal mixtures, Phoneme and Satimage. θ was varied from 0.04 to 0.5 with

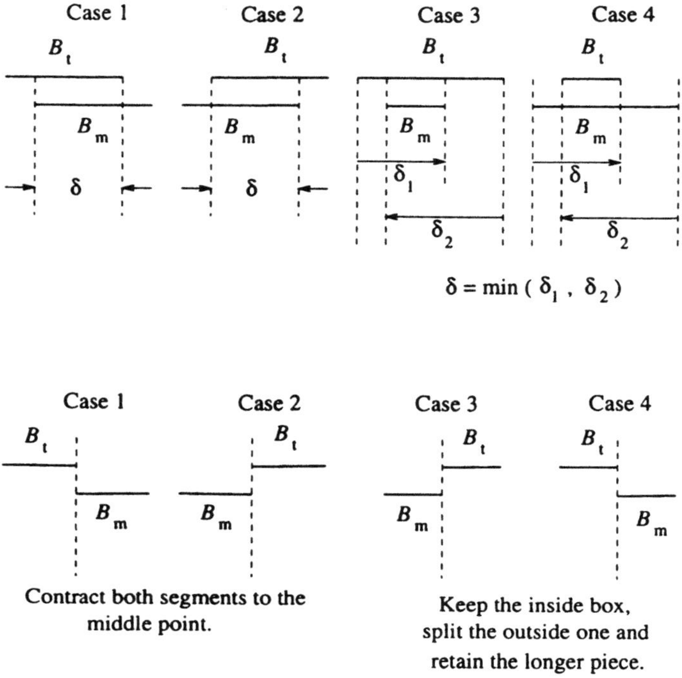

Fig. 6.13. Four cases for the overlap test (upper four diagrams) and hyperbox contraction (lower four diagrams) for the Min-max fuzzy classifier.

step 0.02 and γ was set to 2 and then 10. With all data sets the best training (and test!) results were obtained with the lowest value, $\theta = 0.04$, which is close to the nearest neighbor classifier. Table 6.3 shows the error rate and the number of ties on the class label. The training and testing subsets were used as in the previous experiments. Ties were *excluded* from the calculation of the error rate both in training and testing.

We can see in Figure 6.14 that the match between the boxes and the points in the two classes is far from perfect. Some points are outside the boxes, points with different class labels appear in the same box, and even boxes for class 1, shaded in light grey, are "overtaken" by crosses (class 2). How can this happen? The expansion step of the training process would expand a box to reach a point of the same class regardless of how many points from a different class might fall in the expanded box. Subsequently, the contraction step of the algorithm shrinks the boxes not accounting for how many points fall outside and what their class labels are. Thus, training does not account for the distribution of the points inside the hyperboxes. When θ is small, the Min-max classifier is more likely to resume the shape of the classes than when θ is large. The price to pay for small values of θ is the large number of hyperboxes, many of which will be just points with zero

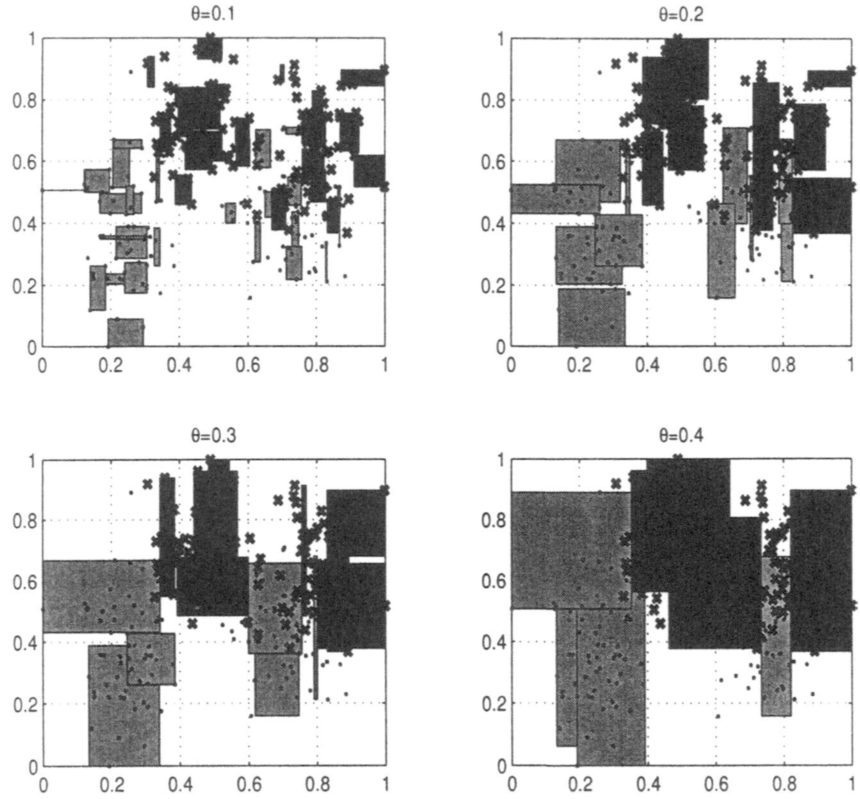

Fig. 6.14. Results from the Min-max fuzzy classifier training with the Normal mixtures data for four values of the maximal box size θ

volume. The results are especially poor in the areas where the classes overlap. The described Min-max training has no mechanism to prevent mislabeling of the boxes in such areas.

Another problem with the Min-max training is that the results depend on the order of submitting the training data and on the order the hyperboxes are checked for overlapping and contracted. To partly eliminate this we can resubmit the (shuffled) data set several times. Although a slightly lower training error rate was achieved with resubmitting the data set, it is not clear how much difference it makes in general.

The Min-max classifier with small θ and high γ (small rim) may leave a significant part of the feature space uncovered, therefore there will be many ties. To overcome this, the parameter values should be selected carefully. In all experiments the overtraining was significant.

Table 6.3. Results from the Min-max fuzzy classifier with the four data sets ($\theta = 0.04$, $T = 5$ epochs through the data set).

Data set	γ	Rules	Training Error [%]	Training Ties	Testing Error [%]	Testing Ties
Cone-torus	2	163	5.32	5	14.54	1
	10	163	5.32	5	14.65	4
Normal mixtures	2	113	2.02	2	13.53	2
	10	113	2.02	2	13.85	11
Phoneme	2	297	0.2	0	17.31	0
	10	297	0.2	1	17.84	874
Satimage	2	218	1.07	33	16.94	351
	10	218	1.07	32	17.16	341

The training algorithm of the Min-max classifier is more sophisticated than those considered so far. The classifier is not linguistically interpretable, it usually requires a significant number of rules, and its performance in our experiments has not been very competitive. However, as with the distributed rules algorithm, the Min-max classifier belongs to the "classical" designs which have inspired further modifications and variations [2, 321, 157, 281].

Adaptive resolution Min-max fuzzy classifier. A modification proposed by Rizzi et al. [281] is called *adaptive resolution min-max classifier*. It improves the Min-max model in two aspects. First, the order of the points in $\mathbf{Z}$ no longer has an effect on the final design, nor does the order of inspecting the hyperboxes for overlap. Second, the parameter θ bounding from above the maximal size of a hyperbox is dropped so that boxes of any size within $[0,1]^T$ can be used where necessary. The training algorithm uses a criterion function

$$F = (1-\lambda)P + \lambda\frac{M}{N}, \tag{6.19}$$

where P is the resubstitution error rate, M is the number of rules (hyperboxes), N is the cardinality of the data set $\mathbf{Z}$ and $\lambda \in [0,1]$ is a regularization parameter. A small λ will lead to many rules and smaller classification error P, whereas a high λ will reduce possible overtraining, at the expense of a higher P. The authors of [281] observe that the choice of λ is not as crucial for the classification performance as the choice of θ in the original Min-max design. Two sets of hyperboxes are kept: $\mathbf{B}_p$, the set of "pure" hyperboxes which contain points from the same class only; and $\mathbf{B}_h$, the set of "hybrid" hyperboxes which contain points from more than one class. The training does one pass through $\mathbf{Z}$ starting with a single hybrid hyperbox containing all of $\mathbf{Z}$, and ends with a set $\mathbf{B}_p$ which covers *the whole* $\mathbf{Z}$ with pure hyperboxes. The new hyperboxes in the two sets are generated by splitting hyperboxes from $\mathbf{B}_h$ taking into account the class labels of the inside points. At each split, the criterion function F is calculated and stored. For calculating the error P, a classifier is built using the current content of $\mathbf{B}_p$. At the end, the design with the smallest F amongst the stored ones is found and taken as the adaptive Min-max classifier. The algorithms is shown in a compact way in Figure 6.15.

ADAPTIVE RESOLUTION MIN-MAX FUZZY CLASSIFIER (TRAINING)
1. Initialize the algorithm with $\mathbf{B}_p = \emptyset$ and $\mathbf{B}_h$ being a hyperbox containing all of $\mathbf{Z}$. 2. While $\mathbf{B}_h$ is not empty, a) Identify the box in $\mathbf{B}_h$ with the largest number of points in it. b) Split that box into two (see the text) and place the two new boxes in the respective sets of pure ($\mathbf{B}_p$) or hybrid ($\mathbf{B}_h$) boxes. c) End % (While) 3. Return $\mathbf{B}_p$ with the labels.

Fig. 6.15. Training of the Adaptive resolution Min-max fuzzy classifier

The split procedure is implemented as follows

1. Identify the hybrid box $B_s \in \mathbf{B}_h$ with the maximal number of points in it.
2. Let $\mathcal{N}(B_s, \omega_i)$ be the number of elements of $\mathbf{Z}$ from class ω_i contained in B_s. Without loss of generality, assume that the two classes with the higher $\mathcal{N}$ are ω_1 and ω_2. Let $\mathbf{m}_{s,1}$ and $\mathbf{m}_{s,2}$ be the centroids of the points from these two classes in B_s. Identify the feature x_j for which the two centroids are at the largest distance from each other.
3. Split B_s into two hyperboxes B_{s1} and B_{s2} with a hyperplane through the middle point of the projection of the segment joining the two centroids onto x_j, parallel to the remaining coordinate axes. Let $m_{s,1,j}$ be the jth component of $\mathbf{m}_{s,1}$, and $m_{s,2,j}$ be the jth component of $\mathbf{m}_{s,2}$, and let $m_{s,1,j} < m_{s,2,j}$. Let

$$mm = \frac{1}{2}(m_{s,1,j} + m_{s,2,j}). \tag{6.20}$$

In terms of lower and upper corners of the hyperboxes, the two new boxes B_{s1} and B_{s2} are formed as follows

$$v_{s1,k} = v_{s,k};\ \forall k = 1,\dots,n;\ \ w_{s1,k} = \begin{cases} w_{s,k},\ k = 1,\dots,n,\ k \neq j \\ mm,\ k = j \end{cases}; \tag{6.21}$$

$$v_{s2,k} = \begin{cases} v_{s,k},\ k = 1,\dots,n,\ k \neq j \\ mm,\ k = j \end{cases};\ w_{s2,k} = w_{s,k};\ \forall k = 1,\dots,n. \tag{6.22}$$

The split is illustrated in Figure 6.16 for $n = 2$ (dimensions), and $c = 3$ (classes). $\mathcal{N}(B_s, \omega_1) = 5$, $\mathcal{N}(B_s, \omega_2) = 7$ and $\mathcal{N}(B_s, \omega_3) = 3$. The split is performed according to ω_1 and ω_2. The two centroids are marked with shaded circles. The split line is parallel to the y axis because the higher distance between the centroids is on the x axis. The two new boxes are again hybrid but eventually, the points will be encapsulated in pure (though possibly small) hyperboxes.

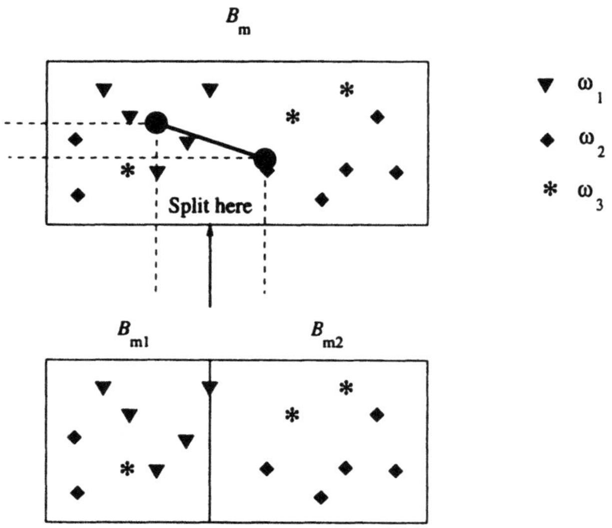

Fig. 6.16. Splitting of hyperboxes for the Adaptive resolution Min-max classifier

For the operation of the Adaptive resolution Min-max classifier we use the same TSK model as in the original Min-max classifier. The parameter γ needs to be specified in advance. The difference lies in the construction of the hyperboxes serving as the rule base.

Example 6.3.2. The Adaptive resolution Min-max classifier was applied to the Normal mixtures data with $\lambda = 0.5$ and $\gamma = 5$. Figures 6.17 and 6.18 show the results from the training. The top of the three plots in Figure 6.17 is the number of hyperboxes in the two sets "pure" ($\mathbf{B}_p$, the increasing curve) and "hybrid" ($\mathbf{B}_h$, the cap-shaped curve) against the number of iterations (splits). Starting from one element in $\mathbf{B}_h$ and none in $\mathbf{B}_p$, the algorithm reaches 67 pure rules in 66 iterations. Figure 6.18 plots the 67 boxes found at the last step. The middle plot in Figure 6.17 shows the training and testing error rates against the number of iterations. When the whole data set $\mathbf{Z}$ is covered in hyperboxes, we achieve zero resubstitution error rate. Clear evidence of overtraining is the declining training error after the 35th iteration and the approximately constant testing error. However, the regularization term in the criterion function F, $\lambda\frac{M}{N}$, is used to decide when the training should be stopped. The bottom plot in Figure 6.17 demonstrates that the criterion function correlates well with the ***testing*** error rate, and therefore can be used to pick a successful model. The minimum of the criterion function for the Normal mixtures data was found at 3 rules, training error 8.29 %, testing error 10.05 % with 69 ties in the training set and 254 ties in the test set. Ties have been excluded when calculating the error rate. ■

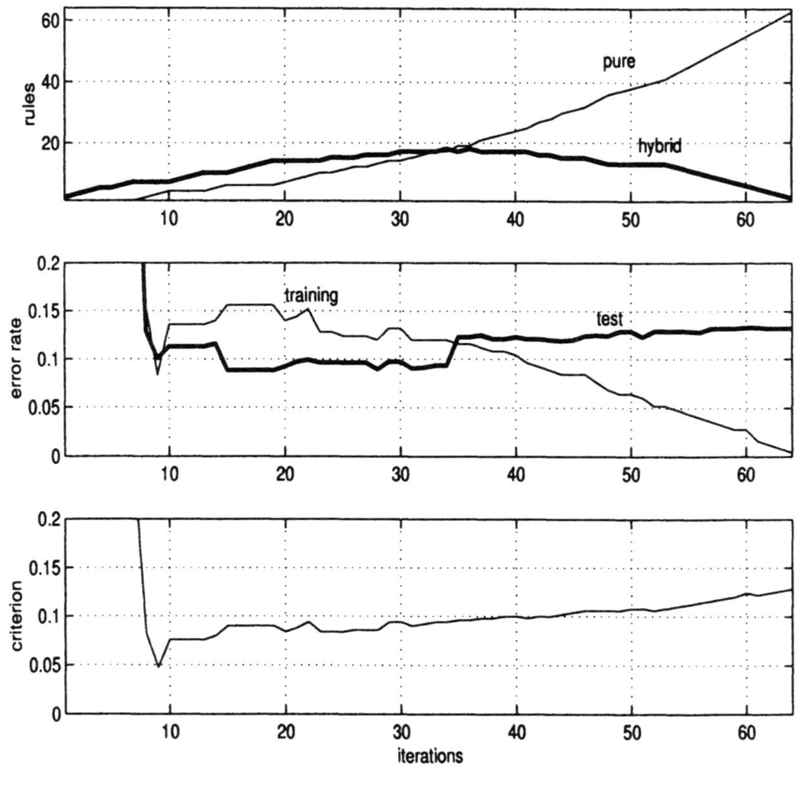

Fig. 6.17. Results from the Adaptive resolution Min-max classifier on Normal mixtures data.

Table 6.4 displays the results from applying the Adaptive resolution Min-max fuzzy classifier to the four data sets. The "winner" was picked according to the value of the criterion function F. Different results have been obtained for different values of the "compromise" coefficient λ. Rizzi et al. use $\lambda = 0.5$ [281].

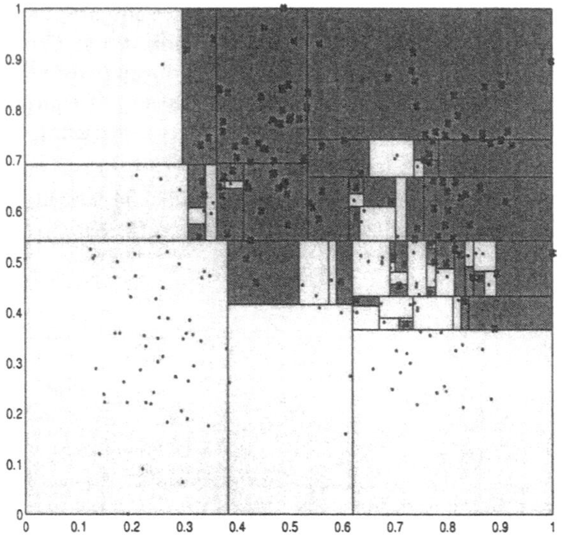

Fig. 6.18. The coverage of the 2-d feature space ($[0,1]^2$) by the Adaptive resolution Min-max fuzzy classifier with the Normal mixtures data (67 rules, 0 resubstitution error, 13.5 % testing error, no ties).

Table 6.4. Results from the Adaptive resolution Min-max fuzzy classifier with the four data sets ($\lambda \in \{0.4, 0.5\}$, $\gamma = 5$).

Data set	λ	Total rules	Rules used	Training Error [%]	Training Ties	Testing Error [%]	Testing Ties
Cone-torus	0.5	109	6	18.11	0	15.53	20
	0.4	109	49	7.75	0	15.00	0
Normal mixtures	0.5	67	3	8.29	0	10.05	254
	0.4	67	3	8.29	0	10.05	254
Phoneme	0.5	144	20	15.03	0	20.49	28
	0.4	144	53	9.0	0	19.13	0
Satimage	0.5	140	25	15.07	11	18.17	19
	0.4	140	71	8.15	11	18.29	24

Abe and Lan [2] propose an Activation-inhibition Min-max classifier. They start with one hyperbox for each class containing all points in **Z** from that class. These boxes are called *activation* hyperboxes. The intersection between boxes of different classes is identified and resolved separately for each pair of classes. The intersection between two hyperboxes is itself a hyperbox which is taken as the new **Z**, and new activation hyperboxes are generated within it. The process continues recursively until all intersections are resolved. The membership function used for the antecedents is trapezoidal as in the original Min-max fuzzy model but the minimum operator is used as the conjunction $\mathcal{A}_t$ (TSK1 model). Thus, the firing strength of a rule is 1 inside the hyperbox and declines to zero within a $1/\gamma$ "rim" around the hyperbox (there are no "stripe tails") such as those shown in Figure 6.11. The idea is very intuitive but the classifier structure becomes more cumbersome than in the above models. This happens because of the hierarchical structure of rules generated for each class. Joshi et al. [157] propose a version of the Min-max model for multiclass membership, i.e., when the classes are not necessarily mutually exclusive.

Identifying class regions by fitting hyperboxes to the data is considered also in [58, 133, 228, 229, 321].

6.4 Tuning antecedents *and* consequents using clustering

This approach for fuzzy systems design has become very popular, especially in fuzzy control. The idea is to identify clusters in the data set **Z**, and generate an if-then rule from each cluster. The membership functions of the antecedents are formed by *projecting* the clusters on the coordinate axes. We usually assume that the centroid of a cluster has the highest degree of membership, and that the membership function decreases monotonically with the distance from the centroid. The most popular fuzzy clustering method, the fuzzy c-means, uses this pattern to calculate the degrees of membership (see Chapter 4, equation (4.92)). Figure 6.19 illustrates the projecting of a cluster in $\Re^2$ back on x and y.

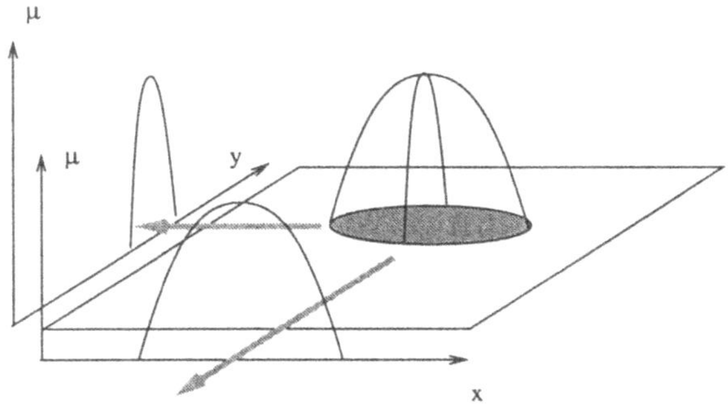

Fig. 6.19. Projection of cluster membership onto feature axes.

Different types of membership functions can be extracted from a cluster. The simplest model is the triangular membership function. The centroid of the cluster is fixed as the (point-) core of the fuzzy set, i.e., the value of x_j with degree of membership 1. The span of the membership function is defined by the smallest and the largest values of x_j amongst all elements of $\mathbf{Z}$ in the respective cluster [133]. An illustration of three clusters of half-torus class from the Cone-torus data with triangular membership functions is shown in Figure 6.20. The three clusters were found by the hard c-means clustering (see Chapter 2. Section 2.3). The centroids are marked with solid circles.

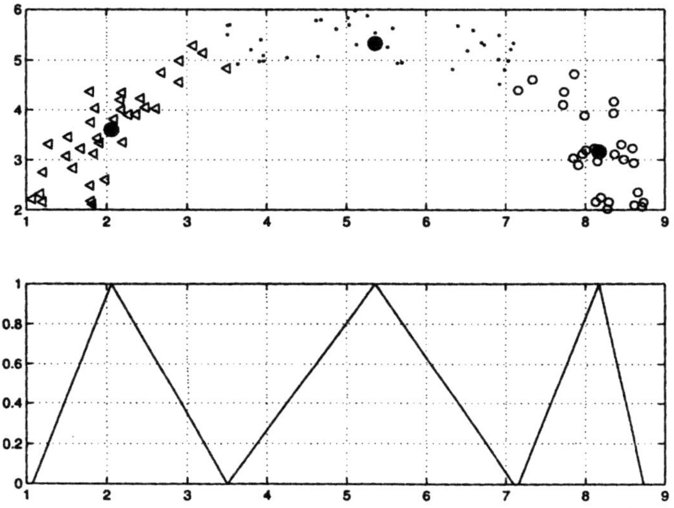

Fig. 6.20. Generating triangular membership functions from clusters

Example 6.4.1. The Cone-torus data was clustered separately into 6 clusters, 2 clusters for each class (presupervised approach). Six triangular membership functions per feature were then generated by projecting the clusters onto the feature axes. Figure 6.21 depicts the data and the membership functions. ■

As with prototype selection by clustering, two approaches are possible: pre-supervised and postsupervised. In the pre-supervised approach, the data for each class is clustered separately. In the postsupervised approach the whole data set is clustered disregarding the class labels.

The membership functions of each cluster are used as the antecedents of a rule corresponding to that cluster. The *consequent* of the rule is a class label (crisp or soft) associated with the cluster. Three labeling variants were applied to find the consequents of the rules when clustering all of $\mathbf{Z}$ (postsupervised approach). Let $\tau_k(\mathbf{z}_t)$ be the firing strength of rule R_k when $\mathbf{z}_t$ ($\in \mathbf{Z}$) is submitted to the fuzzy classifier, $t = 1, \ldots, N$. Let β_i be the sums calculated as in (6.4). The three types of labels are obtained as follows

1. *Crisp labeling*: The rules have as their consequents, vectors in N_{hc} (crisp labels) obtained as

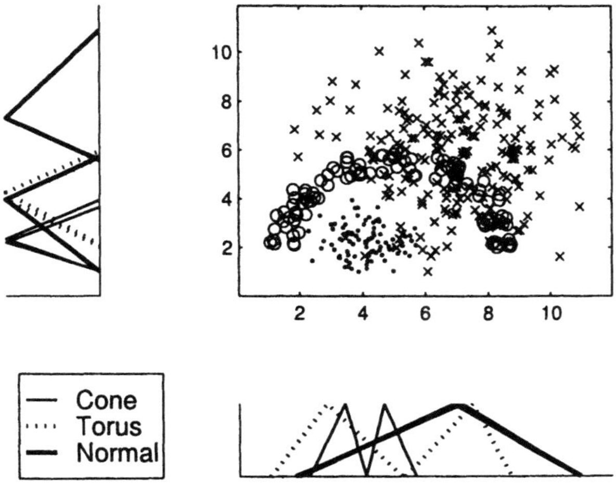

Fig. 6.21. Membership functions on x and y generated by clustering of the Cone-torus data in 6 clusters: 2 per class (presupervised approach)

$$g_{k,i} = \begin{cases} 1, & \text{if } \beta_i = \max_s\{\beta_s\}, \\ 0, & \text{otherwise.} \end{cases} \tag{6.23}$$

Ties are resolved arbitrarily.

2. *Soft singleton labeling*: The labels are in N_{pc} with only one of the c consequent values being greater than zero

$$g_{k,i} = \begin{cases} \beta_i, & \text{if } \beta_i = \max_s\{\beta_s\}, \\ 0, & \text{otherwise.} \end{cases} \tag{6.24}$$

As above, ties are resolved arbitrarily.

3. *Fuzzy labeling*: The consequents are class labels in N_{fc} obtained as

$$g_{k,i} = \frac{\beta_i}{\sum_{s=1}^{c} \beta_s}. \tag{6.25}$$

In the pre-supervised approach the vectors from each class are clustered separately giving rise to two more labeling ideas

4. *Crisp own labeling*: The consequent is the crisp class label of the class whose data was used to find the cluster.
5. *Soft singleton own labeling*: The consequent is a soft label as in *Soft singleton labeling* but the non-zero element is the one corresponding to the class whose data was used to find the cluster. For example, if the elements of $\mathbf{Z}$ from class ω_3 were used to find the cluster for rule R_6, then $g_{6,i} = 0$, $\forall i = 1, \ldots, c, i \neq 3$, and $g_{6,3} = \beta_3$.

Membership functions with bounded support (such as triangular membership functions) may leave a part of the feature space uncovered, i.e., the firing strength of all rules will be zero for the points in such regions. To avoid this situation Gaussian membership functions can be used. The cluster centroid can be picked as the mean, and the standard devisions can be estimated from the data in the cluster. Another option for finding the standard deviation is to assume that the span between the minimum and the maximum (used as the end points of the triangular membership function) is $k\sigma$, where k is a small positive integer.

A series of experiments with the Cone-torus and Normal-mixtures data were carried out using the five labeling modes, triangular or Gaussian membership functions, and a TSK2 classifier. Besides the standard TSK2 output (5.27), we also used

$$g_i(\mathbf{x}) = \frac{\sum_{k=1}^{M} \min\left\{z_{k,i}, \prod_{j=1}^{n} \mu_{j,i(j,k)}(x_j)\right\}}{\sum_{k=1}^{M} \prod_{j=1}^{n} \mu_{j,i(j,k)}(x_j)}. \tag{6.26}$$

Comparing the TSK fuzzy model with the MA models, we regard the minimum in (6.26) and the product in the original formula (5.27) as fuzzy implications associating the firing strength $\tau_k(\mathbf{x})$ and the consequent $z_{k,i}$.

The maximum membership rule was applied at the end to infer the crisp class label. The error rate was measured by the percentage of misclassifications.

All combinations of labeling modes (5 for presupervised and 3 for post-supervised designs), membership function types (2 of them), and the two implications were tried out with both data sets with 4 values of the total number of clusters (rules) M. Thus, for with each data sets we designed $4 \times 5 \times 2 \times 2 = 80$ pre-supervised and $4 \times 3 \times 2 \times 2 = 48$ post-supervised fuzzy classifiers. Since clustering depends on the initialization, it was run 10 times with each of the 256 designs. The best results from the 10 runs are exhibited in the Appendix. "Best result" means the lowest *training* error with the *corresponding* testing error. Wherever training error tied, the one obtained at the earliest time was picked, regardless of whether or not the corresponding testing error was the lowest amongst the tied designs.

All ties resulting from not covering the whole feature space (occurred with triangular membership functions only) were counted as errors thereby giving a somewhat negatively biased error estimates. With the Gaussian membership functions we tried several ways to specify σ. Best results were obtained with $k = 6$ (6σ between the minimum and the maximum of x_j for the respective cluster.

Table 6.5 summarizes the most successful designs with their error rates (in%).

The risk in drawing any conclusion on these results lies in the diversity of the possible fuzzy models. If the designs tried out here did not produce

Table 6.5. Most successful fuzzy TSK classifiers using clustering with the Cone-torus and Normal-mixtures data (full tables with results are given in the Appendix)

	Cone-torus		Normal mixtures	
	Post-supervised	Pre-supervised	Post-supervised	Pre-supervised
Implication	product	minimum	product	product
Membership functions	Gaussian	Gaussian	Gaussian	Gaussian
Labeling mode	3	4	3	1
Number of rules (M)	15	12	6	8
Training error	17.00	16.25	11.60	9.60
Testing error	17.25	16.50	8.90	12.00

any spectacular reduction of the classification error, designs with other combinations of parameters could do so. The problem is to find that winning combination of parameters and set it up as a golden standard for fuzzy classifier design. However, this is not a straightforward problem. Our results show that different combinations can be beneficial for different problems. A general observation can be made that Gaussian membership, pre-supervised designs and product as the implication are the best combination. This combination defines a TSK2 classifier which is closely related to radial basis function (RBF) networks as previously discussed. The linguistic interpretation of models derived from clustering is usually case-based (prototype-based).

A series of elliptical-region based fuzzy classifiers are proposed by Abe and coauthors [1, 3, 291]. The class regions are captured in elliptically shaped subregions and the parameters of the ellipses are used to place and tune the individual membership functions on the feature axes. The membership functions are Gaussians where the covariance matrix is assumed to be diagonal, so that the membership function for each rule on axis x_j has two parameters: the center and the spread (initially the standard deviation). Subsequently, the standard deviation is tuned according to an heuristic procedure until a certain stopping condition on the training error is met.

6.5 Genetic algorithms for tuning fuzzy if-then classifiers

6.5.1 Genetic algorithms (GAs)

Genetic algorithms (GAs) are a popular and widely used optimization paradigm [96, 108]). They take ideas from natural evolution to guide the search algorithm towards good solutions. GAs are not supposed to be an adequate *model* of natural evolution, i.e., all the complexity involved in mating, reproduction, selection, etc., is left behind, and only the basic mechanism is taken. When the basic model is insufficient for some reasons, more sophisticated ideas may come into play, such as niching, migration, sex differentiation, diploidity, dominance, etc [25].

GAs operate on generic structures called **chromosomes**. A chromosome represents a candidate solution. Such candidate solution can be the parameters that we choose to tune by the GA (rules, membership functions, etc.). To run a GA we need to **encode** candidate solutions as chromosomes. Encoding for a subset search is straightforward. The easiest way is to map the set to a binary vector to be used as the chromosome. The ith component of the vector will represent the ith element of the set[1]. A value 1 in position i means that the ith element is in the subset, and 0 means that the element is not in the subset. Thus, every chromosome represents a subset. An illustration is given in Figure 6.22.

$$\mathcal{R} = \{R_1, \ldots, R_{10}\}$$

$$S = \boxed{1\,|\,0\,|\,0\,|\,0\,|\,1\,|\,0\,|\,1\,|\,1\,|\,0\,|\,0}$$

$$S = \{R_1, R_5, R_7, R_8\}$$

Fig. 6.22. Representation of a subset as a chromosome.

A GA optimizes a **fitness function** J. $J(S)$ assesses a candidate solution represented by the chromosome S. In our case, $J(S)$ can be an estimate of the probability of correct classification of the fuzzy model with the parameter values encoded in S. For example, let $\mathcal{R} = \{R_1, \ldots, R_K\}$ be the set of rules of a fuzzy classifier. We can try to select a subset of $\mathcal{R}$ so that the new classifier has the highest possible classification accuracy on some set $\mathbf{Z}$. Finding a subset of $\mathcal{R}$ is an optimization problem in a search space with 2^K elements, the number of all subsets of $\mathcal{R}$. It is usually infeasible to check all subsets. We can randomly pick subsets, evaluate them through the error rate, and select the one that gives the smallest error. By using a genetic algorithm we can *partly* guide the search through the subsets of $\mathcal{R}$ using previous information to construct subsequent candidate solutions.

Given a population of individuals, the GA loops through the following steps:

[1] The elements of the set can be tagged in any order so that each element receives an index.

- mating
- reproduction
- selection

A more detailed description of a GA version for binary chromosomes is given below.

1. *Initialization.* An initial population $\pi = \{S_1, \ldots, S_{ps}\}$ with ps individuals (chromosomes) is generated. Every chromosome is evaluated by the fitness function $J(S_i)$. The higher the fitness, the better the chance for this chromosome or its descendants to survive.
2. *Mating.* A mating set M is selected from π. The most popular way is to use the **roulette wheel**. A probability distribution is set up so that each chromosome in π has a chance, proportional to its fitness value, to be picked in the mating set. The mating set is designed randomly, taking ms chromosomes from π (with replacement) according to the probability distribution. The cardinality of M, ms, is defined in advance. In some GA variants the whole π is used as M.
3. *Crossover.* Parent couples are selected at random from M. According to a prespecified probability of crossover (usually high), each couple of parents produces 2 new chromosomes or identical copies of themselves to form an offspring set O. The number of parent couples is again specified in advance, and reflects how much offspring we would like to create. **Crossover** is used to create two children chromosomes from the two parent ones. Crossover can be performed in different ways, the three most popular ways being: *one-point, two-point* and *uniform*, as illustrated in Figure 6.23. The two parent chromosomes exchange genetic material by either cutting and swapping strings of bits (one and two-point crossover), or by randomly swapping the corresponding bits (uniform crossover).
4. *Mutation.* Each bit of each offspring chromosome is subjected to mutation. Mutation occurs with a (usually small) prespecified probability. For binary chromosomes, mutation just alters the value of the respective bit. After mutation, all chromosomes in O are evaluated by the fitness function.
5. *Selection.* The next population is formed from π and the mutated O. It can be taken at random or part of π can be retained and the other part replaced by elements of O (the percentage replaced is called the **generation gap**). The **elitist strategy** requires that the best chromosome (the one with the highest fitness) always survives in the next population. A simple implementation of the elitist strategy is to pool π and O, rank the chromosomes with respect to their fitness and retain as the new π the best ps of them (this method results in a variable generation gap).

GAs have faced both positive and negative comments in the literature. Below is yet another list of GAs' advantages and limitations

- **Advantages.**

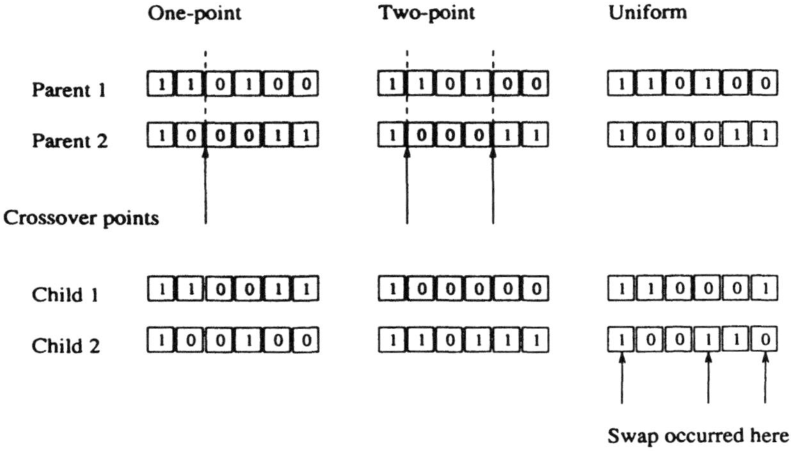

Fig. 6.23. One-point, two-point and uniform crossover.

- Any criterion (fitness) function can be optimized. J does not need to be continuous, smooth, differentiable, etc. The only requirement is that we be able to estimate the value of the function for each chromosome. It is another question how good the GAs are on different types of criterion functions (multimodal, defined over very large search spaces, etc.) The point here is that the GA technically *can* handle any type of criterion function.
- Any information can be encoded into a chromosome. This property is particularly appealing in problems that are analytically intractable. For example, we can encode the feature set and the data set into *one* chromosome and do simultaneously feature selection and data editing [202, 308]. Another example is evolving neural networks by encoding *both* the NN structure and NN parameters in a single chromosome.
- In some problems, such as subset selection, the GA encoding is straightforward. Various integer-valued and real-valued GAs have also been developed with respective definitions of the genetic operators such as crossover and mutation.

Example 6.5.1. Assume that we select fuzzy rules, and each chromosome corresponds to a rule. The rule has n antecedent clauses (one per feature), and each clause is associated with a linguistic value on the respective feature axis. Let k be the number of linguistic values per feature (e.g., $k = 4$ with linguistic values {*small, medium, high, very high*}). The chromosome can be represented by a string of length n where each component takes an integer value between 1 to k. Thus, the string "142" translates into the antecedent "IF x_1 is *small* AND x_2 is *very high* AND x_3 is *medium*". ∎

Contrary to the initial belief that binary encoding is the best one, it has been proven that for search spaces of the same cardinality, the type of encoding is not essential [97], i.e., whatever encoding we use, the GA has the same chance to reach the optimal solution. This result renders a background for using GAs with the most natural encoding for a given problem.

- **Limitations.**
  - GAs are not a panacea. There is a certain balance between how specific and how powerful an algorithm is. In general, all algorithms have the same performance over all possible data sets [349]. However, unlike many optimization methods and algorithms, GAs can be tuned to vary gradually between random search (the most general case) and a variant of hill climbing (the most specific case). In GA terms this is a balance between **exploration** and **exploitation**. Exploration refers to checking new paths that are unseen so far, and exploitation refers to using mostly the information about the good solutions found so far and searching in their vicinity or their breed threads. Random search relies on exploration, whereas hill climbing relies on exploitation. This balance can be tuned by the GA parameters. For example, a small mutation rate plus a high weighting for the best chromosomes in the roulette wheel will put the weight on exploitation. Alternatively, taking the whole of π as the mating set and setting a high mutation rate (e.g., 0.15-0.20) will shift the search towards exploration. Finding a set of parameter values for a good balance for the particular problem is difficult, if at all possible. Sometimes a random search can be a better option (less expensive and more effective) than a badly tuned GA.
  - GAs might take a long time to converge.
  - There is no guarantee that the solution obtained is optimal or even close to an optimal one. Salomon [293] argues that good results with GAs can be expected in simple cases where the fitness function, $J : \Re^q \rightarrow R$, is decomposable, i.e., the optimal solution has as its coordinates in $\Re^q$ the optimal solutions on the individual coordinates. In other words, the problem is solvable as q (easy) one-dimensional optimization problems. For non-decomposable functions, which comprises most of the real life problems, GAs might fail to improve on random search.

6.5.2 GAs for tuning fuzzy classifiers

The publication by Valenzuela-Rendón [329] is perhaps the first work where GAs are used to select fuzzy rules. Since then, GAs have been applied in various studies to design and tune fuzzy if-then systems [132] and in particular fuzzy if-then classifiers. There are three general groups:[2]

[2] Carse et al. [55] consider one more category where both the rule base and the antecedent fuzzy sets are tuned but at separate stages.

1. Fixing and freezing the rule base and tuning the parameters of the membership functions of the antecedents $\mu_{j,k}(x_j)$, $j = 1, \ldots, n$, $k = 1, \ldots, K_j$. The parameters that are tuned are, e.g., the width of the base of the triangle, the width of the plateau of the trapezoidal membership function or the center and the spread parameter of a Gaussian.
2. Fixing and freezing the antecedent membership functions and selecting rules from a rule base of all possible rules.
3. Selecting rules and tuning the membership functions simultaneously.

A series of studies by Ishibuchi and coauthors [148, 149, 142] use GAs for rule selection either from all rules with one resolution or from the extended rule base as in the distributed rules model explained in Section 6.2.3. Fixed and equally spaced triangular membership functions are used for the antecedents. The encoding is straightforward: each bit in the chromosome represents a rule, so that each chromosome represents a subset of the whole rule base. The classification accuracy of the fuzzy if-then classifier with the respective rule base is used as the fitness function. To prevent overtraining, a penalty term is included which penalizes rule subsets of high cardinality. In some studies the penalty term is replaced by an additional fitness functions, so that the GA optimizes two criteria simultaneously.

Carse and coauthors [55] offer an enlightening discussion on two philosophically different GA approaches for rule selection. The **Michigan approach** encodes each if-then rule as a chromosome and the fuzzy system is evolved as the whole population π. In the **Pittsburgh approach** each chromosome encodes the whole rule base, and the best chromosome at the end of the GA determines the winning fuzzy system.[3] Clearly, different fitness functions are needed in each case. The Pittsburgh approach has a straightforwardly measurable fitness: the classification accuracy of the classifier corresponding to the chromosome. In the Michigan approach, the rules have to be evaluated by some indirect fitness function. This can be an asset or a pitfall, and the success is usually determined by the particular application and the experience of the designer. A survey of works on both approaches can be found in [55]. Ishibuchi and coauthors [143] carry out an experimental comparison of the Pittsburgh and Michigan approaches for fuzzy if-then classifier design. Their results are strongly in favor of the Michigan approach. The authors' argument is that more information about good rules can be inherited in the Michigan approach than in the Pittsburgh approach because the latter has one fitness value for the whole set of rules. However, the Pittsburgh approach has been widely used with satisfactory results [132]. A synergy of the two approaches by a hierarchical structuring of the population has also been sought [55].

González and Pérez have developed the SLAVE model (Structural Learning Algorithm in a Vague Environment) and modifications thereof (cf. [109]).

[3] The two approaches are named after the places the work originated from. Both approaches have been originally proposed for crisp rule-based systems.

Their GA evolves one rule at a time. SLAVE starts off with an empty set of rules. The population is formed gradually, through a sequence of GA runs. When the rule satisfies some completeness and consistency criteria (expressed as the fitness function J), the GA is terminated and the rule is added to the current rule base. Part of the data set is associated with the winning rule, and is then excluded from the set to be used in further training. To generate the next rule, the GA is run on the remaining data set. This method is closer to the Michigan approach but differs by the fact that the population π at termination is not used as the final rule base. An interesting feature of their fuzzy if-then classifier is that it uses a conjunction of *disjunctive* compound clauses on each feature. For example, a rule in such fuzzy system can have the form

R_k : IF (x_1 is *small* OR *very large*) AND

x_2 is *medium* AND (x_3 is *medium* OR *large* OR *very large*)

THEN class is ω_1.

The membership functions of the antecedents are specified in advance. The chromosome consists of $K_1 + K_2 + \ldots + K_n$ binary positions (K_j is the number of linguistic terms on axis x_j).

Example 6.5.2. Let $\mathbf{x} \in \Re^3$ be a pixel in an image with x_1 being the gray level intensity {*low, medium, high, very high*}; x_2, the standard deviation of the intensity in a 3 by 3 window {*low, medium, high*}; and x_3, the average of the intensity in a 15 by 15 window {*low, high*}. The chromosome 011001010 in the SLAVE model generates the following rule

R_k : IF (x_1 is *medium* OR *high*) AND

x_2 is *medium* AND x_3 is *low*

THEN class is ...

■

Evolving one rule by one run of the GA underpins the MOGUL algorithm by Cordón et al. [65]. To account for cooperation between the rules already selected by previous GA runs, the authors propose to include more sophisticated fitness functions. Another idea is to break the GA into stages so that at the second stage a subset of rules is evolved amongst the rules selected at the first stage.

Shi et al. [304] encode the whole rule base in a single chromosome (Pittsburgh approach). The maximal number of rules is specified in advance and used in the encoding. For example, if the number of rules is decided to be M, the chromosome contains M strings of length $n + 1$ (one integer per feature and one for an MA type consequent). Each integer specifies which of the linguistic terms is used in the respective clause as in example 6.5.1, and 0 means that that feature is not used in the respective rule. A negation (e.g., *not small*) is also considered as a legitimate antecedent clause. It is encoded in the chromosome by a minus sign. For example, the possible values for an antecedent clause on feature x_2 from example 6.5.2 are $\{-3, -2, \ldots, 2, 3\}$. The type of each membership function and the function parameters are encoded too. Three parameters are used: the type of the function (a total of 5 types are used); "start point" and "end point" on the feature axis (these are encoded as integers from 1 to 10). The chromosome also contains a component for the number of rules. If a total of M rules are allowed and encoded, an integer between 1 and M, called "threshold number of rules" is used to determine how many of the M rules will be used for the chromosome. Thus, if $M = 20$ and the threshold number of rules is 6, the first 6 rules only will be considered. It is possible that not all of these six rules are feasible, i.e., there might be rules with 0's in all the antecedent clauses or a 0 in the consequent clause. Then, for this chromosome, the number of feasible rules will be less than or equal to 6. If there are no feasible rules among the first 6, then the chromosome is considered not viable and its fitness is set to a small value.

Example 6.5.3. Consider a fuzzy if-then classifier designed by the above GA on the Phoneme data ($n = 5$). Let $K = 4$ fuzzy sets be used on each feature axis and let the number of rules M be 10. The chromosome is formed according to the following list (the number of possible integer values for each position is given in the brackets)

1→ threshold number of rules (10)
2→ start point of $\mu_{1,1}(x_1)$ (10)
3→ end point of $\mu_{1,1}(x_1)$ (10)
4→ type of $\mu_{1,1}(x_1)$ (5)
5→ start point of $\mu_{2,1}(x_1)$ (10)
6→ end point of $\mu_{2,1}(x_1)$ (10)
7→ type of $\mu_{2,1}(x_1)$ (5)
...→...
59→ start point of $\mu_{4,5}(x_5)$ (10)
60→ end point of $\mu_{4,5}(x_5)$ (10)
61→ type of $\mu_{4,5}(x_5)$ (5)
62→ clause 1 for rule R_1 (9), see [4]
63→ clause 2 for rule R_1 (9)
64→ clause 3 for rule R_1 (9)
65→ clause 4 for rule R_1 (9)
66→ clause 5 for rule R_1 (9)
67→ consequent for rule R_1 (2), see [5]
68→ clause 1 for rule R_2 (9)

---

[4] The 9 values come from the set $\{-4, -3, \ldots, 3, 4\}$.

[5] The consequent values are modified for the classification purposes. We need two values to encode classes ω_1 and ω_2. In the original model this number was 9.

...→...
120→ clause 4 for rule R_{10} (9)
121→ consequent for rule R_{10} (2)

This *integer-valued* GA has a search space of cardinality

$$M \times (10 \times 10 \times 5)^{nK} \times (2(2K+1)^n)^M$$

which, for this example ($M = 10,\ n = 5,\ K = 4$) is approximately 2^{352}. ∎

The more parameters we tune, the larger the search space becomes (curse of dimensionality). Normally, the fitness function J is multimodal and therefore difficult to optimize. Thus, the chances the GA to converge to an optimal solution for a large search space are not too good.

Russo [290] developed the FuGeNeSys system (FUzzy GEnetic NEural SYStem) for tuning the rule base and the membership functions simultaneously. In his model the number of rules is specified in advance. Each rule (R_k) has n antecedent clauses and a singleton (normal or subnormal) consequent clause. The TSK2 model is used with $z_{k,i} \in [0,1]$. All antecedent membership functions are Gaussians and 2 parameters per function are encoded: the center and the inverse of the standard deviation. Each real value is encoded as a 16-bit binary string. Thus, the chromosome for M rules, n features and c classes (encoding the class label with a binary vector of length c) consists of $(16 \times 2 \times n + c) \times M$ bits. For example, a fuzzy classifier with 10 rules for the Phoneme data will require a chromosome of length 1620, so the search space is 2^{1620}. As reported by Russo [290] a small number of rules is usually sufficient with this model due to the flexibility of the membership function tuning.

There is an interesting difference between the latter two models. In the FuGeNeSys model, each rule has n *rule-specific* membership functions. These functions are not used by any other rule in the rule base (prototype-like model). Thus, each feature axis has M membership functions defined on it regardless of whether or not these correspond to any linguistic labels. The previous model [304] uses the *same* membership functions in all rules. It is then more likely to comply with some linguistic description although the tuning process can distort the membership functions and invalidate the interpretation (see Figure 4.7 in Chapter 4). Carse et al. [55] call this **global** versus **local** fuzzy rules. Then FuGeNeSys evolves local fuzzy rules (rule-specific μ's) whereas the model in [304] evolves global rules.

Carse et al. propose their own method for fuzzy controller design, drawing upon the advantages of both Pittsburgh and Michigan approaches. They choose the basic Pittsburgh representation and call their model Pittsburgh Fuzzy Classifier System # 1 (P-FCS1). They employ a real-valued GA and define appropriate genetic operators for it. Both the rules and the antecedent membership functions are represented in the chromosome. The number of rules is also a parameter tuned together with the rule base. Local membership rules (rule-specific) are used. The P-FCS1 model poses a series of interesting questions and outlines paths for further improvement of training fuzzy systems by GAs.

It is not a priori clear which of the above GA training methods will win a "fair" experimental competition. The leitmotif of this book is that simple models and methods should be tried first and be dismissed only if they do not provide a satisfactory result. Pure rule selection looks simple and effective, and could be the best start-up candidate.

6.6 Fuzzy classifiers and neural networks: hybridization or identity?

We shall try to refrain from calling every classifier a **neural network**. In fact, any classifier can be represented as the good old canonical model in [87], and the appearance of the diagram (Figure 2.2) suggests it be called a "neural network". In this line, the Min-max group of if-then fuzzy models (which "fits neatly into an NN assembly" [157]) was described in the previous section and not under the heading "neuro-fuzzy".

"Neuro-fuzzy" is a huge topic on its own, fashionable and popular, still in its infancy, although the first works were published back in the 70's [212]. It is hardly possible to account for all the diverse and highly heuristic neuro-fuzzy and fuzzy-neuro models (cf. [50, 247, 246, 15]).[6]

Inevitably, the immaturity and the richness of the toolbox of the neuro-fuzzy field generates inconsistency in the definitions, notions and notations. It would not be an exaggeration to say that there are hundreds of models of fuzzy neurons described in the literature. Picking a set of neuro-fuzzy designs and leaving outside hundreds more is not the best option. On the other hand, making an intelligent summary of neuro-fuzzy classifiers and exemplifying each branch seems impossible at this stage. Hence, this section offers some speculations and a few references.

Halgamuge [120] defines "neuro-fuzzy" systems as "*trainable neural networks that can be interpreted as fuzzy systems*". A further specification of this definition is offered by Nauck and Kruse[7]:

> *A neuro-fuzzy system is a fuzzy system that uses a learning algorithm derived from or inspired by neural network theory to determine its parameters (fuzzy sets and fuzzy rules) by processing data samples.*

Buckley and Hayashi [52] distinguish between *hybrid* neural networks and *fuzzy* neural networks. Hybrid NNs are those where the neurons can perform

---

[6] More information can be found on the following WWW addresses: http://www.isis.ecs.soton.ac.uk/resources/nfinfo and http://www.abo.fi/ rfuller/fuzs.html

[7] Found at (http://fuzzy.cs.uni-magdeburg.de/nfdef.html)

various logical operations, and fuzzy NNs are those whose signals or weights or both are fuzzy (e.g., fuzzy numbers). Thus, there can also be a ***hybrid fuzzy*** NN. The pioneering work by Lee and Lee [212] proposed a model of a fuzzy neuron as a generalization of the basic McCulloch and Pitt's model [226]. Fuzzy models of neurons were an important research topic for some time: neurons that perform OR/AND operations [134] and various other fuzzy connectives [54, 245, 162, 274, 354, 358], used as building blocks of neural-network-like schemes able to implement any chain of fuzzy logic expressions [163, 168, 48]. Most of the work in the neuro-fuzzy area is now revolving around solving real-life problems by devising faster and more efficient training algorithms. Three basic tasks relate neural networks and rule-based systems [18]: knowledge initialization, rule extraction and rule refinement. These tasks have inspired many disparate neuro-fuzzy models.

The equivalence between TSK2 type fuzzy models and RBF networks has been noticed and used for training the fuzzy classifier by NN type procedures [154, 155].

Kim and Mendel [169] summarize various kernel-type models and show their general functional equivalence. Among these are designs of probabilistic and non-probabilistic origin such as Parzen density estimator, Specht's probabilistic neural networks [310], generalized regression models, a Kalman filtering models, RBF neural networks, etc. Kim and Mendel argue that by using **fuzzy-based functions** we obtain a richer model which is different from the classical RBF network. The difference comes from the normalization (the denominator of (5.27)) which distinguishes the TSK2 model form the standard RBF output. However, for classification we are only interested in the *ranking* of the output values, most often only in the winner, to assign a class label to $\mathbf{x}$. Therefore, the difference coming from normalization is not essential. This brings all the designs in [169] in the same class of equivalence with regard to classification. The benefit of this equivalence could be in borrowing training ideas from RBF networks and applying them to the fuzzy if-then models. Gradient descent has been one of these [9, 56, 57, 216, 234, 305, 313]. There is a great variety of RBF type neuro-fuzzy models that use other types of learning (mostly heuristic) [58, 62, 121, 122, 123, 120, 178, 208, 239, 370]. A series of MLP (Multi-Layer Perceptron) fuzzy designs has been developed by Mitra and Pal [238, 240, 239, 261]. Nauck and Kruse proposed a neuro-fuzzy classification model NEFCLASS [248] and variants thereof [246, 249, 251, 250, 247], aiming at an accurate and yet transparent fuzzy classifier.

6.7 Forget interpretability and choose a model

This section offers a somewhat heretical conclusion for the Chapter. It looks that the often neglected issue of interpretability is in fact irrelevant. We may think of fuzzy classifiers as black boxes. As long as they have their niche where they work better than any other classification model, they need not be interpretable. So far there is no sound experimental or theoretical evidence as to whether or not fuzzy models can outperform statistical designs. The hope lies in the flexibility of the fuzzy models. Clearly, they can reach small or zero resubstitution error, and handling the generalization problem is just a matter of finding proper regularization strategies and algorithms. The challenge is that there are no paved roads in the realm of fuzzy classifiers. A model has to be chosen from a huge amorphous body of models. Then this model has to be carefully tuned, loads of parameters need to be initialized, etc. This requires skills, intuition, and most of all luck. A comprehensible set of gold-standard fuzzy designs may be distilled in the not very distant future. Neither the curse of dimensionality nor the lack of interpretability is likely to stand in the way.

7. Non if-then fuzzy models

In Chapter 1 we adopted Definition 1 stating that a fuzzy classifier is any classifier which uses fuzzy sets either during its training or during its operation. So, fuzzy classifier modeling stretches beyond fuzzy if-then designs discussed in the previous two chapters. This chapter presents non-if-then fuzzy models. These models can be grouped in different ways (see [39, 81, 115, 118, 273, 320]). However, the boundaries between these groups are not sharp because many of the classification schemes can be assigned to more than one group (see, e.g., [232] where the authors use multiple rule-based prototypes and call their method a knowledge-oriented fuzzy k-nearest neighbor classifier).

7.1 Early ideas

The first ideas about using fuzzy sets in classification emerged soon after fuzzy sets were introduced by Zadeh in 1965. The seminal paper by Bellman, Kalaba and Zadeh [26] defines a fuzzy classification environment where the problem is to design a classifier producing soft labels. They call ***abstraction*** the process of finding a set of discriminant functions $g_i : \Re^n \rightarrow [0,1]$, $i = 1, \ldots, c$, from the labeled data set $\mathbf{Z}$. This corresponds to ***training*** the classifier. Calculating a soft label for a given $\mathbf{x} \in \Re^n$, not in $\mathbf{Z}$, is called *generalization*.

7.1.1 Fuzzy and probabilistic models

Various relationships between fuzziness and probability have been explored. **Combinations** of fuzzy and probabilistic types of uncertainty have been sought at a rather theoretical level. Examples of this are fuzzy probability spaces [160], fuzzy Bayes classification (fuzzy-statistical and statistical-fuzzy models) [86, 236], fuzzy-Bayes decision making [320], etc. **Replacing** the probability density functions (p.d.f's) in the classical statistical pattern recognition model (Chapter 2) by membership functions has appealed to many authors as a logical substitute. This choice has been continuously disputed ever since it emerged. The idea of substitution has given rise to the **possibilistic** models for classification where possibilistic distributions are constructed and used similarly to the probabilistic ones. While there are many methods for estimating probability distributions from a data set, the derivation of possibility distributions is not straightforward.

7.1.2 Fuzzy linear discriminant analysis

Watada and coauthors [343, 344] propose a general linear fuzzy discriminant analysis. The input $\tilde{\mathbf{x}}$ of the classifier is not a point in $\Re^n$ but a set of n *fuzzy numbers* $\tilde{x}_1, \ldots, \tilde{x}_n$, one for each feature. Each element of the data set $\mathbf{Z}$ is also a set of n fuzzy numbers on the feature axes. To calculate the discriminant functions, equations (3.12) are implemented via fuzzy arithmetic. The class means and the entries of the covariance matrices are also fuzzy numbers obtained from $\mathbf{Z}$. In all matrix-type calculations, instead of sum-product, max-min is used. Given the input $\tilde{\mathbf{x}}$, each discriminant function $g_i(\tilde{\mathbf{x}})$ is a fuzzy set defined on the interval [0,1], expressing how certain the decision is for the respective class. Comparing these fuzzy sets in order to assign a class label to $\tilde{\mathbf{x}}$ is not trivial. We either defuzzify $g_1(\tilde{\mathbf{x}}), \ldots, g_c(\tilde{\mathbf{x}})$ and compare the obtained c values, or pick a procedure for comparing fuzzy sets per se. In practice, we hardly know the exact values of the features; measurements are prone to errors; human-defined features are by definition imprecise (e.g., the degree of pain); hence using fuzzy numbers seems a reasonable choice. However, the application of this discriminant model to real problems is difficult because of the implementation of the fuzzy arithmetic involved. Besides, it is not clear whether the fuzzy model is better than the standard analysis in terms of classification performance. A further question is how robust the model is with respect to the choice of membership functions for the fuzzy numbers.

Krusińska [182] suggests using the soft labels of $\mathbf{Z}$, $\mathbf{l}(\mathbf{z}_j) \in [0,1]^c$, $j = 1, \ldots, N$, if available, in computing the means for the c classes and the common covariance matrix. The soft labels are used as the weights in the maximum-likelihood formulas for the means and the covariance matrix. Another option suggested in [183] is to extract "fuzzy" features by grouping a large number of features to form a single score (as in [292], see Example 4.4.4 on page 111), and to use the classical linear discriminant analysis on the new features. Thus, the "fuzzification" of the feature space leads to dimensionality reduction.

A weather forecast by a linear regression is suggested in [373]. The classifier produces a single degree of membership in [0,1] which is compared against $c-1$ thresholds to assign a crisp class label to $\mathbf{x}$. Borodkin and Stadnik [47] also assume that the data set $\mathbf{Z}$ has soft labels in the classes. They develop an algorithm to find a hyperplane in $\Re^n$ such that if we cluster the projections of the data points onto the hyperplane, the similarity between the obtained class (cluster) labels $\hat{\mathbf{l}}(\mathbf{z}_j), j = 1, \ldots, N$ and the original class labels $\mathbf{l}(\mathbf{z}_j), j = 1, \ldots, N$ is high. Stadnik draws an example from a study of lymphogranulomatosis [311].

Like its nonfuzzy counterpart, the **fuzzy perceptron** [167] employs a linear discriminant function $g : \Re^n \rightarrow \Re$ to distinguish between two classes ω_1 and ω_2. The *fuzzy* part of the algorithm is the training, where a soft pre-labeling of the data set $\mathbf{Z}$ is used. An iterative procedure is applied to find $g \equiv \mathbf{w}^T \mathbf{x}^a$, where $\mathbf{w} \in \Re^{n+1}$ is a real-valued vector and $\mathbf{x}^a$ is the augmented vector $[\mathbf{x}^T, 1]^T \in \Re^n \cup \{1\}$. All vectors in $\mathbf{Z}$ are augmented (as $\mathbf{x}$ above) and those from class ω_2 are multiplied by -1. A specially designed pre-labeling procedure assigns to every $\mathbf{z}_j$ a soft label greater than 0.5 in its own class and less than 0.5 in the other class.[1] The training of the perceptron is the same as explained in Chapter 3 (viz. Figure 3.24). The procedure starts with a random $\mathbf{w}$ and updates it by iterating on the elements of $\mathbf{Z}$. If an error occurs when classifying $\mathbf{z}_j$ with the current g, the weight vector $\mathbf{w}$ is updated by

$$\mathbf{w} \leftarrow \mathbf{w} + |l_1(\mathbf{z}_j) - l_2(\mathbf{z}_j)|^m \, \eta \, \mathbf{z}_j \tag{7.1}$$

where $l_i(\mathbf{z}_j)$ is the soft label of $\mathbf{z}_j$ in class $\omega_i, i = 1, 2$, η is a constant, and m is a parameter, usually $m > 1$. This algorithm is shown to converge for linearly separable classes as does the algorithm for the crisp perceptron (Figure 3.24). It is also argued [167] that by discarding borderline vectors from $\mathbf{Z}$ (those with degrees of membership in their own class close to 0.5), the fuzzy perceptron algorithm *may* converge in the case of linearly-nonseparable classes as well.

Fuzzy linear models for classification have not been the most favored. This is probably because fuzzy linear classifiers have not been shown to offer a significant benefit over the nonfuzzy classifiers. The perceptron model is an exception but it is not versatile enough. Since interpretability is not an issue for these models, demonstrating clearly the gain in classification performance is of a primary importance.

7.1.3 Fuzzy relational classifiers

As the name suggests, fuzzy relational classifiers are based on **fuzzy relations** (see Section 4.3.8 on page 107). This model is useful when the features take a small number of discrete values. Therefore, instead of $\Re^n$, a finite feature space $\mathcal{S}$ is considered. An example of a problem generating a discrete feature space is medical diagnosis based on symptoms. The causality connecting the symptoms ($\mathcal{S}$) and the diagnoses (Ω) is sometimes called "medical knowledge" and is represented as the fuzzy relation

$$\mathcal{R} : \mathcal{S} \times \Omega \rightarrow [0, 1], \tag{7.2}$$

or equivalently

$$\mathcal{R}(\mathbf{x}, \omega_i) \in [0, 1]. \tag{7.3}$$

In the simplest case $\mathcal{R}$ is a look-up table.

[1] The pre-labeling procedure is explained in detail in Section 7.2

Example 7.1.1. Let $X = \{X_1, X_2\}$, where X_1 is Symptom1 and X_2 is Symptom2. Let $\Omega = \{\omega_1, \omega_2, \omega_3\}$, such that ω_1 is Disease1, ω_2 is Disease2, ω_3 is Disease3. Each symptom can be either present or absent, therefore $\mathcal{S}$ consists of the following four alternatives $\left\{ \begin{bmatrix} 0 \\ 0 \end{bmatrix}, \begin{bmatrix} 0 \\ 1 \end{bmatrix}, \begin{bmatrix} 1 \\ 0 \end{bmatrix}, \begin{bmatrix} 1 \\ 1 \end{bmatrix} \right\}$. A possible symptom-disease relation $\mathcal{R}$ is expressed by the following table

x^T	ω_1	ω_2	ω_3
0,0	0.3	0.7	0.1
0,1	0.7	0.3	0.0
1,0	0.1	0.4	0.6
1,1	0.4	0.5	0.1

According to $\mathcal{R}$, a patient with Symptom1 present and Symptom2 absent ($[1,0]^T$) will be assigned the soft label $[0.1, 0.4, 0.6]^T$ (third row in the table). ■

The advantage of the fuzzy relational model lies with the "uncertain" $\mathbf{x}$'s, e.g., those patients who have the symptoms to a certain degree. For example, chest pain is associated with ischemic heart disease and can be used for defining the relation $\mathcal{R}$. However, *pain* is not a strictly binary symptom, so there could be degrees of pain. Then $\mathbf{x}$ is expressed as a vector with memberships (possible values for presence or severity of the symptom). To simplify notations we consider here binary features $\{X_1, \ldots, X_n\}$. The extension to multiple-category features is straightforward. Then the feature space $\mathcal{S}$ has 2^n elements as each element is a particular combination of symptom values. Let $\mathbf{f}(\mathbf{x}) = [f_1(\mathbf{x}), \ldots, f_n(\mathbf{x})]^T$ be the vector with degrees of membership showing to what extent feature X_k is *present* in $\mathbf{x}$, $k = 1, \ldots, n$. In the general case we should also have a vector $\bar{\mathbf{f}}(\mathbf{x})$ with degrees of *absence*. For this example we assume that the vector for "absent" is obtained as

$$\bar{\mathbf{f}}(\mathbf{x}) = \mathbf{1} - \mathbf{f}(\mathbf{x}). \tag{7.4}$$

To calculate the output of a relational classifier, we transform the input $\mathbf{x} \in \Re^n$ by defining a fuzzy set f_S over the set of symptoms $\mathcal{S}$, corresponding to that $\mathbf{x}$. The degree of membership $f_S(s)$ is calculated by taking the minimum between the n degrees of membership participating in $\mathcal{S}$, as shown in the following example.

Example 7.1.2. Let $n = 2$ and let s be an element of $\mathcal{S}$, e.g., $s = [0,1]^T$ (Symptom1 absent and Symptom2 present). Assume that $\mathbf{f}(\mathbf{x}) = [0.3, 0.1]^T$, i.e., Symptom1 is present to a mild degree (0.3) and Symptom2 is virtually absent (0.1). Then the degree of membership of $f_S(s)$ is

$$f_S([0,1]^T) = \min\{\bar{f}_1(\mathbf{x})), f_2(\mathbf{x})\} = \min\{(1 - 0.3), 0.1\} = 0.1. \tag{7.5}$$

■

$\mathbf{f}_{\mathcal{S}}$ is called an *input fuzzy set*. As soon as we know the relation $\mathcal{R}$ and the input fuzzy set on $\mathcal{S}$, we can calculate the classifier output for $\mathbf{x}$. We can regard the input fuzzy set $f_{\mathcal{S}}$ as a fuzzy relation $\mathcal{I}$ on a universal set with $1 \times |\mathcal{S}|$ elements $f_{\mathcal{S}}(s)$. Then the soft class label of $\mathbf{x}$ is obtained by **composition** (see Chapter 4, page 107)

$$\mu(\mathbf{x}) = \mathcal{I} \circ \mathcal{R}. \tag{7.6}$$

Example 7.1.3. Let $\mathbf{f}(\mathbf{x}) = [0.3, 0.1]^T$ as in the above example and let $\circ$ be a max-min composition. Then for $\mathcal{S} = \left\{ \begin{bmatrix} 0 \\ 0 \end{bmatrix}, \begin{bmatrix} 0 \\ 1 \end{bmatrix}, \begin{bmatrix} 1 \\ 0 \end{bmatrix}, \begin{bmatrix} 1 \\ 1 \end{bmatrix} \right\}$, we have

$$\mathcal{I} = [0.7, 0.1, 0.3, 0.1]. \tag{7.7}$$

By applying max-min composition in (7.6), we obtain

$$\mu(\mathbf{x}) = [0.7, 0.1, 0.3, 0.1] \circ \begin{bmatrix} 0.3 & 0.7 & 0.1 \\ 0.7 & 0.3 & 0.0 \\ 0.1 & 0.4 & 0.6 \\ 0.4 & 0.5 & 0.1 \end{bmatrix} = [0.3, 0.7, 0.3]. \tag{7.8}$$

Using the maximum membership rule, $\mathbf{x}$ will be assigned class label ω_2. ∎

The entries in the relational table $\mathcal{R}$ can be calculated from data. For example, these can be the relative frequencies of occurrence of each disease given a symptom combination. The table can also be filled in by domain experts. We assume that we have a labeled data set $\mathbf{Z}$ and the following two relations:

- $\mathcal{Q}$: a fuzzy relation on $\mathbf{Z} \times \mathcal{S}$,
- $\mathcal{T}$: a fuzzy relation on $\mathbf{Z} \times \Omega$, i.e., soft class labels for each $\mathbf{z} \in \mathbf{Z}$.

Example 7.1.4. Let $\mathbf{Z}$ be the set {North America, Europe, Africa, Asia, Oceania and Australia, South America}, denoted respectively NA, EU, AF, AS, OC, and SA. Let $\mathcal{S}$ be a set containing the following four elements: {population increase (PI), income per capita (IPC), energy consumption (EC), fresh water availability (FWA)}. A fuzzy relation $\mathcal{Q}$ on $\mathbf{Z} \times \mathcal{S}$ is given in Table 7.1 (the table was designed using data from National Geographic magazine, 1999).

Let $\Omega = \{\omega_1, \omega_2\}$, where the class label expresses the standard of life. Let $\mathcal{R}$ be a fuzzy relation on $\mathcal{S} \times \Omega$, showing the relationship between the elements of $\mathcal{S}$ and the standard of life, and let ω_1 = "good" and ω_2 = "poor". We can design $\mathcal{R}$ by eye, as shown in Table 7.2.

By composition of $\mathcal{Q}$ and $\mathcal{R}$ we can find a class label for each of the elements of $\mathbf{Z}$ as a resultant fuzzy relation $\mathcal{T}$ on $\mathbf{Z} \times \Omega$. Table 7.3 shows $\mathcal{T}$ obtained as $\mathcal{Q} \circ \mathcal{R}$ with max-min and max-product as the composition $\circ$.

Applying the maximum membership rule, out fuzzy relational classifier will put NA, EU, OC and SA in classed in ω_1, and AF and AS in ω_2 with both type composition. ∎

Table 7.1. A fuzzy relation $\mathcal{Q}$ on $\mathbf{Z} \times \mathcal{S}$

	PI	IPC	EC	FWA
NA	0.0490	1.0000	1.0000	0.2947
EU	0.0098	0.5059	0.4034	0.1316
AF	0.3922	0.0240	0.0420	0.0368
AS	1.0000	0.0919	0.0924	0.0737
OC	0.0098	0.5694	0.5070	1.0000
SA	0.1471	0.1369	0.1373	0.5053

Table 7.2. A fuzzy relation $\mathcal{R}$ on $\mathcal{S} \times \Omega$

	Good	Poor
PI	0.1	0.8
IPC	0.8	0.1
EC	0.6	0.3
FWA	0.7	0.1

Table 7.3. A fuzzy relation $\mathcal{T}$ on $\mathbf{Z} \times \Omega$

	max-min		max-product	
	Good	Poor	Good	Poor
NA	0.8000	0.3000	0.8000	0.3000
EU	0.5059	0.3000	0.4047	0.1210
AF	0.1000	0.3922	0.0392	0.3137
AS	0.1000	0.8000	0.1000	0.8000
OC	0.7000	0.3000	0.7000	0.1521
SA	0.5053	0.1471	0.3537	0.1176

One way to calculate $\mathcal{R}$, given $\mathcal{Q}$ and $\mathcal{T}$, is to take the biggest fuzzy relation that satisfies

$$\mathcal{Q} \circ \mathcal{R} \subseteq \mathcal{T}. \tag{7.9}$$

Considering max-min composition as $\circ$, Sanchez [294] gives a method for obtaining $\mathcal{R}$ by solving the relational equation

$$\mathcal{R} = \mathcal{Q}^{-1} \;\; @ \; \mathcal{T}, \tag{7.10}$$

where @ is a composition of fuzzy relations, defined through an α operator explained below (see also [84]).

Let A and B be two fuzzy relations on $U \times V$ and $V \times W$, respectively. The composition and @ is

$$\mu_{A@B}(u, w) = \min_{v \in V} \left\{ \mu_A(u, v) \; \alpha \; \mu_B(v, w)) \right\}, \quad \forall (u, w) \in U \times W, \tag{7.11}$$

where α is the Godelian implication (see Table 4.5)

$$a \; \alpha \; b = \begin{cases} 1, \; a \leq b \\ b, \; \text{otherwise} \end{cases} \quad a, b \in [0, 1]. \tag{7.12}$$

Example 7.1.5. Assume that we have only $\mathcal{Q}$ and $\mathcal{T}$ obtained by max-min composition from Example 7.1.4. Using (7.11) and (7.12), the following $\mathcal{R}$ was found

$$\mathcal{R} = \begin{bmatrix} 0.1000 & 0.8000 \\ 0.5059 & 0.3000 \\ 0.8000 & 0.3000 \\ 0.7000 & 0.1471 \end{bmatrix}. \tag{7.13}$$

This relation is not very different from the one that was used to calculate $\mathcal{T}$ (Table 7.2). ■

A theoretical basis for classification in medicine, based on fuzzy relational equations, is developed by Sanchez, and extended further to interval-valued fuzzy relations [288]. Because of the complexity of the medical diagnostic problems, many authors develop their own domain-specific fuzzy relational classifiers. Adlassnig and coauthors use two different relations for "strength of occurrence" of symptoms and "degree of confirmation" of a diagnosis [4, 5]. They embed these in a medical expert system CADIAG-2 with specific versions for several medical fields. It is argued that along with the degree of "confirmation" we have to derive the degree of "rejection" of a diagnosis, i.e., accounting for how much negative evidence has been accumulated. This is deemed necessary because, unlike in probabilistic systems, here we may get, e.g., "support for diagnosis ω_k" 0.7 and "evidence against diagnosis ω_k" 0.85. This problem is discussed also in [5, 188, 255, 328].

Fuzzy relational classifiers are presented in [76, 104, 105, 211, 271, 272, 320, 332].

An interesting classifier is proposed in [300]. $\mathcal{R}$ is designed not on $\mathcal{S} \times \Omega$ but on $Cl \times \Omega$ where Cl is a set of clusters in $\mathbf{Z}$. These clusters need not be interpretable. Both Cl and $\mathcal{R}$ are obtained from the data set. The classifier is shown to compete well with some non-fuzzy classifiers. Similarly, Mandal and Tanaka [229] consider a relational classifier by partitioning of the feature space and computing $\mathcal{R}$ as a fuzzy relation between the components of the partition and the class labels. These two examples link fuzzy relational classifiers with the clustering-and-relabeling if-then classifiers discussed earlier. The relabeling stage here means to find how each cluster is related to the class labels, i.e., obtaining $\mathcal{R}$.

Pedrycz [273] shows that the fuzzy relational classifier can be made isomorphic to the Bayes classifier under an appropriate definition of $\mathcal{R}$.

We can also note that there is a link between fuzzy relational classifiers and fuzzy if-then classifiers discussed in the previous two chapters. If we design $\mathcal{S}$ to be the set of all possible combinations of linguistic labels, each rule defines c entries in the relation $\mathcal{R}$. Thus, the rule base fills sparsely the $\mathcal{R}$ table.

Example 7.1.6. Consider a classification problem with two features such that there are two labels defined over x_1, {low, high}, and three linguistic labels defined over x_2, {small, medium, big}. Let $\Omega = \{\omega_1, \omega_2, \omega_3\}$ and let the kth rule of the rule base be

R_k : IF x_1 is *low* AND x_2 is *medium*

THEN $g_{k,1} = 0.4$ AND $g_{k,2} = 0.1$ AND $g_{k,3} = 0.7$.

To design a fuzzy relational classifier we form $\mathcal{S}$ to be the set of all six combinations of linguistic labels, i.e., $\mathcal{S} = \{(low,\ small), \ldots, (high,\ big)\}$. The fuzzy relation $\mathcal{R}$ on $\mathcal{S} \times \Omega$ is formed from the rules in the rule base. The rule R_k contributes to $\mathcal{R}$ three degrees of membership, i.e.,

$$\begin{aligned}\mu_{\mathcal{R}}\left((low, med), \omega_1\right) &= 0.4\\ \mu_{\mathcal{R}}\left((low, med), \omega_2\right) &= 0.1\\ \mu_{\mathcal{R}}\left((low, med), \omega_3\right) &= 0.7\end{aligned}$$

Let $\mathbf{x} \in \Re^2$ be submitted to the classifier. Assume that the two antecedent clauses in R_k give $\mu_{1,1}(x_1) = 0.3$ and $\mu_{2,2}(x_2) = 0.6$. Then the degree of membership of the input fuzzy set $f_{\mathcal{S}}$ evaluated at $s = (low, med)$ is

$$f_{\mathcal{S}}((low, med)) = \min\{0.3, 0.6\} = 0.3.$$

Notice that $f_{\mathcal{S}}((low, med))$ is the firing strength $\tau_k(\mathbf{x})$ for minimum used as $\mathcal{A}_t$. Then the relational fuzzy classifier with max-min composition of $\mathcal{I}$ and $\mathcal{R}$ is similar to the TSK models introduced in Chapter 5, and for the max-product composition, the relational classifier is identical with the TSK1 model (page 137) but with soft class labels. ■

Most of these early designs have merely theoretical value. Fuzzy if-then classifiers crystallized as the most practically sound option. A few other models have been retained, e.g., fuzzy k-nearest neighbors and fuzzy prototype classifiers, outlined in the following sections.

7.2 Fuzzy k-nearest neighbors (k-nn) designs

The k-nearest neighbor classifier (k-nn) has many fuzzy and non-fuzzy relatives. There are two main questions:

1. If the reference vectors in $\mathbf{Z}$ are softly labeled in the classes, how do we combine their labels to find the soft output label $\mu(\mathbf{x})$ for $\mathbf{x} \in \Re^n$? Do we use distances in this combination and how?
2. If the reference vectors are not softly labeled in the classes, do we need soft labels at all? If we do, how do we find such soft labels?

7.2.1 Combinations of soft labels

Denote by $\mathbf{Z}^{(\mathbf{x})} = \{\mathbf{z}^{(1)}(\mathbf{x}), \ldots, \mathbf{z}^{(k)}(\mathbf{x})\}$ the set of the k vectors from the reference set $\mathbf{Z}$, which are closest to $\mathbf{x}$ according to some prespecified distance. (To simplify notations we will use $\mathbf{z}^{(k)}$ instead of $\mathbf{z}^{(k)}(\mathbf{x})$.) Let $\mathbf{l}(\mathbf{z}^{(j)}) \in [0,1]^c, j = 1 \ldots, k$, be the *soft* labels of the k nearest neighbors. The most natural way to combine the class labels is to average them as suggested by Jóźwik [159]

$$\mu(\mathbf{x}) = \frac{1}{k} \sum_{j=1}^{k} \mathbf{l}(\mathbf{z}^{(j)}). \tag{7.14}$$

This rule reduces to the classical k-nn rule when the reference vectors are crisply labeled and the classification is done by the maximum membership rule. Averaging seems a logical choice but other operations can also be used in the above equation. For example, we can pick any of the fuzzy aggregation operators from Chapter 4 thereby devising new fuzzy k-nn versions. The choice of an aggregation operation should be guided by some preliminary knowledge about the problem or, empirically, by improvement on classification accuracy.

Many authors include the distances to the nearest neighbors in calculating $\mu(\mathbf{x})$, e.g., [30, 31, 171, 225, 316]. To find the ith component of the soft class label for $\mathbf{x}$, $\mu_i(\mathbf{x})$, Bereau and Dubuisson [30, 31] use

$$\mu_i(\mathbf{x}) = \max_{1 \le j \le k} \left\{ l_i \left(\mathbf{z}^{(j)} \right) \cdot \exp \left[-\lambda \left(\frac{d_j}{d_i^m} \right) \right] \right\} \tag{7.15}$$

where $\lambda > 0$ is a parameter used to control the decrease of the exponent (the authors recommend $\lambda \in [0,1], \lambda \approx 0.5$), d_j is the distance between $\mathbf{x}$ and its jth nearest neighbor $\mathbf{z}^{(j)}$, and d_i^m is the ***average distance*** for class ω_i. In [30, 31] d_i^m is said to be the average distance between the elements of the reference set $\mathbf{Z}$ which have ***high membership*** in ω_i. If $\mathbf{Z}$ is crisply labeled, then d_i^m is the average distance between each pair of vectors from ω_i.

For crisply labeled $\mathbf{Z}$, (7.15) reduces to the single nearest neighbor (1-nn) with a modified distance. The new distance between $\mathbf{x}$ and an element of ω_i is the Euclidean distance d_j divided by d_i^m. To derive this, we start with equation (7.15) and drop the soft membership $l_i\left(\mathbf{z}^{(j)}\right)$. For all neighbors from ω_i, this membership is 1, and for the remaining ones, the membership is 0. Thus, only the neighbors from ω_i will participate in (7.15). Next, since the denominator (d_i^m) of the exponent is the same for all participating neighbors, the maximum will be attained by the neighbor at the smallest Euclidean distance to $\mathbf{x}$. Then $\mu_i(\mathbf{x})$ is determined merely by the closest neighbor from ω_i. The comparison between $\mu_1(\mathbf{x}), \ldots, \mu_c(\mathbf{x})$ in order to determine the class label depends only on d_j/d_i^m. Therefore, if we use the modified distance for the elements of each class, $d_j^* = d_j/d_i^m$, the final (crisp) decision for the class label of $\mathbf{x}$ is determined by the nearest neighbor according to the modified distance. The value of the parameter λ in (7.15) is irrelevant, so is k. (We only assume that k is large enough so that within the set of neighbors $\mathbf{Z}^{(\mathbf{x})}$ there is at least one neighbor from each class.) With crisply labeled $\mathbf{Z}$ this scheme has not been successful on our four data sets. The training and testing error rates in % are summarized in Tables 7.4 to 7.7 (Fuzzy (Béreau and Dubuisson, 1991 [31]), max) together with the results from the crisp 1-nn and other k-nn variants discussed later. Our results showed that scaling the distances by d_i^m did not lead to any improvement over the simple 1-nn rule. It should be noted that Bereau and Dubuisson also use the following (second) membership function [31]. Instead of maximum, they take the summation

$$t_i = \sum_{j=1}^{k} l_i\left(\mathbf{z}^{(j)}\right) \cdot \exp\left[-\lambda\left(\frac{d_j}{d_i^m}\right)\right], \tag{7.16}$$

and then calculate the membership value $\mu_i(\mathbf{x})$ by

$$\mu_i(\mathbf{x}) = \frac{1}{1+\exp\left(\frac{k/2-t_i}{b_i}\right)}, \tag{7.17}$$

where b_i is the *entropy measure of fuzziness* for class ω_i (4.23)

$$b_i = -\sum_{\mathbf{z}\in\mathbf{Z}} \left(l_i(\mathbf{z})\log\left(l_i(\mathbf{z})\right) + \left(1-l_i(\mathbf{z})\right)\log\left(1-l_i(\mathbf{z})\right)\right). \tag{7.18}$$

The classification algorithm in [31] is quite sophisticated. Here we tried just the membership function (7.17) with crisply labeled $\mathbf{Z}$. Tables 7.4 to 7.7 show the *best* results with our four data sets: the lowest training error and the corresponding testing error (Fuzzy (Béreau and Dubuisson, 1991 [31]), $\sum$).

Keller, Gray and Givens [166] propose another formula which is coupled with a procedure for initial fuzzification of the data set discussed in Section 4.4, equation (4.95) on page 115.

$$\mu_i(\mathbf{x}) = \frac{\sum_{j=1}^{k} l_i\left(\mathbf{z}^{(j)}\right)\left(d_j\right)^{-\frac{2}{m-1}}}{\sum_{j=1}^{k}\left(d_j\right)^{-\frac{2}{m-1}}}, \quad i=1,\ldots,c, \tag{7.19}$$

where m is a fuzzification parameter.

Example 7.2.1. Shown in Figure 7.1 is a point $\mathbf{x}$ (the filled square) and its 5 nearest neighbors $\mathbf{z}^{(1)},\ldots,\mathbf{z}^{(5)}$. The neighbors are connected with $\mathbf{x}$ and the distances d_j are marked on the connections.

Let $d_1 = 1$, $d_2 = 1.1$, $d_3 = 1.3$, $d_4 = 1.6$, and $d_5 = 2$ in some units. Let also the soft class labels of $\mathbf{z}^{(1)},\ldots,\mathbf{z}^{(5)}$ be as follows

Table 7.4. Best training (e_{tr}) and the corresponding test error (e_{ts}) of crisp, fuzzy and distance-based k-nn with the **Cone-torus** data. In case of ties on e_{tr}, the smallest number k is taken.

Method	k	e_{tr}	e_{ts}
Crisp	9	15.00	13.00
Fuzzy (Béreau and Dubuisson, 1991 [31]), max	1	24.00	23.25
Fuzzy (Béreau and Dubuisson, 1991 [31]), $\sum$	9	14.75	13.00
Fuzzy (Keller et al., 1985 [166]), $m = 2$	4	15.75	16.50
Fuzzy+pre-labeling ([166]), $m = 2, k^* = 1$	2	13.00	17.25
Theoretical model # 1 (see the text)	5	15.75	16.75
Theoretical model # 2 (see the text)	5	15.00	13.00
Distance-based (Dudani, 1976 [88])	5	15.50	16.00
Distance-based (Macleod et al., 1987 [227]), $\alpha = 0, s = 3k$	9	14.75	12.75
Distance-based (Parthasarathy and Chatterji, 1990 [266])	5	15.75	14.50
Distance-based (Hattori and Takahashi, 1999 [130]),$l = 1$	9	15.25	12.50

Table 7.5. Best training (e_{tr}) and the corresponding test error (e_{ts}) of crisp, fuzzy and distance-based k-nn with the **Normal-mixtures** data. In case of ties on e_{tr}, the smallest number k is taken.

Method	k	e_{tr}	e_{ts}
Crisp	3	14.40	13.40
Fuzzy (Béreau and Dubuisson, 1991 [31]), max	1	15.60	15.60
Fuzzy (Béreau and Dubuisson, 1991 [31]), $\sum$	3	14.40	13.40
Fuzzy (Keller et al., 1985 [166]), $m = 2$	8	14.00	12.30
Fuzzy+pre-labeling ([166]), $m = 2, k^* = 3$	2	10.40	14.40
Theoretical model # 1 (see the text)	1	14.80	15.00
Theoretical model # 2 (see the text)	9	11.60	8.70
Distance-based (Dudani, 1976 [88])	5	12.40	14.00
Distance-based (Macleod et al., 1987 [227]), $\alpha = 0, s = 2k$	4	12.40	13.60
Distance-based (Parthasarathy and Chatterji, 1990 [266])	4	13.20	14.10
Distance-based (Hattori and Takahashi, 1999 [130]),$l = 2$	2	11.60	13.60

Table 7.6. Best training (e_{tr}) and the corresponding test error (e_{ts}) of crisp, fuzzy and distance-based k-nn with the **Phoneme** data. In case of ties on e_{tr}, the smallest number k is taken.

Method	k	e_{tr}	e_{ts}
Crisp	1	17.20	17.70
Fuzzy (Béreau and Dubuisson, 1991 [31]), max	1	17.40	19.15
Fuzzy (Béreau and Dubuisson, 1991 [31]), $\sum$	6	16.00	18.78
Fuzzy (Keller et al., 1985 [166]), $m = 2$	1	17.20	17.70
Theoretical model # 1 (see the text)	1	17.20	17.70
Theoretical model # 2 (see the text)	9	17.00	20.07
Distance-based (Dudani, 1976 [88])	1	17.20	17.70
Distance-based (Parthasarathy and Chatterji, 1990 [266])	4	17.00	17.35
Distance-based (Hattori and Takahashi, 1999 [130]),$l = 2$	4	16.60	17.25

Table 7.7. Best training (e_{tr}) and the corresponding test error (e_{ts}) of crisp, fuzzy and distance-based k-nn with the **Satimage** data. In case of ties on e_{tr}, the smallest number k is taken.

Method	k	e_{tr}	e_{ts}
Crisp	9	15.00	15.05
Fuzzy (Béreau and Dubuisson, 1991 [31]), max	1	22.80	22.59
Fuzzy (Béreau and Dubuisson, 1991 [31]), $\sum$	9	14.60	15.01
Fuzzy (Keller et al., 1985 [166]), $m = 2$	6	14.40	15.92
Theoretical model # 1 (see the text)	5	19.00	18.77
Theoretical model # 2 (see the text)	6	14.20	15.15
Distance-based (Dudani, 1976 [88])	8	15.20	16.76
Distance-based (Parthasarathy and Chatterji, 1990 [266])	9	16.20	17.05
Distance-based (Hattori and Takahashi, 1999 [130]), $l = 1$	5	14.40	15.18

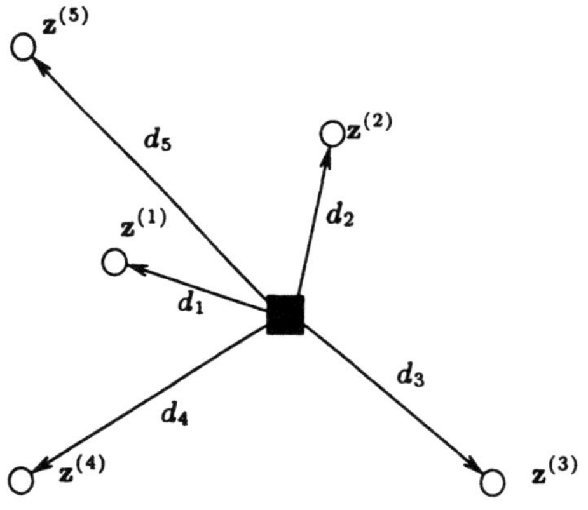

Fig. 7.1. Threshold and sigmoidal activation functions

$$\mathbf{l}(\mathbf{z}^{(1)}) = [0.1, 0.6, 0.8]^T;$$
$$\mathbf{l}(\mathbf{z}^{(2)}) = [0.3, 0.2, 0.1]^T;$$
$$\mathbf{l}(\mathbf{z}^{(3)}) = [1.0, 0.1, 0.0]^T;$$
$$\mathbf{l}(\mathbf{z}^{(4)}) = [0.0, 0.9, 0.5]^T;$$
$$\mathbf{l}(\mathbf{z}^{(5)}) = [0.4, 0.2, 0.2]^T.$$

Below we calculate the the label assigned to $\mathbf{x}$ by two of the above k-nn methods for $k = 5$.

- Jóźwik, (7.14),

$$\mu(\mathbf{x}) = \frac{1}{5}\left(\begin{bmatrix}0.1\\0.6\\0.8\end{bmatrix} + \begin{bmatrix}0.3\\0.2\\0.1\end{bmatrix} + \begin{bmatrix}1.0\\0.1\\0.0\end{bmatrix} + \begin{bmatrix}0.0\\0.9\\0.5\end{bmatrix} + \begin{bmatrix}0.4\\0.2\\0.2\end{bmatrix}\right) = \begin{bmatrix}0.36\\0.42\\0.32\end{bmatrix}.$$

The maximum membership rule will assign $\mathbf{x}$ in ω_2. If crisp k-nn was used, the label of $\mathbf{x}$ would be ω_1 because 3 of the 5 neighbors are from ω_1 (when their labels are "hardened").

- Keller et al., (7.19), for $m = 3$,

$$\mu(\mathbf{x}) = [0.3528, 0.4000, 0.3427]^T,$$

and for $m = 1.1$,

$$\mu(\mathbf{x}) = [0.1299, 0.5462, 0.7062]^T.$$

For $m = 3$, $\mathbf{x}$ is labeled in ω_2 and for $m = 1.1$, in ω_3. The difference occurred because smaller values of m give the closest neighbors higher influence over the final label.

■

The class label of $\mathbf{x}$, according to (7.19), is based on the distance-averaged membership of the k nearest neighbors of $\mathbf{x}$. For labeling $\mathbf{x}$ crisply, we can again drop the denominator which is the same for all classes and serves only to scale the degrees of membership between 0 and 1. For crisply labeled $\mathbf{Z}$, equation (7.19) can be replaced with a set of c discriminant functions (leading to the same crisp class label for $\mathbf{x}$) calculated as follows. Amongst the k neighbors from the set $\mathbf{Z}^{(\mathbf{x})}$, find those that belong to ω_i and calculate

$$g_i(\mathbf{x}) = \sum_{\mathbf{z}^{(j)} \in \omega_i} (d_j)^{-\frac{2}{m-1}}, \quad i = 1, \ldots, c. \tag{7.20}$$

The highest value of $g_i(\mathbf{x})$ indexes the crisp class label for $\mathbf{x}$. The fuzzy k-nn models in [119, 232] reduce to the same set of discriminant functions (7.20) for $m = 3$. Deliberately skipping the pre-labeling procedure, the above method was run for $k = 1, 2, \ldots, 9$, as in the original paper [166] on the four data sets. The results are shown in Tables 7.4 to 7.7 (Fuzzy (Keller et al., 1985 [166]), $m = 2$).

The similarity between various fuzzy k-nn models is discussed by Yang and Chen who propose a **generalized framework** incorporating all the above models, and show some theoretical results about its asymptotical properties [361]. The authors assume that the reference set $\mathbf{Z}$ is pre-labeled and do not consider labeling schemes as part of their generalized model. According to the generalized model [361], the final degree of membership of $\mathbf{x}$ to class ω_i is calculated as

$$t_i = \mathcal{A}_{j=1}^{k}\left\{w_i^{(1)} l_i(\mathbf{z}^{(1)}), \ldots, w_i^{(k)} l_i(\mathbf{z}^{(k)})\right\}, \tag{7.21}$$

$$\mu_i(\mathbf{x}) = \mathcal{F}_i(t_1, \ldots, t_c), \tag{7.22}$$

where

- $w_i^{(j)}$ is a weight, typically accounting for the distance between $\mathbf{x}$ and its jth nearest neighbor $\mathbf{z}^{(j)}$. Yang and Chen propose a weighing function

  $$w_i^{(j)} = \mathcal{W}\left(\mathbf{x}, \mathbf{z}^{(1)}, \ldots, \mathbf{z}^{(k)}\right), \tag{7.23}$$

  so that the weight might depend on all k neighbors. This is needed because often the weight serves as a scaling coefficient, and has as its denominator some "total score" using all k distances. To be most precise, we need to include a set of class-specific parameters as arguments of $\mathcal{W}$ because $w_i^{(j)}$ may depend on the class. For example, d_i^m is used in (7.16), so the weights are class-specific. Often the weighting function is just

  $$w_i^{(j)} = \mathcal{W}\left(\mathbf{x}, \mathbf{z}^{(j)}\right), \tag{7.24}$$

- $\mathcal{A} : \Re^k \rightarrow \Re$ is an aggregation function. Yang and Chen specify $\mathcal{A}$ as summation on the neighbor index j, which is undoubtedly the most natural choice. However, in the model due to Béreau and Dubuisson [30, 31], the *maximum* is taken instead of the sum. Therefore, a more general operation is needed as $\mathcal{A}$. Notice that to make use of the fuzzy aggregation connectives discussed in Chapter 4 (for which $\mathcal{A} : [0,1]^k \rightarrow [0,1]$), we have to scale the k numbers to be aggregated $\left(w_i^{(j)} l_i(\mathbf{z}^{(j)}),\ j = 1, \ldots, k\right)$ in the interval [0,1].
- $\mathcal{F}_i : \Re^c \rightarrow [0,1]$ is an exotic addition to the classical fuzzy k-nn model. It *analyzes* the support for class ω_i, denoted by t_i in (7.21), and determines the final degree of membership $\mu_i(\mathbf{x})$. Most of the models have the identity function as $\mathcal{F}_i, i = 1, \ldots, c$. (The model in [31] using equation (7.17) is rather an exception). The class-specific "rescaling" implemented through $\mathcal{F}_i$ can change the preference induced by t_i's and pick a class label which maximizes $\mu_i(\mathbf{x})$ on i.

Within this generalized framework we can create new fuzzy k-nn schemes by varying $\mathcal{A}$, $\mathcal{W}$, and $\mathcal{F}_i$'s. Since there is no general theory, any heuristic for generating a model is as good as any other one, and the ultimate test is the practical performance.

Yang and Chen [360] proposed an **edited** fuzzy k-nn version based on the generalized model. The scheme is similar to Wilson's editing method defined for crisp k-nn [348] (see Figure 3.14 in Chapter 3). Figure 7.2 shows the algorithm for obtaining the reference set for the fuzzy k-nn rule from $\mathbf{Z}$. Marked for deletion here are all elements of $\mathbf{Z}$ whose soft labels calculated by k_1-nn (denoted $\hat{\mathbf{l}}(\mathbf{z}_j)$) differ significantly from the original (possibly soft) class labels ($\mathbf{l}(\mathbf{z}_j)$). After the training phase, an unlabeled $\mathbf{x}$ is classified by a fuzzy k_2-nn rule, where k_2 is a prespecified integer.

EDITED FUZZY k-NN

1. Pick k_1 to be the number of neighbors for the training stage and the threshold α for the limit discrepancy between soft class labels.
2. For $q = 1$ to N do
   a) Find the k_1 nearest neighbors of $\mathbf{z}_q$ using $\mathbf{Z} - \{\mathbf{z}_q\}$ as the reference set (*leave-one-out*).
   b) Calculate the *soft* class label of $\mathbf{z}_q$, $\hat{\mathbf{l}}(\mathbf{z}_q) = [\hat{l}_1(\mathbf{z}_q), \ldots, \hat{l}_c(\mathbf{z}_q)]^T$ through (7.21) and (7.22) where the weights $w_i^{(j)}$ are calculated by

$$w_i^{(j)} = \exp\left[-\lambda\left(\frac{d_j}{d_i^m}\right)\right] \tag{7.25}$$

   or

$$w_i^{(j)} = \frac{(d_j)^{-\frac{2}{m-1}}}{\sum_{j=1}^{k}(d_j)^{-\frac{2}{m-1}}}. \tag{7.26}$$

   The aggregation $\mathcal{A}$ is summation and $\mathcal{F}_i$ is the identity function.
   c) If $||\mathbf{l}(\mathbf{z}_q) - \hat{\mathbf{l}}(\mathbf{z}_q)|| \geq \alpha$, mark $\mathbf{z}_q$ for deletion.
   d) end % (For q)
3. To obtain $\mathbf{V}$ delete from $\mathbf{Z}$ all marked elements.
4. Return $\mathbf{V}$ with the class labels.

Fig. 7.2. Yang and Chen's fuzzy k-nn editing method for a softly labeled $\mathbf{Z}$. The norm metric $||\cdot||$ is picked in advance.

7.2.2 Pre-labeling the data set

In all fuzzy k-nn models discussed above we allow for *soft* labels of the data set $\mathbf{Z}$ (i.e., $\mathbf{l}(\mathbf{z}_j) \in [0,1]^c, j = 1, \ldots, N$).

Various pre-labeling schemes have been proposed [37, 159, 165, 166, 167, 276, 323]. Not many authors consider a pre-labeling done by an expert who assigns a soft class label to each object in the data set according to his/her opinion about a certain characteristics. For example, the expert might assign to each $\mathbf{z}_j \in \mathbf{Z}$ a degree of typicality or a degree of certainty in the class decision. In some applications fuzzy labeling is naturally enforced, e.g., the severity of cancer metastases in a particular group of lymph nodes [336], acceptability of radiological treatment setups in radiotherapy [215], etc. In most cases only crisp labeling of $\mathbf{Z}$ is available. Then the question is whether *calculating* soft labels will improve the k-nn performance.

A simple (and popular) pre-labeling scheme due to Keller et al. [165, 166] was discussed in Chapter 4, Section 4.4. According to it, $\mathbf{z}_j$ gets a soft class label as in equation (4.95) on page 115. Notice that the scheme guarantees that, if hardened, the *new* soft class label of $\mathbf{z}_j$ will produce the *original* crisp class label. The scheme will affect only those objects which are close to the classification boundaries by diminishing the "certainty" for their own class at the expense of increasing the certainty for the bordering class (or classes). Let k^* be the number of neighbors used in this pre-labeling procedure. We can pick a value of k^* for the pre-labeling that is different from k used for further classification. The higher the value of k^* the wider the borderline region that is modified. Using this pre-labeling scheme with the four data sets, we carried out experiments with the Cone-torus and Normal mixtures data, for $k^* = 1, \ldots, 9$ and $k = 1, \ldots, 9$ (as in [166]). The best training error and the corresponding testing errors are presented in Tables 7.4 and 7.5 (Fuzzy+pre-labeling ([166]), $m = 2$). Keller et al. [166] claim that the improvement on the error rate might not be the main benefit from the fuzzy k-nn model. More importantly, the model offers a degree of certainty which can be used with a refuse-to-decide option. Thus, objects with dubious class labels can be detected and processed separately.

An iterative pre-labeling scheme is proposed by Jóźwik [159] and studied further by Bezdek et al. [37]. First, $\mathbf{Z}$ is split in two parts, which we call here $\mathbf{Z}_{t1}$ with N_{t1} elements and $\mathbf{Z}_{t2}$ with N_{t2} elements ($N_{t1} + N_{t2} = N$). Notice that both parts are used during training. To *test* the fuzzy k-nn with the obtained soft labels, we need a third data set from the same problem. The pre-labeling of $\mathbf{Z}$ is done in three Phases.

Phase 1. During Phase 1 the set $\mathbf{Z}_{t1}$ obtains soft labels. The algorithm starts with the crisp class labels of $\mathbf{Z}_{t1}$. Assume the labels are organized into a label matrix L_0 so that each row is a binary label vector. Using leave-one-out, we calculate the soft class label of each element of $\mathbf{Z}_{t1}$ by the fuzzy k-nn rule (7.14) with $k = 1, \ldots, N_{t1} - 1$. Using the maximum membership rule, crisp labels are assigned and the apparent error rate (misclassification rate) is calculated for each k. (At the very first iteration the results will be the crisp k-nn error rates because for a crisply labeled reference set (7.14) reduces to the crisp k-nn.) The value of k for which the lowest error rate (p_0) is obtained is designated as k_0. The soft class labels of the reference vectors from $\mathbf{Z}_{t1}$, calculated for k_0 are stored as the new matrix with labels L_1 for the reference set. This matrix is then used in (7.14) at the second iteration, where k_1 and p_1 are obtained in the same manner as k_0 and p_0. Thus, a sequence of triplets (L_h, p_h, k_h) is generated, $h = 0, 1, 2, \ldots$ The procedure stops when for some $h+1$, $p_h \le p_{h+1}$. Then the triplet (L_h, p_h, k_h) is taken as the result from Phase 1. We shall denote it as (L', k', p').

Example 7.2.2. For illustration of Phase 1 we consider the synthetic 15-point data set shown in Figure 1.2 and Table 7.9 (see also Chapter 1). The number of errors at the first step (corresponding to the crisp k-nn for $k = 1, \ldots, 14$) is shown in Table 7.8

Choosing $k_0 = 6$ ($p_0 = 0.07$, one misclassification), we obtain the label matrix L_1 shown in Table 7.9. At the next step we use L_1 with (7.14), but the smallest error rate obtained is again 0.07, and the procedure stops. The resultant set of labels at the end of Phase 1 with the 15-point data (all 15 points were used as the first part of the training set $\mathbf{Z}_{t1}$) is in fact L_0, i.e., the crisp labeling with $k_0 = 6$. ■

Table 7.8. Number of neighbors and corresponding errors at the first step of Jóźwik's Phase 1 with the 15-point data

k	1	2	3	4	5	**6**	7	8	9	10	11	12	13	14
e_{tr}	6	5	3	4	2	1	2	2	3	2	4	3	9	15

Table 7.9. The 15-points data set, the initial label set L_0 and the first label set L_1 obtained at Phase 1 of Jóźwik's pre-labeling scheme.

$\mathbf{z}$	x	y	L_0		L_1	
$\mathbf{z}_1$	1.3	3.7	1	0	0.83	0.17
$\mathbf{z}_2$	2.1	4.6	1	0	0.83	0.17
$\mathbf{z}_3$	2.7	6.2	1	0	0.83	0.17
$\mathbf{z}_4$	3.3	4.6	1	0	0.83	0.17
$\mathbf{z}_5$	3.4	2.4	1	0	0.67	0.33
$\mathbf{z}_6$	4.0	1.1	0	1	0.33	0.67
$\mathbf{z}_7$	4.5	3.8	1	0	0.67	0.33
$\mathbf{z}_8$	5.0	6.6	0	1	0.67	0.33
$\mathbf{z}_9$	5.4	1.4	0	1	0.33	0.67
$\mathbf{z}_{10}$	5.7	5.7	1	0	0.50	0.50
$\mathbf{z}_{11}$	6.1	3.9	0	1	0.33	0.67
$\mathbf{z}_{12}$	6.3	1.9	0	1	0.17	0.83
$\mathbf{z}_{13}$	7.4	2.7	0	1	0.17	0.83
$\mathbf{z}_{14}$	7.5	0.9	0	1	0.17	0.83
$\mathbf{z}_{15}$	7.6	5.3	0	1	0.33	0.67

Phase 2. According to the original scheme in [159], at this phase, the second part of the training set, $\mathbf{Z}_{t2}$, is added to the procedure. It is attached to $\mathbf{Z}_{t1}$ and the soft labels for $\mathbf{Z}_{t2}$ are calculated through (7.14) using only L' and k'. The original class labels of $\mathbf{Z}_{t2}$ are disregarded and $\mathbf{Z}_{t2}$ is treated as unlabeled. The resultant label matrix for the whole $\mathbf{Z}$ is used as the initialization of Phase 2 which next operates exactly as Phase 1. However, instead of calculating the error rate p over all of $\mathbf{Z}$, Jóźwik suggests [159] that again only the elements of $\mathbf{Z}_{t1}$ be used, as in Phase 1. The stopping condition is the same as in Phase 1. The result is a labeling matrix L'' for the whole $\mathbf{Z}$ and the corresponding k'' and p''.

In Phase 3, the final reference set and its labeling is chosen. If $p' \geq p''$, choose L' as the labeling of the reference set $\mathbf{Z}_{t1}$ to be used with k'-nn. If $p' < p''$, choose L'' as the labeling of the reference set $\mathbf{Z}$ to be used with k''-nn.

This pre-labeling scheme raises many questions. Is the stopping criterion reasonable (i.e., is the first local minimum of p_h also the global minimum, or, if not, is it a good enough suboptimal solution)? How do we decide on the split of $\mathbf{Z}$ into $\mathbf{Z}_{t1}$ and $\mathbf{Z}_{t2}$? What are the practical benefits of this pre-labeling scheme given the large amount of time it requires for calculating all fuzzy k-nn rules and their scores, $k = 1, \ldots, N_{t1} - 1$, at each step? Bezdek and coauthors examine experimentally the scheme [37] and conclude that Jóźwik's algorithm is well defined in the sense that the stopping rule is consistent. They also find that Phase 2 seldom improves the performance of the Phase 1 design (L', k', p').

7.2.3 *k*-nn variants

Many fuzzy k-nn rules make use of the **distance** between $\mathbf{x}$ and its k neighbors. While fuzzy technology took the usual ad-hoc route, there have been a number of theoretical controversies about the idea of (non-fuzzy) distance-based k-nn rules. The difference between fuzzy and non-fuzzy distance-based k-nn can be summarized as follows. First, in fuzzy k-nn rules, we usually assume that the data points in $\mathbf{Z}$ have either crisp *or soft* labels. This is not the case in distance-based k-nn where $\mathbf{Z}$ is always crisply labeled. Second, the soft class label produced by a fuzzy k-nn design (usually between 0 and 1) can be used as a measure of uncertainty in classification, and can therefore be thresholded and used to create a refuse-to-decide mode. Non-fuzzy distance-based k-nn methods are typically not concerned with degrees of membership. Third, non-fuzzy designs have been studied more rigorously than the fuzzy designs, disclosing some interesting theoretical properties. In terms of classification performance on a finite and usually crisply labeled data set $\mathbf{Z}$, the two groups could be expected to tie.

In this subsection we give a recap of the discussion on some non-fuzzy k-nn variants together with some numerical results on the four data sets used throughout this book.

Patrick summarized using distances in the k-nn model [268]. He details three k-nn rules, of which the third is of interest here as it is related to the distance-based models. Recall the basic nonparametric estimation formula of probability density functions discussed in Chapter 2 (3.20)

$$\hat{p}(\mathbf{x}) = \frac{k}{NV_R}. \tag{7.27}$$

where V_R is the volume of a region R in $\Re^n$ containing the k nearest neighbors of $\mathbf{x}$, and N is the total number of observations (cardinality of $\mathbf{Z}$). Denoting by k_i the number of neighbors from class ω_i amongst the k neighbors, and by N_i the total number of elements of $\mathbf{Z}$ from ω_i, the class-conditional p.d.f's are approximated by

$$\hat{p}(\mathbf{x}|\omega_i) = \frac{k_i}{N_i V_{R_i}}. \tag{7.28}$$

Using the approximation

$$\hat{P}(\omega_i) = \frac{N_i}{N} \tag{7.29}$$

for the prior probabilities, the following estimates of the posterior probabilities are obtained

$$\hat{P}(\omega_i|\mathbf{x}) = \frac{k_i}{N_i V_{R_i}} \cdot \frac{N_i}{N} \cdot \frac{1}{p(\mathbf{x})}$$
$$= \frac{k_i}{N p(\mathbf{x}) V_{R_i}}. \tag{7.30}$$

Assume that R_i is a hypersphere in $\Re^n$ with radius a_i centered at $\mathbf{x}$. Then the volume of R_i can be expressed using the gamma function Γ

$$V_{R_i}(a_i) = \frac{\pi^{n/2}}{\Gamma\left(\frac{1}{2}n+1\right)} a_i^n \tag{7.31}$$

which reduces to[2]

$$V_{R_i}(a_i) = V_n(1)\, a_i^n, \tag{7.32}$$

where $V_n(1)$ is the volume of a hypersphere of radius 1 in $\Re^n$. Substituting (7.32) in (7.30) yields

$$\hat{P}(\omega_i|\mathbf{x}) = \frac{1}{N p(\mathbf{x}) V_n(1)} \cdot \frac{k_i}{a_i^n}. \tag{7.33}$$

We can ignore the first fraction which does not depend on i for a given $\mathbf{x}$, and arrive at the following set of simple discriminant functions

$$g_i(\mathbf{x}) = \frac{k_i}{a_i^n}, \quad i = 1, \ldots, c. \tag{7.34}$$

Different interpretations of the above equation give rise to different k-nn variants.

Theoretical model # 1. Fix $k_i = k$ to be the same for all c classes. Then a_i will be the radius of the hypersphere centered at $\mathbf{x}$ which contains exactly k elements from $\mathbf{Z}$ with class label ω_i. Thus, the radius a_i is the distance between $\mathbf{x}$ and its k-th nearest neighbor from ω_i. Since k is the same for all i's, the largest $g_i(\mathbf{x})$ will be the one with the smallest distance a_i in it. Therefore, according to Theoretical model # 1, $\mathbf{x}$ is assigned to the class of the closest k-th nearest neighbor.

Example 7.2.3. Figure 7.3 illustrates this model. We use $k = 3$ for each of the three classes shown with different markers. The point to be labeled is the boldface "x". The three circles correspond to the regions R, each one containing exactly $k = 3$ elements from the respective class. The third neighbor from class "snowflakes" is closer to $\mathbf{x}$ than the third neighbor from either of the other two classes, therefore $\mathbf{x}$ is labeled as a snowflake. Notice that for $k = 1$, the model coincides with the nearest neighbor (1-nn) design. ■

---

[2] see Calculus books, e.g., [19]

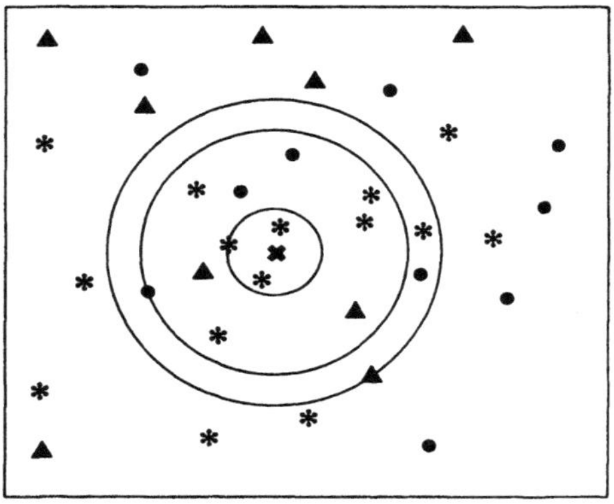

Fig. 7.3. Illustration of k-nn Theoretical model # 1. The point $\mathbf{x}$ is labeled as a "snowflake".

Theoretical model # 2. Fix $k = k_1 + k_2 + \ldots + k_c$. In (7.34) the radius of the hypersphere centered at $\mathbf{x}$ can be taken to be the distance from $\mathbf{x}$ to its k_ith neighbor from ω_i. Then the ratio k_i/a_i^n determines the class label of $\mathbf{x}$.

Example 7.2.4. An illustration of this k-nn model for $k = 11$ is presented in Figure 7.4 on the same data set as in Figure 7.3. The set of 11 nearest neighbors to $\mathbf{x}$, regardless of their class labels, is circled. Within this set, there are $k_1 = 7$ snowflakes, $k_2 = 2$ triangles and $k_3 = 2$ circles. A (hyper-) sphere with radius a encloses the 7 snowflakes, therefore the respective ratio is $g_1(\mathbf{x}) = 7/a^2$. The other two discriminant functions are respectively $g_2(\mathbf{x}) = 2/b^2$ and $g_3(\mathbf{x}) = 2/c^2$. The highest value determines $\mathbf{x}$'s class label. This model also coincides with the baseline 1-nn design for $k = 1$. ■

Theoretical model # 3. This is the classical k-nn model where one hypersphere of radius r contains all k nearest neighbors from $\mathbf{Z}$ (the set $\mathbf{Z}^{(\mathbf{x})}$, regardless of the class labels). Then the value of r becomes irrelevant for the classification decision for a given $\mathbf{x}$, and can be dropped from equation (7.34). The remaining discriminant functions are the number of neighbors k_i. In the above 11-nn example, $\mathbf{x}$ will be labeled as a snowflake.

We can also vary k_i and plug into (7.34) the respective radius a_i. Results from experiments with Theoretical models #1 and # 2 are shown in Tables 7.4 to 7.7 (Theoretical model # 1 and # 2). Given are the best training error rates and the respective testing error rates.

All these variants are ***asymptotically*** Bayes-optimal, i.e., for $N \to \infty$, $k \to \infty$ and $k/N \to 0$, equation (7.33) produces the true posterior probability, therefore (7.34) ensures the minimal-error classification (minimal risk for 0-1 loss matrix, see Chapter 2). In the finite-sample case, however, there could be better k-nn models. In search of such a model, Dudani proposed a distance-based k-nn [88]. The discriminant functions are

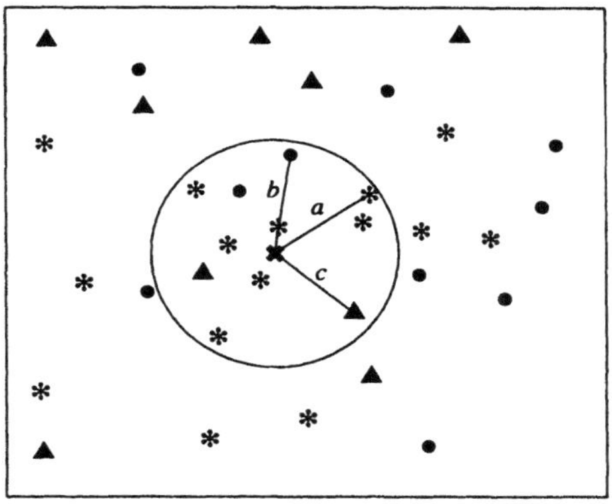

Fig. 7.4. Illustration of k-nn Theoretical model # 2. The three relevant distances are denoted a, b and c.

$$g_i(\mathbf{x}) = \sum_{j=1}^{k} w_j(\mathbf{x}) l_i(\mathbf{z}^{(j)}), \quad i = 1, \ldots, c, \tag{7.35}$$

where $w_j(\mathbf{x})$ are weights calculated by

$$w_j(\mathbf{x}) = \begin{cases} \frac{d_k - d_j}{d_k - d_1}, & d_k \neq d_1, \\ 1, & d_k = d_1, \end{cases} \tag{7.36}$$

where d_i is the distance between $\mathbf{x}$ and its ith neighbor from the reference set. Thus, the nearest neighbor is awarded a weight of 1, and the k-th one, the weight of 0 (practically not used in the equation). The author believed that the finite-sample accuracy of the weighted model will surpass that of the unweighted one, and supported this claim with two experimental examples. We applied Dudani's weighted k-nn with $k = 1, \ldots, 9$, to the four data sets: Cone-torus, Normal mixtures, Phoneme and Satimage. The best training accuracy and the respective test accuracy are displayed in Tables 7.4 to 7.7 (Distance based (Dudani, 1976 [88])).

Dudani's experiments have been criticized for taking all k-nn ties as errors [22, 242]. Bailey and Jain [22] propose three ways of tie-breaking:

- random
- using fewer neighbors
- using more neighbors

and show on the Dudani's experimental setup that the distance-weighted and the unweighted k-nn are practically indistinguishable. They prove the following theorem

Theorem 7.2.1. *(Bailey and Jain, [22]) In the infinite sample case ($N \to \infty$) the probability of error of the majority k-nearest neighbor rule is minimum among all weighted k-nearest neighbor rules (ties are resolved randomly).*

The authors form the difference between the probability of committing an error by any weighted rule T and the majority rule M, given $\mathbf{x}$ and the set of its nearest neighbors $\mathbf{Z}^{(\mathbf{x})}$ from $\mathbf{Z}$ (of cardinality N)

$$\Delta_N = P_N(e|\mathbf{x}, \mathbf{Z}^{(\mathbf{x})}, T) - P_N(e|\mathbf{x}, \mathbf{Z}^{(\mathbf{x})}, M) \tag{7.37}$$

and prove that for $N \to \infty$, Δ_N is always nonnegative. Thus, the k-nn rule for a fixed k may not be Bayes-optimal, but it is *asymptotically* the best one among the rules using the same k neighbors. In fact, the "consistency" of the generalized fuzzy k-nn model claimed by Yang and Chen [361] verifies that in the infinite-sample case the fuzzy and the non-fuzzy k-nn are the same.

In the finite-sample case, however, it is not clear which of the distance-based or fuzzy models is the best. Macleod et al. [227] state the following (presumably false!) hypothesis

> *Hypothesis: The error rate of the unweighted k-nn rule is lower than that of any weighted k-nn rule even when the number of training samples is finite.*

The authors disprove the hypothesis by an example showing analytically that a weighted 2-nn rule ($x \in \Re$, $\Omega = \{\omega_1, \omega_2\}$) gives a lower overall error rate than the corresponding unweighted rule for *any* training set generated from the specified p.d.f's. This result opens up the door for new distance-based k-nn models. Macleod et al. [227] generalize Dudani's weighting scheme by introducing a parameter α and using the s-th nearest neighbor for scaling the distances. The overall formula is the same as (7.35) but the weights are obtained by

$$w_j(\mathbf{x}) = \begin{cases} \frac{d_s - d_j + \alpha(d_s - d_1)}{(1+\alpha)(d_s - d_1)}, & d_s \neq d_1, \\ 1, & d_s = d_1. \end{cases} \tag{7.38}$$

For $\alpha = 0$ and $s = k$, (7.38) becomes the original formula due to Dudani. The values of these parameters used in the numerical experiments carried out in [227] were:

- Case 1: $\alpha = 0$, $s = \{2k, 3k\}$.
- Case 2: $\alpha = 0$, $s = \{N, [N/c]\}$.
- Case 3: $s = k$, $\alpha = \{1, 2\}$.

The weighted k-nn by Macleod et al. was applied to the Cone-torus and Normal mixtures data. The same three cases of parameter values were reproduced. The best training and the corresponding testing errors are given in Tables 7.4 and 7.5 (Distance-based (Macleod et al., 1987 [227])).

Parthasarathy and Chatterji [266] propose to estimate the posterior probabilities as an average of k estimates of type (7.33). They argue that if the individual estimates are independent and of the same variance θ, the variance of the averaged estimate is θ/k. Neglecting the terms which do not depend on i in the equation proposed by the authors, the discriminant function for class ω_i is

$$g_i(\mathbf{x}) = \sum_{j=1}^{k} \frac{k(\omega_i, j)}{d_j^n}, \tag{7.39}$$

where $k(\omega_i, j)$ is the number of neighbors from class ω_i among the nearest j neighbors of $\mathbf{x}$. For example, let $k = 6$, and neighbors $\mathbf{z}^{(2)}, \mathbf{z}^{(3)}$ and $\mathbf{z}^{(5)}$ have class label ω_i. Then

$$g_i(\mathbf{x}) = \frac{0}{d_1^n} + \frac{1}{d_2^n} + \frac{2}{d_3^n} + \frac{2}{d_4^n} + \frac{3}{d_5^n} + \frac{3}{d_6^n}. \tag{7.40}$$

It is not clear whether the individual estimates $\frac{k(\omega_i,j)}{d_j^n}$ are independent. More likely, they are not because each set of j neighbors of $\mathbf{x}$, $j = 2, \ldots, k$ already contains the nearest $j-1$ neighbors used for calculating a previous estimate. The authors assume that the set of discriminant functions (7.39) should be used for classes with equal prior probabilities and suggest another formula where each g_i is multiplied by $\hat{P}(\omega_i)$. However, the estimates of posterior probabilities (7.33) already account for the $P(\omega_i)$'s, and so does the averaging formula (7.39). Therefore, multiplying the discriminant functions by the prior probabilities only moves the classification boundary towards the more probable class and is theoretically unnecessary as long as our aim is to minimize the error rate. The results of our experiments using (7.39) with the four data sets are shown in Tables 7.4 to 7.7 (Distance-based (Parthasarathy and Chatterji, 1990, [266])).

A distance-based k-nn design which is similar in spirit is proposed by Hattori and Takahashi [130]. Similarly to Theoretical model # 1 the authors use k neighbors from each class. Instead of comparing the distances to the k-th neighbor, they sum up all k distances (each one raised to its l-th power) and assign $\mathbf{x}$ to the class of the smallest total distance. Hence, the following discriminant functions are used

$$g_i(\mathbf{x}) = -\sum_{j_i=1}^{k} d_{j_i}^l, \tag{7.41}$$

where j_i is the index of the nearest neighbors of $\mathbf{x}$ whose class label is ω_i.

This method was also applied to the four data sets for $l = 1$ and $l = 2$. The best training and the corresponding testing error rates are shown in Tables 7.4 to 7.7 (Distance-based (Hattori and Takahashi, 1999 [130])).

A Dempster-Shafer k-nn version is proposed by Denœux [74]. Each neighbor from $\mathbf{Z}^{(\mathbf{x})}$ and its label are considered as an "atomic" piece of evidence. All pieces are pooled by means of Dempster's rule of combination (belief theory) producing two components for each discriminant function: $g_i^{bel}(\mathbf{x})$ and $g_i^{pl}(\mathbf{x})$. The first component accounts for the *belief* that the true class of $\mathbf{x}$ is ω_i, and the second component accounts for the *plausibility* for that. The second component, $g_i^{pl}(\mathbf{x})$, is calculated using the evidence that $\mathbf{x}$ is *not* from ω_i, i.e., it reflects the degree of "doubt". Usually both components take a value in the interval [0,1] but do not necessarily sum up to 1. We can use the maximum membership rule considering only the "support" part of the discriminant function (the first component, $g_i^{bel}(\mathbf{x})$) thereby selecting the *most credible* class label. Alternatively, we can pick the label corresponding to the *least doubt* (highest degree of plausibility $g_i^{pl}(\mathbf{x})$). The two class labels may not coincide which calls for further analysis of $\mathbf{x}$.

Joussellin and Dubuisson [158] analyze the sequence of the class labels of the nearest neighbors to detect doubtful cases which need a second look by an expert or another classification procedure. Rovatti et al [287] go further to propose a weighted (but not distance-weighted!) k-nn model. The weights are used as in (7.35) and a constant is added for each class. The other differences are that: (1) the weights are class-specific (w_{ij}); and (2) they do not depend on the distances d_j between $\mathbf{x}$ and its jth neighbor, $j = 1, \ldots, k$. The weights are calculated in advance and account for how "reliable" the j-th neighbor is (in general) for class ω_i.

Example 7.2.5. Consider a two-class problem and pick $k = 3$. Assume the following weight vectors have been obtained at the training stage of the Rovatti et al. model: $[0.6, 0.4, -0.1, \mathbf{0.3}]^T$ for class ω_1 and $[0.8, 0.9, -0.3, \mathbf{0.1}]^T$ for class ω_2. The boldface entry is the constant. These weights will be used for calculating the discriminant functions by (7.35). Let, for some $\mathbf{x}$, the set of class labels of the first tree neighbors be $\{\omega_1, \omega_2, \omega_1\}$. The two discriminant functions are then

$$g_1(\mathbf{x}) = (1) \times 0.6 + (0) \times 0.4 + (1) \times (-0.1) + 0.3 = 0.8 \tag{7.42}$$

and

$$g_2(\mathbf{x}) = (0) \times 0.8 + (1) \times 0.9 + (0) \times (-0.3) + 0.1 = 1.0 \tag{7.43}$$

In this case $\mathbf{x}$ will be assigned to ω_2 although neither the majority 3-nn nor the 1-nn rule agree with this. The reason for this result in our example is that during training the second neighbor has proven to be very indicative for ω_2 and is respectively praised by a high weight (0.9). ■

Rovatti et al. [287] derive the equations for calculating the weights $w_{ij}, i = 1, \ldots, c, j = 1, \ldots, k+1$, using the least squares method.

Here we mention in chronological order a few application of various fuzzy k-nn models. Hussein et al. [141] use fuzzy perceptron and fuzzy k-nn for user identification in computer systems. The features are based on the time duration between keystrokes when a user's password is entered. Cabello et al. [53] apply fuzzy k-nn for detecting and classifying arrhythmia in five classes: four diagnostic classes and a class of artifacts. Wang et al. [336] construct a mathematical model using fuzzy k-nn for estimating the lymph node metastases in gastric cancer patients. Kim and Han [170] apply fuzzy k-nn to distinguish between 32 classes of texture images. Two industrial applications using fuzzy k-nn are developed by Liao and Li [217]. The authors analyze radiographic images of welded structures for flaws and also tool failure in milling surfaces of high chromium materials. Several fuzzy k-nn and nearest prototype schemes have been designed and applied for grading fish products [137]. Leszczynski et al. [215] propose a special weighting scheme for the fuzzy k-nn rule. They use the model to distinguish between acceptable and unacceptable setups during a session of radiological treatment for patients.

7.2.4 Fuzzy k-nn or non-fuzzy k-nn?

We introduced a number of k-nn models which all can easily be called *fuzzy*. Suffice it to scale the discriminant functions between 0 and 1 and call them "degrees of membership", and practically, there is no difference between the fuzzy and non-fuzzy groups. None of the *fuzzy* k-nn models actually exploits fuzzy set theory beyond the simple notion of a fuzzy set. The main difference between fuzzy and nonfuzzy k-nn models is the pre-labeling of $\mathbf{Z}$ in the fuzzy models.

The single best-training error and its corresponding testing error give an idea about the training and generalization properties of the models. An eyeball comparison of the error rates in Tables 7.4 to 7.7 suggests that there is no clear winner amongst the fuzzy and non-fuzzy k-nn models. It should be mentioned that the number of neighbors k is not *exactly* what is shown in the tables. For example, Theoretical method # 2 uses k neighbors from each class. Indeed, the rule uses only k of those (the farthest) but all $c \times k$ neighbors have to be found. The same holds for Hattori and Takahashi's method, where effectively the number of neighbors used in the calculations is $c \times k$. Macleod's design needs the distance from x to its s-th neighbor, hence all s neighbors must be identified even though only $k + 1$ are used in the calculations. This "unequal start" makes the comparison of the methods difficult.

Fuzzy k-nn is a popular choice for applications because it is simple and also gives some information about the certainty (in a broad sense) of the classification decision. However, as the results in this subsection suggest, the benefit from using a fuzzy version instead of the classical k-nn needs to be explored individually for each problem. For some problems fuzzy (or distance-based) k-nn might be the better choice but this has to be experimentally verified.

7.3 Generalized nearest prototype classifier (GNPC)

7.3.1 Fuzzy prototype classification, 1-nn, and nearest mean

Although in Chapter 3 we introduced prototypes as points in $\Re^n$ (page 44), the general concept of "prototype" is much wider

- a point in the feature space $\Re^n$;
- a fuzzy set on the feature space $\Re^n$ or a discrete subset $\mathcal{S} \subset \Re^n$;
- a list of linguistic values;
- a set of prescribed rules or intervals; etc.

The classification decision is based on a comparison of $\mathbf{x}$ to the respective prototype. Measures used for this comparison can be: proximity, similarity, coincidence, degree of equivalence, matching, energy to make a "deformable" prototype as close as possible to the input pattern, etc. [80, 81, 207, 275, 285]. We owe this variety to the rich definition of "prototype". In this subsection we consider **pointwise prototypes**, $\mathbf{v}_j \in \mathbf{V}, j = 1, \ldots, v$. As with the crisp prototypes, there can be

- one prototype per class ($v = c$ prototypes);
- more than one prototype per class ($v > c$ prototypes). Note that the number of prototypes can be greater than N (the number of elements in the data set $\mathbf{Z}$.

A great advantage of fuzzy prototype classifiers is that we can have a good classifier with **less than one prototype per class**. This is possible because we can use soft labels of the prototypes to the classes, so that each class is taken into account even with fewer than c prototypes.

The **nearest prototype classifier** (1-NP) is defined as

$$D_{1NP}(\mathbf{x}) = l(\mathbf{v}_k) \in \Omega \iff$$

$$||\mathbf{x} - \mathbf{v}_k|| \leq ||\mathbf{x} - \mathbf{v}_i||;\ \forall i = 1, \ldots, v;\ i \neq k, \tag{7.44}$$

where $||.||$ is a pre-specified norm. Ties are broken randomly. When $\mathbf{V} = \mathbf{Z}$ and $v = N$, then 1-NP is the 1-nn classifier. When $v = c$ and $\mathbf{V}$ are obtained as the sample means for each class in $\mathbf{Z}$, 1-NP is often called the *minimum-distance* classifier [87, 324] (we denote it by 1-NM for "nearest mean"). Since the 1-NP definition does not specify how many prototypes are used and how they are found, both 1-nn and 1-NM are special cases of the 1-NP classifier.

7.3.2 GNPC definition

There are many classification techniques that are based explicitly or implicitly on similarity to point prototypes. A scheme that unifies many disparate techniques and provides a common framework for comparative analysis is the **Generalized Nearest Prototype Classifier** (GNPC) [199].

Let $\mathbf{l}(\mathbf{v}_i) = [l_{1i}, \ldots, l_{ci}]^T \in [0,1]^c$ be the soft label vector of the prototypes $\mathbf{v}_i \in \mathbf{V}$. The value l_{ji} can be regarded as the strength of the association of $\mathbf{v}_i$ with class ω_j. The common characteristic of nearest prototype classifiers is that they calculate ***similarities*** $s(\mathbf{x}, \mathbf{v}_i)$ of $\mathbf{x}$ to each of the prototypes $\mathbf{v}_i$ and then combine the $s(\mathbf{x}, \mathbf{v}_i)$, with $\mathbf{l}(\mathbf{v}_i), i = 1, \ldots, v$, to label $\mathbf{x}$. Intuitively, the closer $\mathbf{x}$ is to $\mathbf{v}_i$ *and* the higher the association weight l_{ji} is, the greater the plausibility that $\mathbf{x}$ comes from ω_j. Figure 7.5 shows an example with $v = 4$ prototypes in $\Re^2$ that are associated with $c = 3$ classes by crisp label vectors. For convenience we shall denote the *crisp* class labels by $\mathbf{e}_j$, where $I_{crisp} = \{\mathbf{e}_1, \ldots, \mathbf{e}_c\}$ is the canonical basis of $\Re^c$. That is, objects from class i are labeled by $\mathbf{e}_i = [e_{1i}, \ldots, e_{ci}]^T$, $e_{ji} = 1$ if $j = i$, and 0, otherwise. We denote by $\mathbf{L}_V$ the matrix of crisp label vectors $[\mathbf{l}(\mathbf{v}_1), \ldots, \mathbf{l}(\mathbf{v}_v)], \mathbf{l}(\mathbf{v}_j) \in I_{crisp}$. For the example in Figure 7.5, $\mathbf{L}_V = [\mathbf{e}_1, \mathbf{e}_2, \mathbf{e}_2, \mathbf{e}_3]$. The solid, dashed, and dotted lines indicate the crisp relationship of prototypes to classes 1, 2, and 3, respectively.

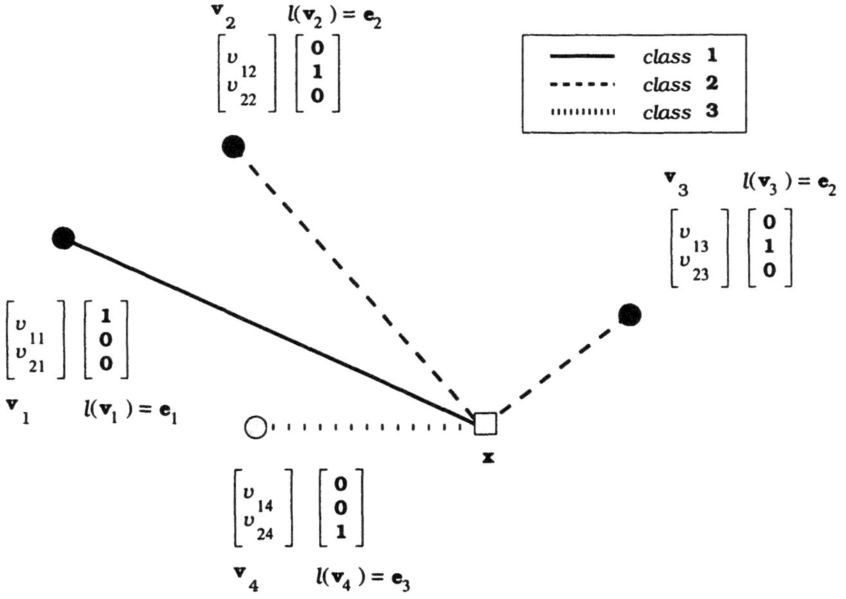

Fig. 7.5. Nearest prototype classifier with crisp labels

Using Euclidean distance (i.e., $d_E(\mathbf{x}, \mathbf{v}_i) = \sqrt{(\mathbf{x} - \mathbf{v}_i)^T(\mathbf{x} - \mathbf{v}_i)}$), the 1-NP classifier [324] will assign label ω_2 to $\mathbf{x}$ in because $d(\mathbf{x}, \mathbf{v}_3) < d(\mathbf{x}, \mathbf{v}_i)$, $i =$

$1, 2, 4$, and $l(\mathbf{v}_3) = \mathbf{e}_2$. Since $v > c$ in Figure 7.5, some authors call this a nearest multiple prototype design [40, 324].

It is natural to assume that the similarity s between $\mathbf{x}$ and $\mathbf{v}_i$ has maximal value when $\mathbf{x} = \mathbf{v}_i$ and decreases with the distance between the two vectors in $\Re^n$. In general, similarity measures are required only to be reflexive and symmetric. Here we restrict similarity a little more than this.

Definition 7.3.1. *A norm-induced similarity function*

$$s(\mathbf{x}, \mathbf{v}_i; \theta) \equiv s(\Delta(\mathbf{x}, \mathbf{v}_i); \theta), \tag{7.45}$$

where θ is a set of parameters of s, is any monotonically decreasing function $s : \Re^+ \to [0, 1]$ of any norm metric Δ on $\Re^n$.

For example, s might be

$$s(\mathbf{x}, \mathbf{v}_i; \theta) \equiv s(d_E(\mathbf{x}, \mathbf{v}_i); \emptyset) = \exp\left(-\frac{1}{2} d_E(\mathbf{x}, \mathbf{v}_i)^2\right). \tag{7.46}$$

Thus, assigning to $\mathbf{x}$ the label of the *most similar* prototype is equivalent to assigning to $\mathbf{x}$ the label of the *nearest* prototype. Another widely used example is based on the Mahalanobis norm,

$$s(\mathbf{x}, \mathbf{v}_i; \theta) \equiv s(d_M(\mathbf{x}, \mathbf{v}_i); S) = \exp\left(-\frac{1}{2} d_M(\mathbf{x}, \mathbf{v}_i)^2\right). \tag{7.47}$$

where

$$d_M(\mathbf{x}, \mathbf{v}_i) = \sqrt{(\mathbf{x} - \mathbf{v}_i)^T S^{-1} (\mathbf{x} - \mathbf{v}_i)}, \tag{7.48}$$

where S is a covariance matrix. Using d_M and $h_i \in \Re^+$ as a smoothing parameter associated with prototype $\mathbf{v}_i$, s can be defined as the Gaussian kernel centered at $\mathbf{v}_i$ (3.35) [100]. To simplify notation we write $s(\mathbf{x}, \mathbf{v}_i; \theta_i) = s_i$ and call s a similarity function, meaning always the *norm-induced similarity* as in Definition 7.3.1.

Figure 7.6 shows the example in Figure 7.5 but with possibilistic labels $\mathbf{l}(\mathbf{v}_i)$. The soft labeling of prototypes is indicated in Figure 7.6 by using all three types of lines (solid, dashed, and dotted) to link each prototype to $\mathbf{x}$.

We need to compute the membership of $\mathbf{x}$ in each class using $\mathbf{l}(\mathbf{v}_i)$ and $s(\mathbf{x}, \mathbf{v}_i)$, $i = 1, \ldots, v$. The label matrix in this case is $\mathbf{L}_V = [\mathbf{l}(\mathbf{v}_1), \ldots, \mathbf{l}(\mathbf{v}_v)]$ and can be viewed as a fuzzy relation on $\Omega \times \mathbf{V}$ expressing the association between the classes and the prototypes. For example, the label matrix for the four prototypes in Figure 7.6 is

$$\mathbf{L}_V = \begin{bmatrix} 0.7 & 0.0 & 0.2 & 0.1 \\ 0.6 & 0.8 & 0.4 & 0.3 \\ 0.1 & 0.3 & 0.1 & 0.4 \end{bmatrix}.$$

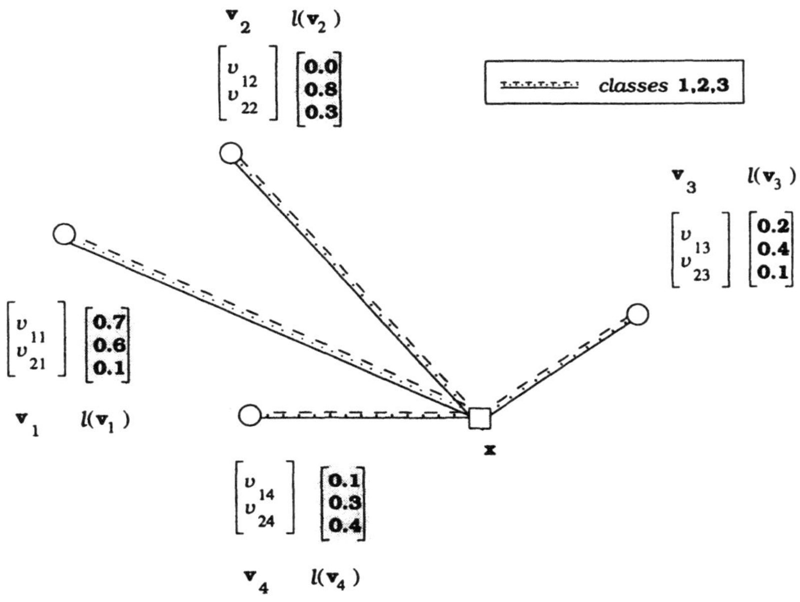

Fig. 7.6. Nearest prototype classifier with possibilistic labels. The three types of lines (solid, dashed, and dotted) show that each prototype links $\mathbf{x}$ to all three class labels

We can also define an input fuzzy set $\mathcal{I}$ on $\mathbf{V}$ expressing the similarity between the input $\mathbf{x}$ and the prototypes, $s(\mathbf{x}, \mathbf{v}_1), \ldots, s(\mathbf{x}, \mathbf{v}_v)$. The class label of $\mathbf{x}$ is calculated by the composition

$$\mu(\mathbf{x}) = \mathbf{L}_V \circ \mathcal{I}. \tag{7.49}$$

We assume that the composition $\circ$ consists of two fuzzy aggregation connectives, $\mathcal{S}$ and $\mathcal{T}$ (refer to page 107). Typically $\mathcal{T}$ is an intersection operation (a t-norm) while $\mathcal{S}$ is either union or mean n-place operation.

Definition 7.3.2. *[199, 198] The Generalized Nearest Prototypes Classifier (GNPC) is the 5-tuple $(V, \mathbf{L}_V, s, \mathcal{T}, \mathcal{S})$ where*

- *$V = \{\mathbf{v}_1, \ldots, \mathbf{v}_v\}$, $\mathbf{v}_i \in \Re^n$ is the set of prototypes;*
- *$\mathbf{L}_V \in [0,1]^{c \times v}$ is the label matrix for the prototypes in c classes;*
- *$s(\Delta(\mathbf{x}, \mathbf{v}_i); \theta)$ is a similarity function as in Definition 7.3.1;*
- *$\mathcal{T}$ is a t-norm defined over fuzzy sets, and $\mathcal{S}$ is an aggregation operator. (Thus the composition $\circ$ is completely specified.)*

For an input $\mathbf{x} \in \Re^n$, the GNPC calculates the similarity vector $[s_1, \ldots, s_v]^T$, produces the label vector $\mu(\mathbf{x})$ using (7.49), and assigns the crisp class label $\omega_k \in \Omega$ to $\mathbf{x}$ if

$$\mu_k(\mathbf{x}) = \max_{i=1,\ldots,c} \{\mu_i(\mathbf{x})\}. \tag{7.50}$$

Ties are broken randomly.

Note especially that the GNPC is a crisp classifier, no matter what type of labels are in the columns of $\mathbf{L}_V$. Generally, a GNPC representation can be built by answering five basic questions:

- **Q1.** *How many* prototypes do we need? (v = ?)
- **Q2.** Given ($\mathbf{Z}$), how do we *find* the prototypes $\{\mathbf{v}_i\}$?
- **Q3.** How do we *create the prototype label matrix* $\mathbf{L}_V$?
- **Q3.** How do we *choose the similarity function* s and how do we obtain θ_i?
- **Q4.** How do we *combine* the similarities with the labels (specify $\mathcal{S}$ and $\mathcal{T}$).

For the example in Figures 7.5 we can design the following GNPC: $V = \{\mathbf{v}_1, \mathbf{v}_2, \mathbf{v}_3, \mathbf{v}_4\}$, $\mathbf{L}_V = [\mathbf{e}_1, \mathbf{e}_2, \mathbf{e}_2, \mathbf{e}_3]$, $s(\mathbf{x}, \mathbf{v}_i; \emptyset) = \exp\left(-(\mathbf{x} - \mathbf{v}_i)^T(\mathbf{x} - \mathbf{v}_i)\right)$, $\mathcal{T}$ is **product**, $\mathcal{S}$ is **max**). The class label for $\mathbf{x}$ is

$$\mu(\mathbf{x}) = \begin{bmatrix} 1\,0\,0\,0 \\ 0\,1\,1\,0 \\ 0\,0\,0\,1 \end{bmatrix} \begin{bmatrix} 0.0002 \\ 0.0011 \\ 0.0369 \\ 0.0111 \end{bmatrix} = \begin{bmatrix} 0.0002 \\ 0.0369 \\ 0.0111 \end{bmatrix}.$$

Consequently, GNPC assigns $\mathbf{x}$ to ω_2. For the example in Figure 7.6,

$$\mu(\mathbf{x}) = \begin{bmatrix} 0.7\;0.0\;0.2\;0.1 \\ 0.6\;0.8\;0.4\;0.3 \\ 0.1\;0.3\;0.1\;0.4 \end{bmatrix} \begin{bmatrix} 0.0002 \\ 0.0011 \\ 0.0369 \\ 0.0111 \end{bmatrix} = \begin{bmatrix} 0.0074 \\ 0.0148 \\ 0.0044 \end{bmatrix}.$$

According to this result $\mathbf{x}$ will be given label ω_2.

The GNPC model encapsulates five intersecting (and non-exhaustive) families:

- **F1.** Clustering-and-relabeling [40];
- **F2.** Parzen classifier [100];
- **F3.** RBF networks [43, 131];
- **F4.** LVQ-type classifiers [175];
- **F5.** Edited nearest neighbor rules [69].

As we show next, the only operation $\mathcal{T}$ and $\mathcal{S}$ that are needed for the five families of classifiers $\mathbf{F}_1, \ldots, \mathbf{F}_5$, are:

- $\mathcal{T}$ operations: **product**
- $\mathcal{S}$ operations: **maximum, average**

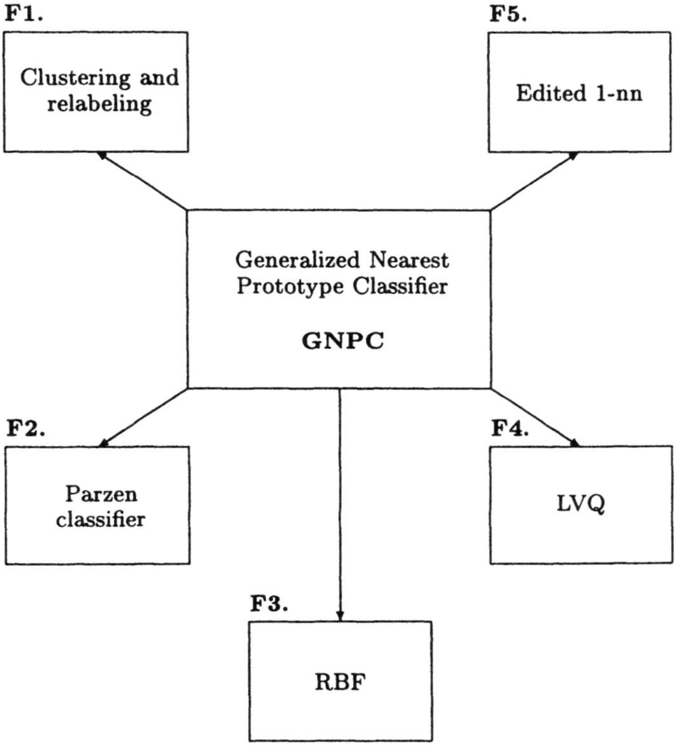

Fig. 7.7. Five groups of GNPC's.

7.3.3 The five families of GNPCs

Figure 7.7 shows the five groups of classifiers.

In the sequel, **A1-A5** stand for the answers to questions **Q1-Q5** shown above.

Baseline designs. Table 7.10 gives the **GNPC** representation of three *baseline* designs: 1-nn, 1-NP, and the linear discriminant classifier (LDC) for equiprobable classes [87] (see Chapter 3). Notation $\bar{\mathbf{V}}$ in Table 7.10 stands for the sample means $\bar{\mathbf{V}} = \{\bar{\mathbf{v}}_1, \ldots, \bar{\mathbf{v}}_c\}$, which are used as the prototypes here

$$\bar{\mathbf{v}}_i = \frac{1}{N_i} \sum_{l(\mathbf{z}_j)=\omega_i} \mathbf{z}_j, \quad 1 \le i \le c. \tag{7.51}$$

Table 7.10. Baseline GNPC designs. $\mathbf{L}_Z$ denotes the $c \times N$ label matrix of $\mathbf{Z}$ constructed as $[\mathbf{l}(\mathbf{z}_1), \ldots, \mathbf{l}(\mathbf{z}_N)]$.

Classifier →	1-nn	1-NP	LDC
A1. (v)	$v = n$	$v = c$	$v = c$
A2. $(\mathbf{V})$	$\mathbf{V} = \mathbf{Z}$	$\mathbf{V} = \bar{\mathbf{V}}$	$\mathbf{V} = \bar{\mathbf{V}}$
A3. $(\mathbf{L}_V)$	$\mathbf{L}_V = \mathbf{L}_Z$	$\mathbf{L}_V = [\mathbf{e}_1, \ldots, \mathbf{e}_c]$	$\mathbf{L}_V = [\mathbf{e}_1, \ldots, \mathbf{e}_c]$
A4. $(s; \theta)$	(7.46), $d_E; \emptyset$	(7.46), $d_E; \emptyset$,	(7.47), $d_M; S$
A5. $(\mathcal{T}, \mathcal{S})$	(prod, max)	(prod, max)	(prod, max)

Note that the representation of a classifier or a family of classifiers in terms of GNPC *is not unique*. We may construct $\mathbf{L}_V$ in many ways, and choose $s, \mathcal{T}$, and $\mathcal{S}$ accordingly. Therefore, the rest of this section provides *a* GNPC representation of each of the five families of classifiers.

Clustering and relabeling (F1). An easy scheme to build a GNPC is clustering followed by relabeling. First, data vectors are pooled disregarding the labels $\mathbf{L}_Z$ and clustered into $v \geq c$ clusters. We call this *unsupervised design*. Each cluster is represented by its centroid as a single prototype.

The set of prototypes is then *relabeled* as discussed in the previous chapters, and used with the 1-NP.

GNPC representation:

- **A1.** $v \geq c$, the number of clusters v can be either fixed or decided empirically, e.g., by a cluster validity measure;
- **A2.** $\mathbf{V}$ is found by clustering: any procedure that produces a set of point prototypes using a measure of similarity between $\mathbf{z}_j$ and $\mathbf{v}_i$ can be used.
- **A3.** $\mathbf{L}_V$ is crisp and the columns are obtained by the relabeling procedure.
- **A4.** s is a similarity function (it may or may not require θ_i);
- **A5.** $\mathcal{T}$ is **product**, $\mathcal{S}$ is **max**.

Some authors use this type of GNPC to compare clustering procedures by counting the number of misclassification after the prototype relabeling. This, however can hardly be any indication of how good a *clustering* procedure is because the classes represented in $\mathbf{Z}$ might not be conveniently shaped into clusters, and the match could be spurious. Generally, the classification accuracy of GNPCs in this family will not be as high as if we used the class labels $\mathbf{L}_Z$ to guide the procedure towards prototypes $\mathbf{V}$ [199].

An advantage can be gained by making a soft relabeling, i.e., $l(\mathbf{v}_i) \in [0, 1]^c$, thereby accounting for all of the class labels represented within the i-th crisp cluster, $C_i, i = 1, \ldots, v$. For example, let C be a cluster containing m_1 samples from class ω_1, m_2 samples from class ω_2, and m_3 samples from class ω_3, $m_1 + m_2 + m_3 = m$. Then the prototype of cluster C can be labeled by the vector $\left[\frac{m_1}{m}, \frac{m_2}{m}, \frac{m_3}{m}\right]^T$. Other relabeling schemes were discussed in Section 6.4.

Alternatively, we may cluster the data vectors from $\mathbf{Z}$ separately for each class (***supervised design***). A special case of this design is the 1-NM classifier where we assume that the points of a class *are* one single cluster, and hence, the prototypes are computed as the sample means. Generally, more than one prototype per class can be used, the number of prototypes may differ from class to class, and the labels can be either crisp or fuzzy.

Parzen classifier (F2). Using the kernel estimates of the class-conditional p.d.f's (e.g., the Gaussian kernel (3.35) with identity covariance matrix) and the expression N_i/N as an estimate of the prior probability for class ω_i, we obtain a Parzen classifier with the following GNPC representation

GNPC representation:

- **A1.** $v = n$;
- **A2.** $\mathbf{V} = \mathbf{Z}$;
- **A3.** $\mathbf{L}_V = \mathbf{L}_Z$;
- **A4.**
$$s_i = \frac{1}{n\theta_i^n} K\left(\frac{(\mathbf{x} - \mathbf{v}_i)}{\theta_i}\right),$$
where K is a kernel function (viz. Chapter 3), $\theta_i \in \Re^+$ is the smoothing parameter for class ω_i;
- **A5.** $\mathcal{T}$ **is product,** $\mathcal{S}$ **is average.**

In the classical Parzen model the smoothing parameter θ_i is the same for all prototypes of class ω_i. If we attach a specific θ_j to each prototype, $j = 1, \ldots, v$, and use some training procedure to tune the value, the model is called *Kernel Discriminant Analysis* (KDA) [136]. Parzen model is also isomorphic and isofunctional to *Probabilistic Neural Nets* (PNN) [310].

Using all the vectors in $\mathbf{Z}$ as prototypes might be computationally infeasible. Therefore *reduced* versions have been proposed where $v < n$. The main questions then become how to find these v "most relevant" prototypes and how to tune the parameter values. This relates the reduced Parzen (KDA, PNN) model to RBF networks where selecting/tuning the prototypes is a central question. The methods for finding prototypes discussed in 3.3 can be used.

Radial Basis Function Networks (F3). RBF networks were introduced in Chapter 3. The RBF output is calculated through (3.70). The crisp class label of $\mathbf{x}$ is decided by the index of the maximal $\mu_k(\mathbf{x}), k = 1, \ldots, c$, as in equation (7.50). Because of the restriction at Definition 7.3.1, we consider only RBF networks whose radial-basis function co-domain is [0,1], and whose hidden-output connections are numbers in the interval [0,1]. Thus, the GNPC definition embeds a class of RBF networks based on "semiparametric" mixture modeling [326].

GNPC representation:

- **A1.** $v < n$;
- **A2.** $\mathbf{V}$ is either selected and fixed in advance or tuned during the training stage;
- **A3.** $\mathbf{l}(\mathbf{v}_i) = [w_{1i}, w_{2i}, \ldots, w_{ci}]^T$ (the RBF NN weights);
- **A4.** $s_i = \psi(\mathbf{x}, \mathbf{v}_i; \theta_i)$; θ_i consists of the parameter values of the radial basis function associated with prototype $\mathbf{v}_i$.
- **A5.** $\mathcal{T}$ is **product**, $\mathcal{S}$ is **average**.

LVQ-type procedures (F4). LVQ classifiers compute $\mathbf{V}$ by sequential competitive learning [131, 175] as discussed in Chapter 3. Although LVQ classifiers are usually regarded as neural network classifiers, their representation is closer to the (non-neural-network) clustering and relabeling family **F1** than to the RBF networks family **F3**. Unsupervised LVQ classifiers differ from **F1** only in the way of *finding* the prototypes, and therefore can be regarded as a special case of **F1**. We consider **F4** separately because of the large number of LVQ models that have been developed. In contrast to the broad GNPC representation of **F1**, here we *specify* LVQ as the prototype extraction procedure in answer **A2**.

GNPC representation:

- **A1.** The number of prototypes v is usually fixed in advance;
- **A2.** $\mathbf{V}$ is found by an LVQ algorithm (there is a great variety of such algorithms);
- **A3.** $\mathbf{L}_V$ is crisp, and is fixed prior to prototype tuning;
- **A4.** s is a similarity function;
- **A5.** $\mathcal{T}$ is **product**, $\mathcal{S}$ is **max**.

Edited 1-Nearest Neighbor (F5). In numerous papers comparing pattern classifiers, the classical nearest neighbor (1-nn) method is recognized as a good competitor to many neural networks and other classification paradigms. It is usually pointed out that, although one of the best, the 1-nn method requires large memory and time, and therefore other classifiers are recommended. Edited versions of 1-nn design (some of them discussed in Chapter 3) are excellent examples of flexible, small, and accurate GNPCs.

GNPC representation:

- **A1.** v depends on the editing technique. In some designs it can be specified in advance, or a desirable upper limit can be set up (e.g., when using genetic algorithms [200]);
- **A2.** $\mathbf{V}$ is usually a subset of $\mathbf{Z}$, but can be also a subset of $\Re^n$ in the case of prototype extraction;
- **A3.** $\mathbf{L}_V$ is constituted from the crisp label vectors of the selected points from $\mathbf{Z}$;

- **A4.** s is a similarity function;
- **A5.** $\mathcal{T}$ is **product**, $\mathcal{S}$ is **max**.

Related to this group are the fuzzy k-nearest neighbor classifiers discussed in the previous section. They use fuzzy or possibilistic labels $\mathbf{l}(\mathbf{z}_j) \in [0,1]^c$ and combine s with the labels of the k nearest neighbors of $\mathbf{x}$.

Other related designs. There is a group of GNPCs that are not a coherent family and therefore are not treated as such here. This group comprises various heuristic schemes whose unifying concept is the connection between the 1-nn rule and non-competitive-learning type neural networks such as *multi-layer perceptrons* (MLP) and RBF networks. Two topics have been studied:

- constructing neural networks that implement *exactly* the 1-NN rule [49, 103, 179, 243];
- optimizing the 1-NN (GNPC) performance by designing a neural network (of MLP or RBF type) that provides a close or exact representation of the GNPC, and tuning V (and θ_i, where applicable) [71, 318, 359, 362, 371].

The second option looks the more promising because the GNPC parameters are derived by some NN training procedure while "exact implementation" means that we either already have the prototypes $\mathbf{V}$ or use all of $\mathbf{Z}$. Each of the neural network designs is very specific and includes many heuristics. Therefore it is difficult to distill a *common* representation. Since each architecture and training scheme in this group stems from the 1-nn rule, *functionally* the resultant classifiers are GNPCs.

Fuzzy prototype classifiers have been applied to medical diagnosis [92, 174, 181] showing that they compete well with statistical classifiers. Another largely studied application field is character recognition [60, 177, 230, 302, 340, 372].

8. Combinations of multiple classifiers using fuzzy sets

8.1 Combining classifiers: the variety of paradigms

Different classifiers can be built using the labeled data set $\mathbf{Z}$. Instead of choosing for further use the classifier with the best accuracy, we can keep a set of them. Let $\mathcal{D} = \{D_1, \ldots, D_L\}$ be a set of L classifiers designed on the data set $\mathbf{Z}$. The idea is to combine their outputs hoping to increase the accuracy beyond that of the best classifier in the pool $\mathcal{D}$. This is a theoretically justified hope as we show later but there is no guarantee that picking an arbitrary set of classifiers will render a successful team. Combining classifiers has been an important research topic coming under different names in the literature:

- combination of multiple classifiers [172, 209, 282, 350, 352];
- classifier fusion [62, 102, 115, 164];
- mixture of experts [150, 151, 156, 256];
- committees of neural networks [43, 79];
- consensus aggregation [28, 29, 252];
- voting pool of classifiers [24];
- dynamic classifier selection [350];
- composite classifier system [70];
- classifier ensembles [79, 95];
- divide-and-conquer classifiers [61];
- pandemonium system of reflective agents [309];
- change-glasses approach to classifier selection [189], etc.

The paradigms of these models differ on the assumptions about classifier dependencies; the type of classifier outputs; aggregation strategy (global or local); aggregation procedure (a function, a neural network, an algorithm), etc. Apart from the variety of combination paradigms, the individual classifiers in the pool can be built using:

1. *Different subsets of features* [352]. This is especially convenient when a large number of features are available, naturally grouped into sets, e.g., in text and speech recognition [58, 263, 330]
2. *Different subsets of the data set* $\mathbf{Z}$. We can pick random subsets of $\mathbf{Z}$, take bootstrap samples or use nearest neighbor editing (see Chapter 3) [11]. In the boosting model by Drucker et al. [79], a cascade of three classifiers

is trained using parts of **Z**, "filtered" out by previous constituents of the cascade.

3. *Different classifier models (or initializations).* Using the same **Z**, we can derive different classifier outputs by choosing different classifier models (e.g., [350]) or by initializing differently the same classifier model. (Most often the classifier is a neural network, typically an MLP.)
4. *Any combination of the above three items.*

Example 8.1.1. Let $\mathcal{D} = \{D_1, D_2\}$. Figure 8.1 shows three ways of building D_1 and D_2. The training data set **Z** consists of N vectors, $\mathbf{z}_1, \ldots, \mathbf{z}_N$. The training set for D_1 is denoted by Z_1, and for D_2 by Z_2. In Cases 1 and 2, the two classifiers could be of the same type, e.g., QDC. In Case 3, either the two models should be different, or the chosen model should allow for changing its parameters so that D_1 differs from D_2. For example, in Case 3, D_1 could be an RBF and D_2 an MLP NN. ■

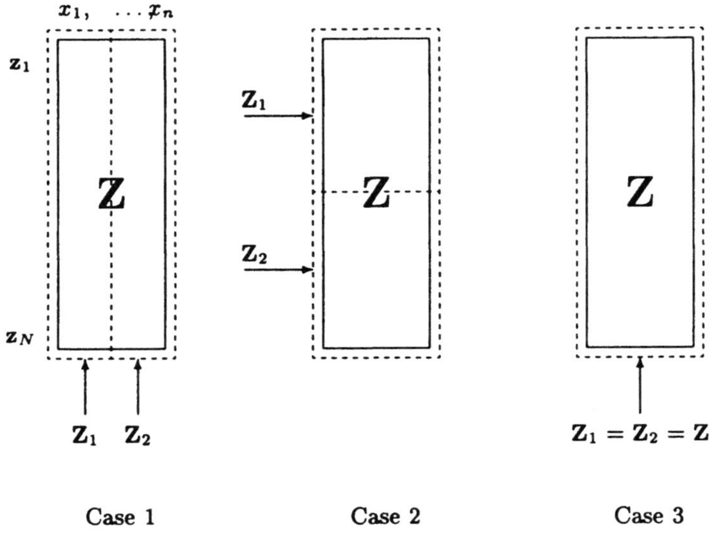

Fig. 8.1. Three ways of building individual classifiers in multiple classifier systems

According to the type of the classifier output, Xu et al. [352] divide classifiers into

1. Type 1, *abstract level* classifiers. The classifier output is a crisp class label, i.e.,

$$D : \Re^n \to \Omega. \tag{8.1}$$

2. Type 2, rank level classifiers. The classifier output is an ordered subset of Ω.
3. Type 3, *measurement level* classifiers. The output is a c dimensional vector with "degrees of support" for each class. Most often this is a soft label, i.e.,

$$D : \Re^n \rightarrow [0,1]^c. \tag{8.2}$$

Type 2 contains Type 1, and Type 3 contains both Type 1 and Type 2 as special cases. We shall denote the output of the combination scheme by the label vector for $\mathbf{x}$

$$\mu(\mathbf{x}) = [\mu_1(\mathbf{x}), \ldots, \mu_c(\mathbf{x})]^T \in [0,1]^c. \tag{8.3}$$

Practically, this is the standard classifier output expressed in terms of the values of the c discriminant functions $g_1(\mathbf{x}), \ldots, g_c(\mathbf{x})$, but scaled in $[0,1]^c$. Again, maximum membership rule can be applied to label $\mathbf{x}$ into only one of the classes.

Here we deal with measurement level classifiers (Type 3). Classifier outputs are usually made comparable by scaling them to the interval [0,1]. For some classifiers these values can be treated as classifier-conditional posterior probabilities for the classes [90]. However, a probabilistic interpretation of the output is not always adequate. Even when it is, many times it does not lead very far without some assumptions, which may appear unrealistic and restrictive. Such assumptions are that the individual classifiers use mutually independent subsets of features or commit independent misclassification errors.

The more general interpretation of classifier outputs as the *support* for the classes is the basis of fuzzy aggregation methods, examples of which are simple connectives between fuzzy sets, the fuzzy integral [62, 63, 102, 117, 164, 335], and Dempster-Shafer fusion [224, 282, 352].

There is a "pattern recognition" way to look at the classifier combination problem: we can treat the classifier outputs as the input to a second-level classifier in some **intermediate feature space**, and design a classifier for the second level. The use of classifiers discussed so far over the intermediate feature space is not always appropriate because the class distributions might not be well-behaved as shown later in Example 8.3.7.

An analogy can be drawn between classifier combinations and team decision making. Each classifier is an "expert" (member of the team), and the team decision is expected to be of a higher value (according to a certain criterion) than the decision of the "best" individual member.

There are generally two types of combination: **classifier selection** and **classifier fusion** ([350]) . The presumption in classifier *selection* is that each classifier is "an expert" in some local area of the feature space. When a feature vector $\mathbf{x} \in \Re^n$ is submitted for classification, the classifier responsible for the vicinity of $\mathbf{x}$ is given the highest credit when assigning the class label to $\mathbf{x}$. We can nominate exactly one classifier to make the decision, as in [278], or more than one "local expert", as in [14, 151]. Classifier ***fusion*** assumes that ***all*** classifiers are trained over the whole feature space, and are thereby considered as **competitive** rather than **complementary** [252, 352]. Figures 8.2 and 8.3 illustrate the two approaches in layman terms.

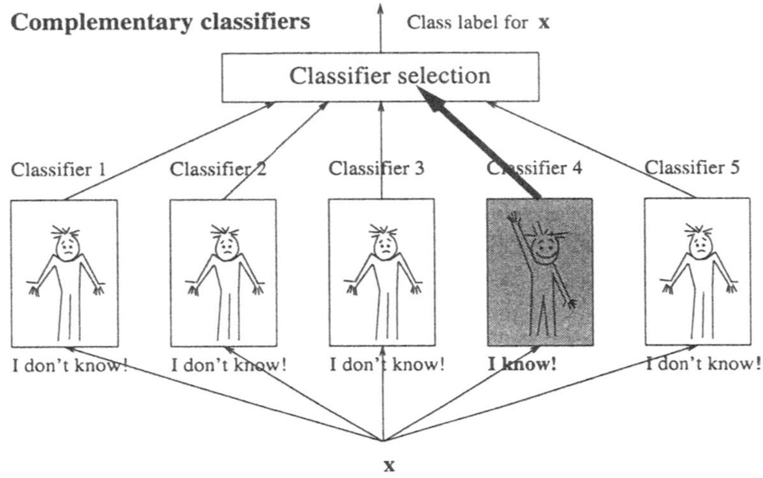

Fig. 8.2. Classifier selection

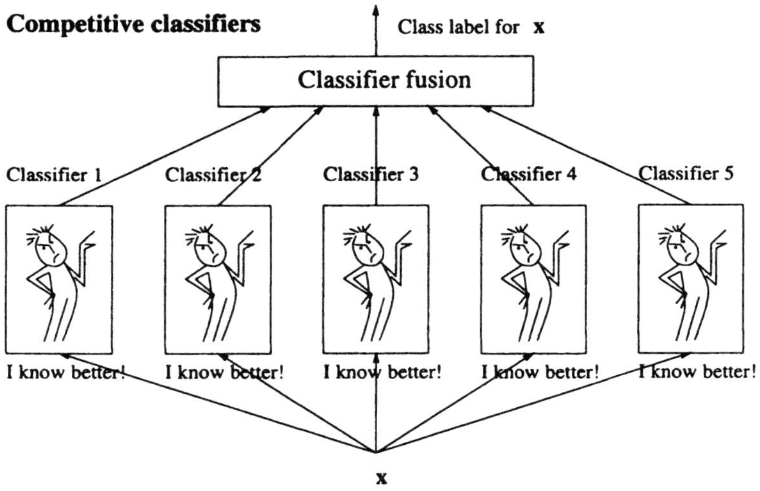

Fig. 8.3. Classifier fusion

It is curious how these two ideas have been merged. Instead of nominating one "expert" we can nominate a small group of them. We can then take their judgements and weight them (where appropriate) by the level of expertise they have on $\mathbf{x}$. We can also assume that each expert has the same level of expertise over the whole feature space (wrong but useful!) and assign a fixed "competence" coefficient to each expert. Thus, the classifier with the highest individual accuracy could be made the "leading expert" in the team. When many classifiers become involved, the scheme is shifted from classifier selection towards classifier fusion. The assumption that a classifier has the same accuracy over the whole feature space is unrealistic. Therefore, ideally, we should be able to combine classifier votes with regard to the particular $\mathbf{x}$ to be classified. The two approaches are discussed in more detail in the next two sections.

8.2 Classifier selection

Classifier selection has been proposed in the form of a composite classifier system by Dasarathy and Sheela in 1978 [70]. They combine a linear classifier and a k-nearest neighbor (k-nn) classifier. The authors suggest to identify a conflict domain in the feature space and use k-nn in that domain while using the linear classifier elsewhere. A similar approach is proposed in [189] where the classification procedure switches between a set of possible classifiers, based on the *certainty* of the current classification decision. This idea mimics the decision making in real life situations, e.g., in medical diagnostics, where help is sought if the confidence of the current decision-maker is not high enough. Rastrigin [278] proposes to use a set of L classifiers $\mathcal{D} = \{D_1, \ldots, D_L\}$ and a *meta-classifier* which decides in whose region of competence the input $\mathbf{x}$ falls. The nominated classifier is then responsible for the decision. Below we show why this idea works.

We denote the output of the ith classifier by $D_i(\mathbf{x}) = [d_{i,1}(\mathbf{x}), \ldots, d_{i,c}(\mathbf{x})]^T$, where $d_{i,j}(\mathbf{x})$ is the degree of "support" given by classifier D_i to the hypothesis that $\mathbf{x}$ comes from class ω_j. Let D^* be the classifier with the highest overall accuracy among the elements of $\mathcal{D}$. Let the feature space $\Re^n$ be divided into K regions of competence, $K > 1$. Denote the regions by $R_1, \ldots, R_K$. Note that these regions are not related with the classification regions, nor do they need to have a specific shape or size.

Example 8.2.1. Shown in Figure 8.4 is a partition of the feature space for the 15-point data set into four regions regardless of the classification boundaries. The boundaries of the nearest neighbor classifier are overlayed. The four regions are used in classifier selection. We nominate a classifier from $\mathcal{D} = \{D_1, \ldots, D_L\}$ for each region R_i, and use it to classify all $\mathbf{x}$'s that fall inside R_i. ∎

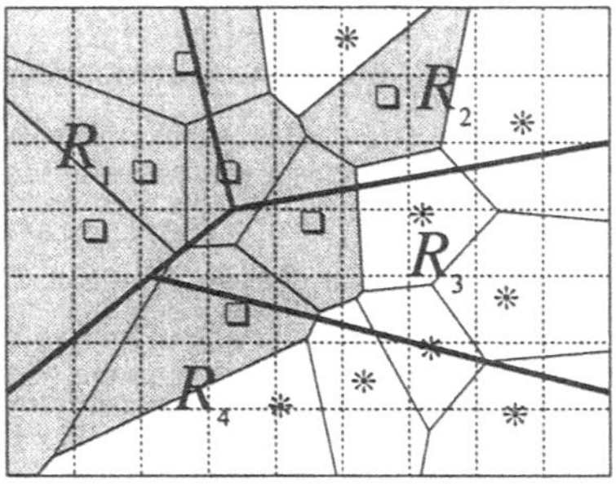

Fig. 8.4. An example of dividing the feature space into four regions

Denote by $P(D_i|R_j)$ the probability of correct classification of D_i in region R_j. Let $D_{i(j)} \in \mathcal{D}$ be the classifier responsible for region $R_j, j = 1, \ldots, K$. The overall probability of correct classification is

$$P(correct) = \sum_{j=1}^{K} P(R_j)P(correct|R_j), \tag{8.4}$$

or equivalently,

$$P(correct) = \sum_{j=1}^{K} P(R_j)P(D_{i(j)}|R_j). \tag{8.5}$$

To maximize $P(correct)$, we assign $D_{i(j)}$ so that

$$P(D_{i(j)}|R_j) \geq P(D_t|R_j), \forall\, t = 1, \ldots, L. \tag{8.6}$$

Hence

$$P(correct) \geq \sum_{j=1}^{K} P(R_j)P(D^*|R_j) = P(D^*). \tag{8.7}$$

The above equation shows that the combined scheme performs better than the best classifier D^* in the pool $\mathcal{D}$, *regardless of the way the feature space has been partitioned.* The only condition (and, of course, the trickiest one) is to ensure that $D_{i(j)}$ *is* the best amongst the L classifiers in $\mathcal{D}$ for region R_j. The extent to which this is satisfied determines the success of our classifier selection model.

There are several ways to design a classifier selection system

1. **Static classifier selection** [278]. The regions are specified prior to classifying the unlabeled vector $\mathbf{x}$. Two approaches are possible: (a) specify the regions and then assign a responsible classifier for each region (e.g.,

the model in [331]), or (b) given $\mathcal{D}$, find a region (possibly a set of regions) where each classifier is the best one (e.g., in [278]). Although probably more efficient, the second approach is difficult to implement. Rastrigin [278] suggests using a meta-classifier which "recognizes" the region of $\mathbf{x}$, say, R_j, and then $D_{i(j)}$ labels $\mathbf{x}$. Approach (a) can be implemented in different ways

- Assuming that the feature space is limited (it can always be scaled to satisfy this), partition it into regularly shaped regions, e.g., by a very coarse grid. This will define "bins" in the feature space similarly to the histogram methods. Use these bins as R_j's. Estimate the classification accuracy of the L classifiers on each bin R_j, and nominate as $D_{i(j)}$ the classifier with the best accuracy.
- Cluster the data in $\mathbf{Z}$ into K clusters, disregarding the class labels. Each cluster defines a region corresponding to the Voronoi cell of the cluster centroid. In other words, each point in the feature space is first labeled in the region with the closest centroid, e.g., R_j, and processed further by the respective classifier $D_{i(j)}$. This method ensures that reasonable amount of points fall in the regions $R_j, j = 1, \dots, K$.

2. **Dynamic classifier selection**. The choice of the classifier to label $\mathbf{x}$ is based on the certainty of the decision. Preference is given to the more certain classifiers. For example, if the 5-nn rule is being used in a 2-class problem, and 3 of the neighbors vote for class ω_1 and 2 for class ω_2, we can resort to, say, 3-nn or 1-nn, thereby changing the classification rule *dynamically* [189]. Woods and coauthors [350] consider a dynamic selection scheme. Given $\mathbf{x}$, all classifiers in $\mathcal{D}$ produce class labels. If they all agree, then $\mathbf{x}$ is given that label, else, the accuracy of each classifier is estimated *in the vicinity of* $\mathbf{x}$. The classifier with the highest accuracy is authorized to label $\mathbf{x}$. Thus, the regions are estimated during the classification process.

Within the static classifier selection approach, Verikas et al.[331] suggest using a weighted combination of classifier outputs in each region. Assuming that $\mathbf{x}$ falls in region R_j, the support for class ω_i is

$$\mu_i(\mathbf{x}) = \sum_{k=1}^{L} w_{k,j} d_{k,i}(\mathbf{x}), \; \mathbf{x} \in R_j \tag{8.8}$$

The weights $w_{k,j}, k = 1, \dots, L, j = 1, \dots, K$ are region-specific but not class-specific. Verikas et al. suggest

$$w_{k,j} = \frac{\hat{P}(D_k|R_j)}{\sum_{t=1}^{L} \hat{P}(D_t|R_j)} \tag{8.9}$$

where $\hat{P}$ stands for an estimate of P. The denominator of (8.9) is the same for all classes and can be dropped to arrive at the following (equivalent) set of discriminant functions

$$g_i(\mathbf{x}) = \sum_{k=1}^{L} \hat{P}(D_k|R_j) d_{i,k}(\mathbf{x}), \ \mathbf{x} \in R_j, \ \ i = 1, \ldots, c. \tag{8.10}$$

Assume we use the estimate

$$\hat{P}(D_k|R_j) = N_{D_k}/N_{R_j}, \tag{8.11}$$

where N_{D_k} is the number of correctly classified elements from $\mathbf{Z}$ in region R_j, and N_{R_j} is the total number of members of that region. Then we can also drop N_{R_j} from the equation and arrive at an algorithmically simple classification rule. Let $R_1, \ldots, R_K$ be found by clustering of $\mathbf{Z}$ into K clusters, so that each region is the Voronoi cell of a cluster centroid. Therefore we call this method **clustering and weighted sum**:

Given $\mathbf{x}$, calculate classifier outputs $d_{i,t}(\mathbf{x}), i = 1, \ldots, L, t = 1, \ldots, c$. Find the closest cluster centroid to specify R_j. Assign $\mathbf{x}$ to class ω_s iff, within R_j,

$$\sum_{k=1}^{L} N_{D_k} d_{k,s}(\mathbf{x}) \geq \sum_{k=1}^{L} N_{D_k} d_{k,t}(\mathbf{x}), \forall t = 1, \ldots, c. \tag{8.12}$$

Ties are resolved arbitrarily.

The "clustering and weighted sum" rule was used in the experiments reported at the end of this section. For comparison we also ran the following "pure" selection method, called **clustering and selection** in which, after determining the region of $\mathbf{x}$, e.g., R_j, we just accept the class label given by $D_{i(j)}$.

8.3 Classifier fusion

Consider again the set of classifiers $\mathcal{D} = \{D_1, \ldots, D_L\}$. We construct $\hat{D}$, the fused output of the L classifiers as

$$\hat{D}(\mathbf{x}) = \mathcal{F}(D_1(\mathbf{x}), \ldots, D_L(\mathbf{x})), \tag{8.13}$$

where $\mathcal{F}$ is an *aggregation rule.*

The classifier outputs can be organized in a **decision profile** (DP) as the matrix

Output of classifier $D_i(\mathbf{x})$

$$DP(\mathbf{x}) = \begin{bmatrix} d_{1,1}(\mathbf{x}) & \dots & d_{1,j}(\mathbf{x}) & \dots & d_{1,c}(\mathbf{x}) \\ \dots & & & & \\ d_{i,1}(\mathbf{x}) & \dots & d_{i,j}(\mathbf{x}) & \dots & d_{i,c}(\mathbf{x}) \\ \dots & & & & \\ d_{L,1}(\mathbf{x}) & \dots & d_{L,j}(\mathbf{x}) & \dots & d_{L,c}(\mathbf{x}) \end{bmatrix}. \qquad (8.14)$$

Support from classifiers $D_1 \dots D_L$ for class j

We can rewrite (8.13) as

$$\hat{D}(\mathbf{x}) = \mathcal{F}(DP(\mathbf{x})). \qquad (8.15)$$

Various groups of fusion methods can be defined on the basis of the aggregation function $\mathcal{F}$. Some methods calculate the support for class ω_i using only the ith column of $DP(\mathbf{x})$, i.e., the individual support for ω_i given by $D_1, \dots, D_L$, regardless of what the support for the other classes is. Fusion methods that use the DP class-by-class will be called **class-conscious (CC)** [201]. We refer to the alternative group as **class-indifferent (CI)** methods. Within this group are fusion by fuzzy integral, the Dempster-Shafer type of fusion and the Decision Template model, all discussed later. Notice the difference between the class-conscious and class-indifferent groups of methods. The former use the *context* of the DP, i.e., recognizing that a column corresponds to a class, but disregard part of the information. Class-indifferent methods use the whole DP but disregard the context which could be useful. This grouping makes sense only for fusion schemes that use the decision profile DP. Aggregation schemes which use a single class label as individual classifier output will be denoted as **crisp (C)**. We start with the group of crisp fusion methods.

8.3.1 Majority vote

The majority vote is a popular classifier fusion technique [23, 24, 209, 210, 352]. Type 1 (crisp) classifier outputs are considered, i.e., each individual classifier assigns a single class label to $\mathbf{x}$ (abstract level classifiers), i.e., the classifier **votes** for this class. The final class label of $\mathbf{x}$ is the one that is voted for by the most classifiers in the pool. Sometimes a *majority* is required, i.e., a class label is assigned if *more than a half* of the classifiers vote for that class [352].

In terms of the decision profile DP, class label ω_k is assigned to $\mathbf{x}$ iff

$$\sum_{j=1}^{L} d_{j,k}(\mathbf{x}) \geq \sum_{j=1}^{L} d_{j,i}(\mathbf{x}), i = 1, \ldots, c, \quad d_{j,i}(\mathbf{x}) \in \{0, 1\}. \tag{8.16}$$

Ties are broken randomly. Since each classifier output is a binary label vector, the summation gives the number of votes for the respective class.

There is an analogy between majority vote and the k-nearest neighbor classifier, so results about k-nn are directly applicable for majority vote. Also, k-nn variants which are not based on *distance* can be applied as majority vote schemes (e.g., the model due to Rovatti et al. [287]).

How much accuracy can we gain using majority vote compared to the accuracy of the single best classifier in the pool $\mathcal{D}$? To start with, we can calculate the added value for the special case of L *independent* classifiers with *the same* classification accuracy, where L is *odd.* Assume that the probability that each classifier gives the correct class label is p, for any $\mathbf{x} \in \Re^n$. The majority vote will give an accurate class label if at least $\lceil L/2 \rceil$ classifiers give correct answers ($\lceil a \rceil$ denotes the *ceiling*, i.e., the nearest integer greater than a). The overall probability for correct classification is given by the Binomial formula

$$P(correct) = \sum_{m=0}^{\lfloor L/2 \rfloor} \binom{L}{m} p^{L-m}(1-p)^m, \tag{8.17}$$

where $\lfloor a \rfloor$ denotes the *floor*, i.e., the nearest integer smaller than a. Lam and Suen [210] acknowledge that Marquis de Condorcet, the French philosopher, mathematician, and politician, is usually credited with using this formula in a decision-making context. The probability of correct classification of the pool for $p = 0.6, 0.7, 0.8$ and 0.9, and $L = 3, 5, 7$ and 9, is tabulated in 8.1.

Table 8.1. Tabulated values of the majority vote accuracy of L independent classifiers with individual accuracy p

	$L = 3$	$L = 5$	$L = 7$	$L = 9$
$p = 0.6$	0.6480	0.6826	0.7102	0.7334
$p = 0.7$	0.7840	0.8369	0.8740	0.9012
$p = 0.8$	0.8960	0.9421	0.9667	0.9804
$p = 0.9$	0.9720	0.9914	0.9973	0.9991

For both odd and even L's, Lam and Suen [210] prove that

1. If $p > 0.5$, then $P(correct)$ in (8.17) is monotonically increasing and

$$P(correct) \rightarrow 1 \quad \text{as} \quad L \rightarrow \infty. \tag{8.18}$$

2. If $p < 0.5$, then $P(correct)$ at (8.17) is monotonically decreasing and

$$P(correct) \to 0 \quad \text{as} \quad L \to \infty. \tag{8.19}$$

3. If $p = 0.5$, then $P(correct) = 0.5$ for any L.

This result shows (what is intuitively acceptable) that we can expect improvement over the individual accuracy p only when p is higher than 0.5.

Above we assumed that half of the group plus one must be correct in order to incur the correct class label. All other cases are counted as misclassifications. However, this is not always the case, and the correct class label can be assigned even when the number of correct votes is smaller than the simple majority. Let our pattern recognition problem be to distinguish between the 26 classes of images of handwritten letters. Let $L = 5$ and let class labels for a given $\mathbf{x}$ be $\{'A','H','B','R','B'\}$. Let the true class label of $\mathbf{x}$ be $'B'$. Although only 2 (less than $\lfloor L/2 \rfloor + 1$) classifiers produce the correct class label, $\mathbf{x}$ will be still correctly classified by (8.16). Since the misclassification error can only be reduced as the example shows, we can expect an even bigger improvement over the individual error rate than that shown in Table 8.1.

It is interesting to find out what happens if the assumption of independence is dropped. Given the data set $\mathbf{Z}$, for each classifier D_i, we design an N-dimensional output vector $\mathbf{y}_i = [y_{1,i}, \dots, y_{N,i}]^T$ of *correct classification*, such that

$$y_{j,i} = \begin{cases} 1, & \text{if } D_i \text{ correctly recognizes } \mathbf{z}_j \\ 0, & \text{if } \mathbf{z}_j \text{ is misclassified by } D_i \end{cases}, \tag{8.20}$$

$j = 1, \dots, N$. The L classification accuracies are then estimated as

$$P_c(D_i) = \frac{1}{N} \sum_{j=1}^{N} y_{j,i}, \quad i = 1, \dots, L. \tag{8.21}$$

The results of a pair of classifiers D_i and D_k can be summarized in two equivalent ways as shown in Table 8.2. (See also Chapter 2, Table 2.2.)

We can employ various statistics to assess how "similar" the two classifiers D_i and D_k are [6]. We choose the Q statistic

$$Q_{i,k} = \frac{N^{11}N^{00} - N^{01}N^{10}}{N^{11}N^{00} + N^{01}N^{10}}, \tag{8.22}$$

where N^{ab} is the number of elements $\mathbf{z}_j$ of $\mathbf{Z}$ for which $y_{j,i} = a$ and $y_{j,k} = b$ (see Table 8.2). It can be easily verified that for statistically **independent** classifiers, $Q_{i,k} = 0$. However, $Q = 0$ does not imply the independence of the classifiers. Clearly, Q varies between -1 and 1. Classifiers that tend to recognize *the same* objects correctly will have values of Q close to 1, and those which commit errors on different objects will render Q negative.

Table 8.2. Two equivalent ways of representing the output of a pair of classifiers D_i and D_k. The entries in the tables are the number of times the respective combination occurs amongst the N elements of $\mathbf{Z}$

Enumerating all possibilities

Combination $\mathbf{y}_i\mathbf{y}_k$	Number of occurrences
00	N^{00}
01	N^{01}
10	N^{10}
11	N^{11}

A 2 × 2 table

	D_k correct (1)	D_k wrong (0)
D_i correct (1)	N^{11}	N^{10}
D_i wrong (0)	N^{01}	N^{00}

Total, $N = N^{00} + N^{01} + N^{10} + N^{11}$.

Example 8.3.1. Shown below are five 2 × 2 tables and the respective values of Q for $N = 100$ objects in $\mathbf{Z}$. Note that the pair of classifiers in each table have the same overall accuracy.

$Q = -1$

0	50
50	0

$Q = -0.5$

30	30
30	10

$Q = 0$

36	24
24	16

$Q = 0.5193$

32	18
18	32

$Q = 1$

50	0
0	50

■

A similar characteristic is the **correlation coefficient**, which in the above notation is

$$\rho_{i,k} = \frac{N^{11}N^{00} - N^{01}N^{10}}{\sqrt{(N^{11} + N^{10})(N^{01} + N^{00})(N^{11} + N^{01})(N^{10} + N^{00})}}. \tag{8.23}$$

It can be shown that Q and ρ have the same sign, and $|\rho| \le |Q|$. Assume all classifiers in $\mathcal{D}$ have the same individual accuracy. Intuitively, a higher majority vote accuracy can be expected if D_i's have negative values of Q, i.e., the classifiers commit their errors on *different* objects. On the contrary, classifiers which are positively related (make the same mistakes) will hardly improve on the individual accuracy. By a synthetic example, we show that the intuition is partly correct but the dependence between the majority vote accuracy and the measure of relationship Q is ambivalent.

Let $\mathcal{D} = \{D_1, D_2, D_3\}$ and $N = |\mathbf{Z}| = 10$. Assume that all classifiers have the same individual accuracy of correct classification, $p = 0.6$. This is manifested by each classifier labeling correctly 6 of the 10 elements of $\mathbf{Z}$. Given these requirements, *all* possible combinations of distributing 10 elements into the 8 combinations of outputs of the three classifiers are shown in Table 8.3. For a correct overall decision by majority vote for some $\mathbf{z}_j \in \mathbf{Z}$, at least two of the three outputs $\mathbf{y}_i$ should be 1. The last column of Table 8.3 shows the majority vote accuracy of each of the 28 possible combinations. It is obtained as the proportion (out of 10 elements) of the sum of the entries in columns '111', '101', '011' and '110' (two or more correct votes). The best and the worst cases are highlighted in the table.

Table 8.3. All possible combinations of correct/incorrect classification of 10 objects by three classifiers so that each classifier recognizes exactly 6 objects. The entries in the table are the number of occurrences of the specific binary output of the three classifiers in the particular combination. The majority vote accuracy P_{maj} is shown in the last column.

No	111	101	011	001	110	100	010	000	P_{maj}	
1	0	2	2	2	4	0	0	0	0.8	
2	0	2	3	1	3	1	0	0	0.8	
3	0	3	3	0	3	0	0	1	**0.9**	← best case
4	1	1	1	3	4	0	0	0	0.7	
5	1	1	2	2	3	1	0	0	0.7	
6	1	2	2	1	2	1	1	0	0.7	
7	1	2	2	1	3	0	0	1	0.8	
8	2	0	0	4	4	0	0	0	0.6	
9	2	0	1	3	3	1	0	0	0.6	
10	2	0	2	2	2	2	0	0	0.6	
11	2	1	1	2	2	1	1	0	0.6	
12	2	1	1	2	3	0	0	1	0.7	
13	2	1	2	1	2	1	0	1	0.7	
14	2	2	2	0	2	0	0	2	0.8	
15	3	0	0	3	2	1	1	0	0.5	
16	3	0	0	3	3	0	0	1	0.6	
17	3	0	1	2	1	2	1	0	0.5	
18	3	0	1	2	2	1	0	1	0.6	
19	3	1	1	1	1	1	1	1	0.6	
20	3	1	1	1	2	0	0	2	0.7	
21	4	0	0	2	0	2	2	0	**0.4**	← worst case
22	4	0	0	2	1	1	1	1	0.5	
23	4	0	0	2	2	0	0	2	0.6	
24	4	0	1	1	1	1	0	2	0.6	
25	4	1	1	0	1	0	0	3	0.7	
26	5	0	0	1	0	1	1	2	0.5	
27	5	0	0	1	1	0	0	3	0.6	
28	6	0	0	0	0	0	0	4	0.6	

The table offers at least two interesting facts

- We can gain up to 30 % increase in the classification accuracy over the individual rate (best case in Table 8.3). This is a substantial improvement, especially noticing that the accuracy of the majority vote of 3 *independent* classifiers, each one of accuracy 0.6, is 0.648 (Table 8.1).
- Combining classifiers using the majority vote is beneficial or "neutral" in a great deal of the cases. In this example, in 12 of the 28 possible cases (42.9 %) the accuracy is greater than the theoretical value for independent classifiers ($P_{maj} \geq 0.7 > 0.648$). For another 11 cases (39.3 %), the accuracy did not improve on the individual rate ($P_{maj} = p = 0.6$). In the remaining 5 cases (17.8 %) the overall accuracy was below the individual error rate ($P_{maj} < 0.6$).

The Q statistic was calculated for each pair of classifiers for each of the 28 combinations. For the winning combination ($P_{maj} = 0.9$), $Q_{1,2} = Q_{2,3} = Q_{1,3} = -0.5$. For the worst case ($P_{maj} = 0.4$), $Q_{1,2} = Q_{2,3} = Q_{1,3} = 0.333$. Although supporting intuition, this result appears to be not very indicative. Table 8.4 shows the sorted P_{maj} and the corresponding $Q_{1,2}, Q_{2,3}$ and $Q_{1,3}$. As seen in the table, there is no pattern for the relationship between P_{maj} and the Q's. For a general observation, we averaged separately the Q's for all 12 combinations for which $P_{maj} > 0.6$ (favorable) and the 16 combination for which $P_{maj} \leq 0.6$ (unfavorable). The averaged Q of the favorable combinations is -0.1227, and that of the unfavorable combinations is 0.2873. However, the values of the Q's are scattered in the whole range from -1 to 1, and extracting a consistent relationship seems impossible.

The same type of synthetic experiment was carried out for $N = 100$ objects. A total of 36151 distributions of the 100 objects from **Z** were possible, so that each of the classifiers recognizes exactly 60 of them. From these combinations, 14941 (41.3 %) have $P_{maj} > 0.648$ (favorable group), and from the unfavorable group, 11270 (31.2 %) have $P_{maj} < 0.6$. The averaged values of Q for the two groups are similar to the values in our previous example, -0.1109 for the favorable group and 0.2320 for the unfavorable one. Figure 8.5 shows the histograms of all Q's for the favorable and unfavorable groups of combinations. Generally, the favorable Q's tend to be more on the negative side but, in fact, the Q's for each group span the whole interval $[-1, 1]$. Theoretical limits on the majority vote accuracy are obtained in [206]. Equations relating P_{maj}, p, Q, and L are derived for the best possible and the worst possible scenario.

The majority vote is used in the experiments described at the end of this chapter.

8.3.2 "Naive"-Bayes combination

This scheme assumes that the classifiers are mutually independent (this is the reason we use the name "naive"); Xu et al. [352] and others call it *Bayes* combination. For each classifier D_j, a $c \times c$ confusion matrix CM^j is calculated

Table 8.4. Sorted by P_{maj} combination from Table 8.3 and the corresponding Q's

No	P_{maj}	$Q_{1,2}$	$Q_{1,3}$	$Q_{2,3}$
21	0.4	0.33	0.33	0.33
15	0.5	0.88	-0.50	-0.50
17	0.5	0.33	-0.50	0.33
22	0.5	0.88	0.33	0.33
26	0.5	0.88	0.88	0.88
8	0.6	1.00	-1.00	-1.00
9	0.6	0.88	-1.00	-0.50
10	0.6	0.33	-1.00	0.33
11	0.6	0.33	-0.50	-0.50
16	0.6	1.00	-0.50	-0.50
18	0.6	0.88	-0.50	0.33
19	0.6	0.33	0.33	0.33
23	0.6	1.00	0.33	0.33
24	0.6	0.88	0.33	0.88
27	0.6	1.00	0.88	0.88
28	0.6	1.00	1.00	1.00
4	0.7	0.88	-1.00	-1.00
5	0.7	0.33	-1.00	-0.50
6	0.7	-0.50	-0.50	-0.50
12	0.7	0.88	-0.50	-0.50
13	0.7	0.33	-0.50	0.33
20	0.7	0.88	0.33	0.33
25	0.7	0.88	0.88	0.88
1	0.8	0.33	-1.00	-1.00
2	0.8	-0.50	-1.00	-0.50
7	0.8	0.33	-0.50	-0.50
14	0.8	0.33	0.33	0.33
3	0.9	-0.50	-0.50	-0.50

by applying D_j to the training data set. The (k, s)th entry of this matrix, $cm^j_{k,s}$ is the number of elements of the data set whose true class label was ω_k, and were assigned by D_j to class ω_s. By $cm^j_{.,s}$ we denote the total number of elements labeled by D_j into class ω_s (this is calculated as the sum of the sth column of CM^j). Using $cm^j_{.,s}$, a $c \times c$ label matrix LM^j is computed, whose (k, s)th entry $lm^j_{k,s}$ is an estimate of the probability that the true label is ω_k given that D_j assigns crisp class label s.

$$lm^j_{k,s} = \hat{P}\left(\omega_k | D_j(\mathbf{x}) = \omega_s\right) = \frac{cm^j_{k,s}}{cm^j_{.,s}}, \tag{8.24}$$

For every $\mathbf{x} \in \Re^n$, D_j yields a crisp label vector $D_j(\mathbf{x})$ pointing at one of the classes, say, $\omega_s \in \Omega$. Considering the label matrix for D_j, LM^j, associated with ω_s is a *soft label vector* $[\hat{P}\left(\omega_1 | D_j(\mathbf{x}) = \omega_s\right), \ldots, \hat{P}\left(\omega_c | D_j(\mathbf{x}) = \omega_s\right)]^T$, which is the sth column of the matrix. Let $s_1, \ldots, s_L$ be the crisp class labels assigned to $\mathbf{x}$ by classifiers $D_1, \ldots, D_L$, respectively. Then, by the indepen-

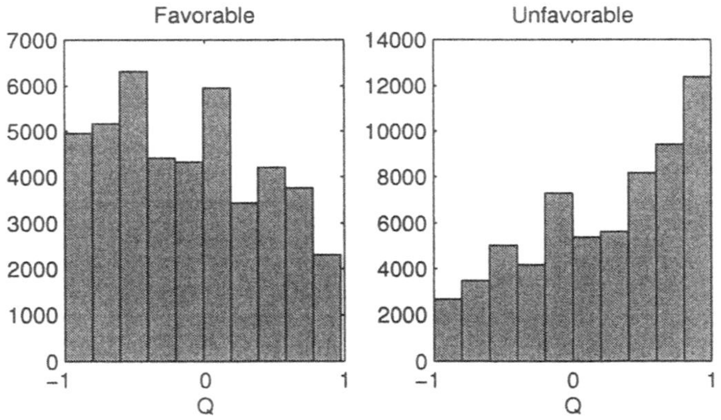

Fig. 8.5. Histograms of the Q statistic for the "favorable" ($P_{maj} > 0.648$) and "unfavorable" ($P_{maj} \leq 0.648$) combinations of classifier outputs, $N = 100$.

dence assumption, the estimate of the probability that the true class label is ω_i, is calculated by

$$\mu_i(\mathbf{x}) = \prod_{j=1}^{L} \hat{P}\left(\omega_i | D_j(\mathbf{x}) = s_j\right) = \prod_{j=1}^{L} lm^j_{i,s_j}, \quad i = 1, \ldots, c. \tag{8.25}$$

Example 8.3.2. Consider a problem with L=2 classifiers, D_1 and D_2, and $c = 3$ classes. Let the number of training data points be $N = 20$. ¿From these, let 8 be from ω_1, 9 from ω_2 and 3 from ω_3. Suppose the following confusion matrices have been obtained for the two classifiers

$$CM^1 = \begin{bmatrix} 6 & 2 & 0 \\ 1 & 8 & 0 \\ 1 & 0 & 2 \end{bmatrix}, \quad \text{and} \quad CM^2 = \begin{bmatrix} 4 & 3 & 1 \\ 3 & 5 & 1 \\ 0 & 0 & 3 \end{bmatrix}. \tag{8.26}$$

The two label matrices obtained from CM^1 and CM^2 are

$$LM^1 = \begin{bmatrix} 6/8 & 2/10 & 0 \\ 1/8 & 8/10 & 0 \\ 1/8 & 0 & 1 \end{bmatrix}, \quad \text{and} \quad LM^2 = \begin{bmatrix} 4/7 & 3/8 & 1/5 \\ 3/7 & 5/8 & 1/5 \\ 0 & 0 & 3/5 \end{bmatrix}. \tag{8.27}$$

Assume $D_1(\mathbf{x}) = \omega_2$ and $D_2(\mathbf{x}) = \omega_1$ for the input $\mathbf{x} \in \Re^n$. Using the second column of LM^1 and the first column of LM^2, we calculate the output of the Naive Bayes classifier fusion scheme as follows

$$\mu_1(\mathbf{x}) = \hat{P}(\omega_1|D_1(\mathbf{x}) = \omega_2)\ \hat{P}(\omega_1|D_2(\mathbf{x}) = \omega_1)$$
$$= \frac{2}{10} \times \frac{4}{7} = \frac{4}{35}\,;$$

$$\mu_2(\mathbf{x}) = \hat{P}(\omega_2|D_1(\mathbf{x}) = \omega_2)\ \hat{P}(\omega_2|D_2(\mathbf{x}) = \omega_1) \tag{8.28}$$
$$= \frac{8}{10} \times \frac{3}{7} = \frac{12}{35}\,;$$

$$\mu_3(\mathbf{x}) = \hat{P}(\omega_3|D_1(\mathbf{x}) = \omega_2)\ \hat{P}(\omega_3|D_2(\mathbf{x}) = \omega_1)$$
$$= 0$$

The maximum membership rule will class $\mathbf{x}$ in ω_2. ■

8.3.3 Behavior-knowledge space (BKS)

Let again $(s_1, \ldots, s_L) \in \Omega^L$ be the crisp class labels assigned to $\mathbf{x}$ by classifiers $D_1, \ldots, D_L$, respectively. Every possible combination of class labels is an index regarded as a cell in a look-up table (BKS table)[139]. The table is designed using the data set $\mathbf{Z}$. Each $\mathbf{z}_j$ is placed in the cell indexed by $D_1(\mathbf{z}_j), \ldots, D_L(\mathbf{z}_j)$. The elements in each cell are tallied and the most representative class label is selected for this cell. Thus, each entry in the BKS table is one of the following: a single class label (the one that is most often encountered amongst the elements of $\mathbf{Z}$ in this cell); no label (the cell is empty because no element of $\mathbf{Z}$ had the respective combination of class labels); or a set of tied class labels (if more than one class have the same highest number of elements in this cell).

Example 8.3.3. Let $c = 3$, $L = 2$, $N = 100$. The design of a BKS look-up table is displayed in Table 8.5.

Table 8.5. Design of a BKS table. The entries are the number of elements with the particular label combination from the two classifiers

$s_1, s_2 \rightarrow$	1,1	1,2	1,3	2,1	2,2	2,3	3,1	3,2	3,3
ω_1	10	3	5	0	1	4	7	0	0
ω_2	3	0	4	0	16	4	2	2	0
ω_3	3	6	5	0	6	4	4	5	6
Cell label	ω_1	ω_3	tie ω_1, ω_3	None	ω_2	tie $\omega_1, \omega_2, \omega_3$	ω_1	ω_3	ω_3

■

The decision for an $\mathbf{x} \in \Re^n$ is made according to the class label of the cell indexed by $D_1(\mathbf{x}), \ldots, D_L(\mathbf{x})$. Ties are broken randomly. If an empty cell is hit, the class label is chosen at random from Ω. The operation of BKS is illustrated in Figure 8.6.

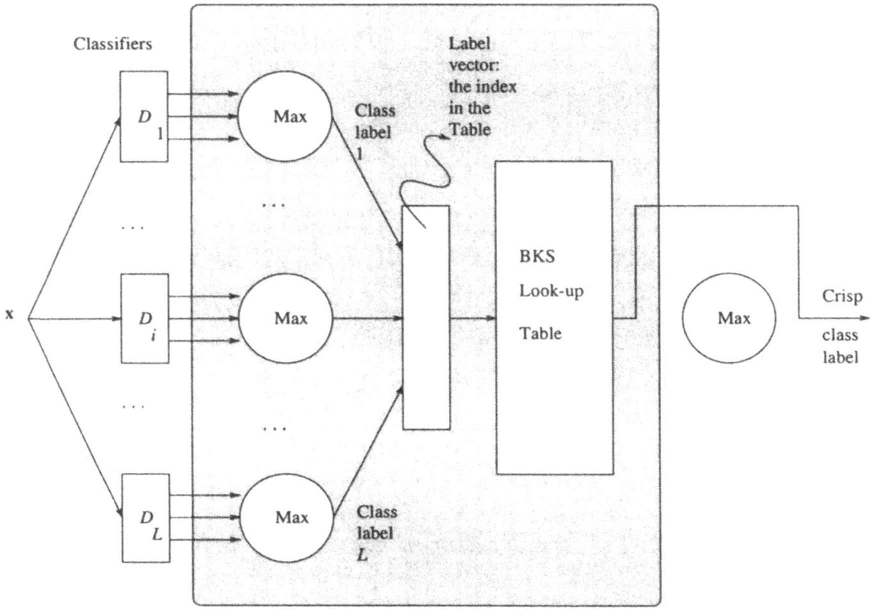

Fig. 8.6. Operation of BKS method for classifier fusion

8.3.4 Simple fuzzy connectives for classifier fusion

This set of fusion designs belong to the **class-conscious** group because each $\mu_i(\mathbf{x})$ is calculated using ***only*** the ith column of the decision profile $DP(\mathbf{x})$. We use the L-place operators ***minimum, maximum, average and product*** (cf Chapter 4) as the aggregation rule $\mathcal{F}$ in (8.15):

$$\mu_i(\mathbf{x}) = \mathcal{F}\ (d_{1,i}(\mathbf{x}), \ldots, d_{L,i}(\mathbf{x}))\,, \quad i = 1, \ldots, c. \tag{8.29}$$

Figure 8.7 shows the operation of simple aggregation rules.

Example 8.3.4. The following example helps to clarify the four fusion methods. Let $c = 3$ and $L = 5$. Assume that for a certain $\mathbf{x}$,

$$DP(\mathbf{x}) = \begin{bmatrix} 0.1 & 0.5 & 0.4 \\ 0.0 & 0.0 & 1.0 \\ 0.4 & 0.3 & 0.4 \\ 0.2 & 0.7 & 0.1 \\ 0.1 & 0.8 & 0.2 \end{bmatrix}.$$

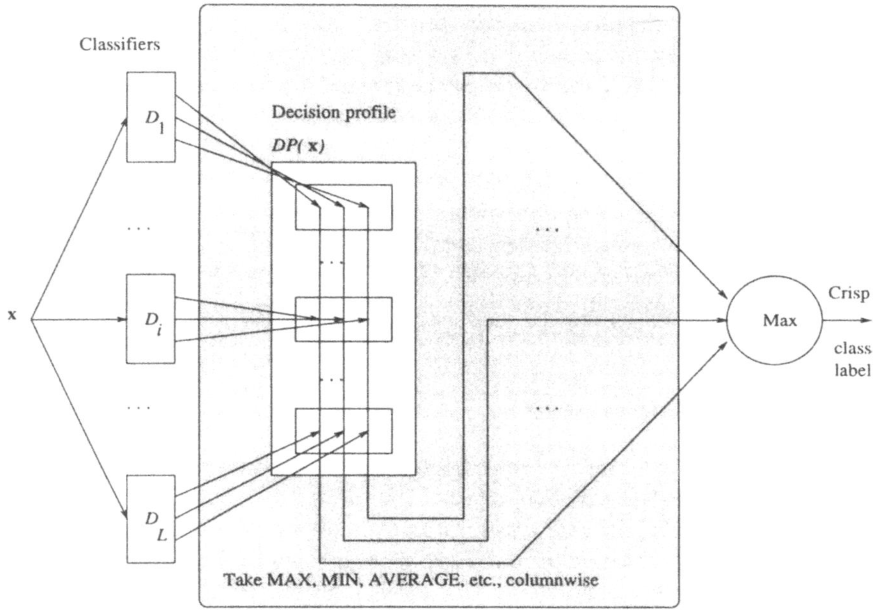

Fig. 8.7. Operation of the simple aggregation rules

Applying each of the operators columnwise, we obtain as the final soft class label $\mu(\mathbf{x})$

$$\begin{aligned} \text{Minimum} &= [\,0.0,\ 0.0,\ 0.1\,]^T; \\ \text{Maximum} &= [\,0.4,\ 0.8,\ 1.0\,]^T; \\ \text{Average} &= [\,0.16,\ 0.46,\ 0.42\,]^T; \\ \text{Product} &= [\,0.0,\ 0.0,\ 0.0032\,]^T. \end{aligned}$$

If hardened, minimum, maximum, and product will label $\mathbf{x}$ in class ω_3, whereas the average will put $\mathbf{x}$ in class ω_2. ■

Ordered Weighting Average (OWA) operators (see Chapter 4) can also be applied as $\mathcal{F}$ [194].

Example 8.3.5. Using the data from Example 8.3.4, and applying OWA with the coefficient vector $[0.1, 0.2, 0.4, 0.2, 0.1]^T$, we obtain $\mu(\mathbf{x}) = [0.22, 0.39, 0.44]^T$, and assign $\mathbf{x}$ to ω_3. ■

Recall that the OWA coefficients are not associated with a particular classifier D_i but with the ***places*** in the ordered outputs. Take, for example, class ω_j. The OWA coefficients are associated with the ordered entries of the jth column of the decision profile $DP(\mathbf{x})$. This prevents crediting one particular "expert" with the highest competence over the whole feature space, as it would be the case if we assigned fixed weights to the classifiers. If the favourite expert (classifier) has received the credit because of overfitting the training data, then by praising it, we can face poor generalization. Thus, classifier fusion by OWA seems more robust than the weighted average, where the coefficients are derived on the basis of classifier performance. It is worth noticing that the fuzzy integral for classifier fusion takes this idea further so that OWA aggregation is a special case of it. We can either pick the set of OWA coefficients according to some assumptions about the classifier outputs or can obtain them using the data set **Z**.

The model in Example 8.3.5 reflects the case where we believe that very high or very low support given by any of the L classifiers could be biased, and the medium values are more likely to be precise.

8.3.5 Probabilistic product

The probabilistic product is a simple aggregation formula [197, 319] (derived in [46]) in the *class-conscious* group. The probabilistic product gives the Bayes decision if the classifiers use mutually independent subsets of features and yield the true posterior probability, on their respective feature subspaces.

Let the set of features $X = \{X_1, \ldots, X_n\}$ be partitioned into L disjoint subsets $\{X^{(1)}, \ldots, X^{(L)}\}$ such that

$$\bigcup_{i=1}^{L} X^{(i)} = X, \tag{8.30}$$

and

$$X^{(i)} \cap X^{(j)} = \emptyset, \ \forall\, i, j, \ 1 \le i, j \le L, i \neq j. \tag{8.31}$$

Denote by $\mathbf{x}^{(i)}$ a vector in the subspace $\Re^{|X^{(i)}|}$ spanned by the features in $X^{(i)}$. Thus, the feature vector $\mathbf{x}$ is represented as $\mathbf{x} = [\mathbf{x}^{(1)}, \ldots, \mathbf{x}^{(L)}]^T$. We assume that the feature subsets are **conditionally independent**, i.e.,

$$p(\mathbf{x}|\omega_j) = \prod_{i=1}^{L} p(\mathbf{x}^{(i)}|\omega_j), \ \ \mathbf{x}^{(i)} \in \Re^{|X^{(i)}|}, \ \ j = 1, \ldots, c, \tag{8.32}$$

Let $P(\omega_j)$ denote the prior probability for class ω_j. Assume that all individual classifiers produce $d_{i,j}(\mathbf{x}^{(i)}) = P(\omega_j|\mathbf{x}^{(i)})$ (the true posterior probability for class ω_j on the feature set $X^{(i)}$). For classifier D_i,

$$d_{i,j}(\mathbf{x}^{(i)}) = P(\omega_j|\mathbf{x}^{(i)}) = \frac{P(\omega_j)p(\mathbf{x}^{(i)}|\omega_j)}{p(\mathbf{x}^{(i)})} \tag{8.33}$$

The overall posterior probability is

$$P(\omega_j|\mathbf{x}) = \frac{P(\omega_j)\, p(\mathbf{x}|\omega_j)}{p(\mathbf{x})}$$
$$= \frac{P(\omega_j)\, \prod_{i=1}^{L} p(\mathbf{x}^{(i)}|\omega_j)}{p(\mathbf{x})}. \tag{8.34}$$

Multiplying the fraction by

$$\frac{P(\omega_j)^L \prod_{i=1}^{L} p(\mathbf{x}^{(i)})}{P(\omega_j)^L \prod_{i=1}^{L} p(\mathbf{x}^{(i)})}, \tag{8.35}$$

we obtain

$$P(\omega_j|\mathbf{x}) = \frac{\prod_{i=1}^{L} p(\mathbf{x}^{(i)})}{p(\mathbf{x})P(\omega_j)^{L-1}} \prod_{i=1}^{L} \frac{P(\omega_j)p(\omega_j|\mathbf{x}^{(i)})}{p(\mathbf{x}^{(i)})}$$
$$= \frac{\prod_{i=1}^{L} p(\mathbf{x}^{(i)})}{p(\mathbf{x})} \frac{1}{P(\omega_j)^{L-1}} P(\omega_j|\mathbf{x}^{(i)}). \tag{8.36}$$

Noticing that the first multiplier does not depend on the class label, we can design a set of discriminant functions proportional to the true posterior probabilities,

$$g_j(\mathbf{x}) = \frac{\prod_{i=1}^{L} d_{i,j}(\mathbf{x})}{P(\omega_j)^{L-1}}, \quad j = 1, \ldots, c, \tag{8.37}$$

and derive $\mu_j(\mathbf{x})$, $j = 1, \ldots, c$, by scaling them into $[0, 1]$.

Even if the classifier outputs are not the true probabilities but some estimates thereof, and the independence is not guaranteed, the probabilistic product has been found to work well for classifier fusion [201]. (The estimates of $P(\omega_j)$ are found on the training sample as $\frac{N_j}{N}$.)

8.3.6 Fuzzy integral

Fuzzy integral was introduced in Chapter 4. Here we use it as a **class-indifferent** model for classifier fusion, as in [62, 102, 164, 335]. For an input $\mathbf{x}$, we calculate c vectors of length L. Each such vector corresponds to a class, and contains L values of a *fuzzy measure*. Then the ith column of the decision profile, containing the L values of support for class ω_i, is sorted and *fused* with the fuzzy measure for that class to get $\mu_i(\mathbf{x})$. Thus, fuzzy integration is interpreted as searching for the maximal grade of agreement between the objective evidence (provided by the sorted classifier outputs for class ω_i) and the expectation (the L fuzzy measure values) [62]. The problem is how to find the fuzzy measure vectors. Most authors that use the fuzzy integral suggest computing a λ-fuzzy measure g_λ. To find $\mu_k(\mathbf{x})$, the support for class ω_k, the following procedure is applied [1]

1. Fix the L **fuzzy densities** $g^1, \ldots, g^L$. These densities can be interpreted as the *importance* of the classifiers. Wang et al. [335] report that good results have been obtained with $g^i = \hat{P}(D_i)/2$, i.e., half of the estimated probability of correct classification of classifier D_i; Cho and Kim [62] use $g^i = \hat{P}(D_i)$, $i = 1, \ldots, L$.
2. Calculate $\lambda > -1$ as the only real root greater than -1 of the equation (see (4.59))
$$\lambda + 1 = \prod_{i=1}^{L}(1 + \lambda g^i). \tag{8.38}$$
3. For a given $\mathbf{x}$ sort the kth column of $DP(\mathbf{x})$ to obtain $[d_{i_1,k}(\mathbf{x}), d_{i_2,k}(\mathbf{x}), \ldots, d_{i_L,k}(\mathbf{x})]^T$, $d_{i_1,k}(\mathbf{x})$ being the highest degree of support, and $d_{i_L,k}(\mathbf{x})$, the lowest.
4. Sort the densities correspondingly, i.e., $g^{i_1}, \ldots, g^{i_L}$.
5. Set $g(1) = g^{i_1}$.
6. For $t = 2$ to L, calculate recursively
$$g(t) = g^{i_t} + g(t-1) + \lambda g^{i_t} g(t-1).$$
7. Calculate the final degree of support for class ω_k as in (4.62)
$$\mu_k(\mathbf{x}) = \max_{t=1}^{L} \left\{\min\{d_{i_t,k}, g(t)\}\right\}. \tag{8.39}$$

Notice that the fuzzy measure vector might be different for each class, and is also specific for the current $\mathbf{x}$. Two fuzzy measure vectors will be the same only if the ordering of the classifier support is the same.

Example 8.3.6. Let $c = 4$ and $L = 3$. Consider the DP and classifier accuracies used as the fuzzy densities g^1, g^2 and g^3, shown in Table 8.6. Solving equation (4.59) we get $\lambda = -0.94977$. Applying the above procedure separately to each column of the table, we get the results shown in Table 8.7.

Table 8.6. Decision profile for x

$DP(\mathbf{x}) =$

class →	1	2	3	4
$D_1(\mathbf{x})$	0.2	0.1	0.4	0.3
$D_2(\mathbf{x})$	0.7	0.0	0.1	0.2
$D_3(\mathbf{x})$	0.1	0.1	0.6	0.2

$g^1 = \hat{P}(D_1) = 0.63,$
$g^2 = \hat{P}(D_2) = 0.70,$
$g^3 = \hat{P}(D_3) = 0.66,$

Thus, the class label for $\mathbf{x}$ is ω_1. ■

[1] The procedure is given in Chapter 4, Section 4.3.5. Here we repeat it briefly for completeness of the text.

Table 8.7. Applying the FI procedure to the columns of $DP(\mathbf{x})$ in Table 8.6

column 1	g
0.7	0.70
0.2	0.91
0.1	1.00

column 2	g
0.1	0.66
0.1	0.90
0.0	1.00

column 3	g
0.6	0.66
0.4	0.90
0.1	1.00

column 4	g
0.3	0.63
0.2	0.90
0.2	1.00

$\mu_1(\mathbf{x}) = 0.7$ $\mu_2(\mathbf{x}) = 0.1$ $\mu_3(\mathbf{x}) = 0.6$ $\mu_4(\mathbf{x}) = 0.3$

The FI scheme is illustrated in Figure 8.8.

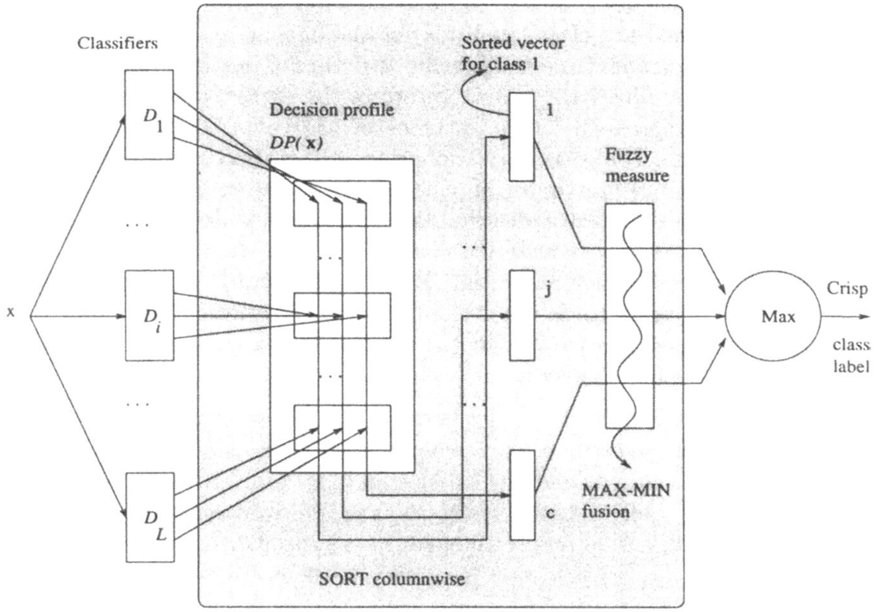

Fig. 8.8. Operation of fuzzy integral for classifier fusion

8.3.7 A brute-force approach to classifier fusion

The choice of an aggregation operator $\mathcal{F}$ depends on the interpretation of $d_{i,j}(\mathbf{x}), i = 1, \ldots, L,\ j = 1, \ldots, c$. We can regard $d_{i,j}(\mathbf{x})$ as an estimate of the posterior probability $P(\omega_j|\mathbf{x})$ produced by classifier D_i (denoted

$\hat{P}_i(\omega_j|\mathbf{x}), i = 1,\ldots,L,\ j = 1,\ldots,c$). Optimal (in the Bayesian sense) combination of these estimates is not straightforward. Kittler et al. [172] show two (different!) ways of combining such estimates starting from the same assumption of independence of individual classifiers. For many classifiers, the estimates $\hat{P}_i(\omega_j|\mathbf{x})$ can have both large bias and variance, which, together with a faulty independence assumption, can invalidate the probabilistic approach.

In the brute-force approach we use *all of* $DP(\mathbf{x})$ (**class-indifferent** methods) to calculate the support for each class. We call the space spanned by the classifier outputs, $[0,1]^{(L\cdot c)} \subset \Re^{(L\cdot c)}$, the **intermediate feature space**. Each vector in this space is an "expanded" version of the DP matrix obtained by concatenating its L columns. The problem now is to design the second (fusion) stage of the classifier using the intermediate features, disregarding the matrix context. (That is why this group of methods is called **class-indifferent**). Here we can use *any* classifier with the intermediate features as inputs and the class label $\mu(\mathbf{x})$ as the output.

The difficulty comes from the specific structure of the intermediate feature space. If all L classifiers are perfect (produce the correct crisp class label for every $\mathbf{x}$), then there will be no variance of the values of $DP(\mathbf{x})$ over the subset of the data from class ω_i. The covariance matrices for the classes (or the single covariance matrix for all classes) are therefore singular. Classifiers such as linear and quadratic discriminant classifiers, which are based on the assumption of normally distributed classes, will fail when trying to estimate and invert the covariance matrices. To get high overall accuracy, we try to use the most accurate individual classifiers. The higher the accuracy of all classifiers, the more likely it is that the covariance matrix of the intermediate features will be close to singular.

Example 8.3.7. Consider two classifiers D_1 and D_2 giving the values in Table 8.8 for 10 objects in $c = 2$ classes from a certain data set. Let all data points that we consider be labeled in class ω_1. The values were generated at random and independently, so that the expected support for class 1 from D_1 is in [0.9, 1.0], and from classifier D_2, in [0.8, 1.0]. Each four-tuple $[d_{1,1}(\mathbf{x}), d_{1,2}(\mathbf{x}), d_{2,1}(\mathbf{x}), d_{2,1}(\mathbf{x})]^T$ is a point in the intermediate feature space. The mean is

$$\mathbf{m} = (0.9475, 0.0525, 0.8830, 0.1170)^T,$$

and the covariance matrix is

$$S = \begin{bmatrix} 0.0007 & -0.0007 & -0.0004 & 0.0004 \\ -0.0007 & 0.0007 & 0.0004 & -0.0004 \\ -0.0004 & 0.0004 & 0.0019 & -0.0019 \\ 0.0004 & -0.0004 & -0.0019 & 0.0019 \end{bmatrix}.$$

To calculate a linear or quadratic discriminant function we have to invert S. In Matlab, a warning is displayed that S is close to singular or badly scaled and that the results may be inaccurate. The following matrix results:

Table 8.8. Intermediate feature values for class i from classifiers D_1 and D_2

D_1		D_2	
Class 1	Class 2	Class 1	Class 2
0.9807	0.0193	0.9007	0.0993
0.9318	0.0682	0.8200	0.1800
0.9697	0.0303	0.8357	0.1643
0.9458	0.0542	0.8710	0.1290
0.9849	0.0151	0.8364	0.1636
0.9302	0.0698	0.8680	0.1320
0.9622	0.0378	0.9316	0.0684
0.9140	0.0860	0.9421	0.0579
0.9146	0.0854	0.9318	0.0682
0.9406	0.0594	0.8932	0.1068

$$S^{-1} = \begin{bmatrix} 4.6117 & 4.6117 & 0.0000 & 0.0000 \\ 4.6117 & 4.6117 & -0.0000 & 0 \\ 0.0000 & 0 & 4.6117 & 4.6117 \\ 0.0000 & 0.0000 & 4.6117 & 4.6117 \end{bmatrix} \quad \underline{\mathbf{1.0e+18}}.$$

■

As the example shows, parametric models such as linear and quadratic discriminant classifiers (LDC and QDC) assuming normal densities [87], might not be a good brute-force option. The (semiparametric) logistic classifier (LOG) [16], and Fisher's discriminant [87] could be a better choice.

8.3.8 Decision templates

Assume that we know the desirable DP for each class in advance.

Example 8.3.8. Consider an example with $L = 3$ and $c = 4$. The most desirable (presumably) decision profile for class ω_3 is the "crisp" decision profile shown in Table 8.9. Then the aggregation rule $\mathcal{F}$ in (8.13) can be designed to measure the correspondence of the current $DP(\mathbf{x})$ to the "model" for class ω_i, $i = 1, \ldots, c$.

Table 8.9. Most desirable (presumably) decision profile for class ω_3

class →	1	2	3	4
$D_1(\mathbf{x})$	0	0	1	0
$D_2(\mathbf{x})$	0	0	1	0
$D_3(\mathbf{x})$	0	0	1	0

■

Definition 8.3.1. *The* **decision template** *[203] $DT_i(\mathbf{Z})$ of class ω_i is the $L \times c$ matrix $DT_i(\mathbf{Z}) = [dt_i(k,s)(\mathbf{Z})]$ whose (k,s)th element is computed by*

$$dt_i(k,s)(\mathbf{Z}) = \frac{\sum_{j=1}^{N} Ind(\mathbf{z}_j, \omega_i)\, d_{k,s}(\mathbf{z}_j)}{\sum_{j=1}^{N} Ind(\mathbf{z}_j, \omega_i)}, \quad k = 1, \ldots, L, \quad s = 1, \ldots, c, \tag{8.40}$$

where $Ind(\mathbf{z}_j, \omega_i)$ is the indicator function (3.2). To simplify notation, $DT_i(\mathbf{Z})$ will be denoted by DT_i.

Thus, the decision template DT_i for class ω_i is the average of the decision profiles of the elements of the training set $\mathbf{Z}$ from ω_i. When $\mathbf{x} \in \Re^n$ is submitted for classification, the DT scheme matches $DP(\mathbf{x})$ to $DT_i, i = 1, \ldots, c$, and produces the soft class labels

$$\mu_i(\mathbf{x}) = \mathcal{S}(DT_i, DP(\mathbf{x})), \quad i = 1, \ldots, c, \tag{8.41}$$

where $\mathcal{S}$ is interpreted as a *similarity* measure. The higher the similarity between the decision profile of the current $\mathbf{x}$ and the decision template for class ω_i (DT_i), the higher the support for ω_i ($\mu_i(\mathbf{x})$). Notice that we use the word "similarity" in a broad sense, meaning "degree of match" or "likeness", etc. Among the measures of similarity that we consider in the experiment are the four (proper) measures of similarity, the five inclusion indices, and the consistency measure defined via equations (4.75) - (4.88) in Chapter 4 (page 106). For intersection and union we use the minimum and maximum, respectively, and for complement, $\mu_{\bar{A}}(u) = 1 - \mu_A(u)$. However, there is no reason to prefer these measures. Since the general idea is to compare the matrix $DP(\mathbf{x})$ to c template matrices $(DT_1, \ldots, DT_c)$, any measure that does this might be appropriate.

The next decision template model is based on the normalized squared Euclidean distance between matrices DP and DT_i,

$$N(DP, DT_i) = \mu_i(\mathbf{x}) = 1 - \frac{1}{L\,c} \sum_{k=1}^{L} \sum_{s=1}^{c} (dt_i(k,s) - d_{k,s}(\mathbf{x}))^2. \tag{8.42}$$

While we use only the Euclidean norm in this study, there is no reason to stop at this choice. Any norm could be used in (8.42), e.g., the Minkowski norms or the Mahalanobis norm. It is important to notice the difference between *integral* and *point-wise* measures. Integral measures are based on cardinality $(S_1, S_2, S_3, I_1, I_2, I_3, N)$, while pointwise measures use a single degree of membership to determine their value (S_4, I_4, I_5, C). Therefore, point-wise measures tend to be more sensitive to outliers and prone to errors than integral measures. Figure 8.9 illustrates how the DT scheme operates.

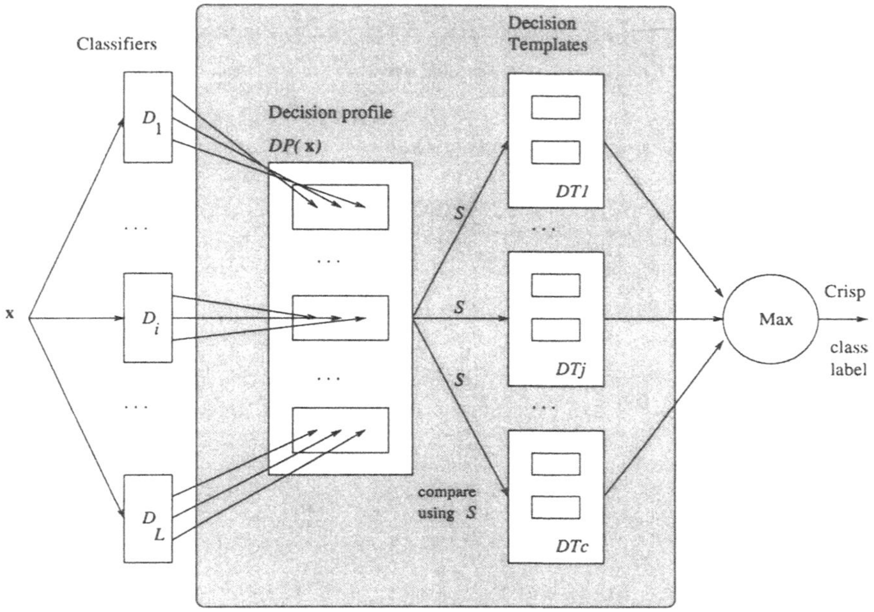

Fig. 8.9. Architecture of the decision templates classifier fusion scheme

Example 8.3.9. Let $c = 3$ and $L = 2$. Assume that the three decision templates are found to be

$$DT_1 = \begin{bmatrix} 0.7 & 0.2 & 0.4 \\ 0.4 & 0.3 & 0.3 \end{bmatrix}, \quad DT_2 = \begin{bmatrix} 0.4 & 0.6 & 0.1 \\ 0.5 & 0.5 & 0.1 \end{bmatrix}, \quad DT_3 = \begin{bmatrix} 0.3 & 0.4 & 0.7 \\ 0.1 & 0.5 & 0.8 \end{bmatrix}. \tag{8.43}$$

Let the decision profile for $\mathbf{x}$ be

$$DP(\mathbf{x}) = \begin{bmatrix} 0.1 & 0.2 & 0.6 \\ 0.2 & 0.8 & 0.4 \end{bmatrix}. \tag{8.44}$$

To find the class label for $\mathbf{x}$ we compare $DP(\mathbf{x})$ with each decision template. Choosing the consistency index C (4.88) to measure the similarity between the matrices (the maximum of the pairwise minimums of the corresponding elements), we obtain

$$\mu_1(\mathbf{x}) = 0.4, \quad \mu_2(\mathbf{x}) = 0.5, \quad \mu_3(\mathbf{x}) = 0.6. \tag{8.45}$$

The maximum membership rule will label $\mathbf{x}$ in ω_3. ■

8.3.9 Dempster-Shafer combination

The Dempster-Shafer fusion model [282] also belongs to the **class-indifferent** group as it uses the whole DP to calculate a single component of the output vector $\mu_i(\mathbf{x})$. The classifier outputs $D_i(\mathbf{x}), i = 1, \ldots, L$ are possibilistic labels. The following steps are performed:

1. Let DT_j^i denote the ith row of the decision template for class ω_j. We calculate the "proximity" Φ between DT_j^i and $D_i(\mathbf{x})$ for every class $j = 1, \ldots, c$ and for every classifier $i = 1, \ldots, L$. As recommended in [282], this proximity is calculated as

$$\Phi_{j,i}(\mathbf{x}) = \frac{\left(1 + ||DT_j^i - D_i(\mathbf{x})||^2\right)^{-1}}{\sum_{k=1}^{c} \left(1 + ||DT_k^i - D_i(\mathbf{x})||^2\right)^{-1}}, \tag{8.46}$$

   where $|| * ||$ is any matrix norm.
2. Using (8.46), we calculate for every class, $j = 1, \ldots, c$; and for every classifier, $i = 1, \ldots, L$, the following *belief* degrees

$$b_j\left(D_i(\mathbf{x})\right) = \frac{\Phi_{j,i}(\mathbf{x}) \prod_{k \neq j}(1 - \Phi_{k,i}(\mathbf{x}))}{1 - \Phi_{j,i}(\mathbf{x}) \left[1 - \prod_{k \neq j}(1 - \Phi_{k,i}(\mathbf{x}))\right]}.$$

3. The final DS label vector with membership degrees has the components

$$\mu_{\hat{D}}^j(\mathbf{x}) = K \prod_{i=1}^{L} b_j\left(D_i(\mathbf{x})\right), \quad j = 1, \ldots, c,$$

   where K is a normalizing constant.

8.4 Experimental results

8.4.1 Methods and experimental setup

As the first level classifiers we trained 10 multilayer perceptron (MLP) neural networks (see Chapter 3) with 10 hidden nodes at the single hidden layer. To get different classifiers, MLP training was started from a different initialization each time. Three of the four data sets were used: Cone-torus, Normal-mixtures and Phoneme. The Satimage data was left out because of the large amount of time and memory required to train a proper MLP for that data set. Besides, the current design (10 hidden nodes) failed to produce any reasonable classification accuracy for the Satimage data. Along with the class-conscious/class-indifferent grouping, we also distinguish between methods that do not need training at the second (aggregation) level and methods that need such training. The fusion models in the first group, denoted by (1), have no parameters that need to be tuned whereas the methods in the second group, denoted by (2), do. The fusion schemes studied here were as follows:

- **Group C(1)**
  1. MAJ. Majority voting[2].
- **Group C(2)**
  2. NB. Naive Bayes.
  3. BKS. Behavior Knowledge Space method.
- **Group CC(1)**
  4. MAX. Maximum aggregation rule.
  5. MIN. Minimum aggregation rule.
  6. AVR. Average aggregation rule.
  7. PRO. Product aggregation rule.
- **Group CC(2)**
  8. PPR. Probabilistic product.
  9. FI. Fuzzy integral.
- **Group CI(2)**
  10. DS. Dempster-Shafer.
  11. LDC. Linear discriminant classifier on the intermediate-output space.
  12. QDC. Quadratic discriminant classifier on the intermediate-output space.
  13. LOG. Logistic classifier on the intermediate-output space.
- **Group CI(2) (DTs)** Each DT scheme is denoted by "DT:ss", where "ss" stands for the respective similarity measure, e.g., DT:I5. By DT:NM we denote the DT scheme based on the Euclidean distance. "NM" stands for the "nearest mean". Thus, there are a total of 11 DT schemes.
- **Classifier selection**
  25. CS. Clustering and selection. ($K = 8$ for the Cone-torus and Phoneme data, and 4 for the Normal-mixtures data)
  26. CWS. Clustering and weighted sum. ($K = 8$ for all experiments)
- **Reference models**
  27. SB. The single best (individual) accuracy.
  28. OR. The oracle. The **oracle** works as follows: assign the *correct* class label to **x** iff at least one individual classifier produces the correct class label of **x** (when its decision is hardened).

Tables 8.10, 8.11 and 8.12 show the classification accuracy for the three data sets, respectively. Not surprisingly, the fusion schemes have approximately the same performance. To find out which were consistently better (even a little better) than the others, we sorted the testing accuracies and calculated their ranks. To the right of classification accuracy in each column

[2] We did not consider the reject option – all ties were broken randomly.

is the ***rank*** of the fusion scheme, based on that column. For an individual test, the ranks range from 1 (poorest) to 26 (best). The last column in each table is the total rank (the sum of the two) for the respective data set. The overall ranks from the training and testing experiments with the three data sets are shown in Table 8.13. Table 8.14 shows the 28 schemes sorted by the sum of their ranks (on test error only). Since there were three data sets, the maximum possible score is $28 \times 3 = 84$, which is achieved by the oracle. If the same classifier had ranked lowest at all 3 tests, the minimum cumulative value of 3 would be attained, but this did not happen. The single best accuracy (SB) is taken to be the benchmark because there is no point applying combination schemes that perform less accurately than the best individual classifier in the pool $\mathcal{D}$.

Table 8.10. Error rates and ranks for the **Cone-torus** data

Method	Training		Test		Total
	error	rank	error	rank	rank
Maj	0.1650	8	0.1350	12.5	20.5
NB	0.1575	17.5	0.1275	22.5	40.0
BKS	0.1025	26	0.1325	18	44.0
Max	0.1675	6	0.1200	24	30.0
Min	0.1625	9.5	0.1450	5.5	15.0
Avr	0.1675	6	0.1325	18	24.0
Pr	0.1700	4	0.1400	7.5	11.5
PPr	0.1600	12	0.1600	3	15.0
FI	0.1350	23	0.1100	25.5	48.5
DS	0.1575	17.5	0.1350	12.5	30.0
LDC	0.1250	24	0.1050	27	51.0
QDC	0.0975	27	0.1100	25.5	52.5
LOG	0.4150	2	0.5225	1	3.0
DT:I1	0.1575	17.5	0.1350	12.5	30.0
DT:I2	0.1575	17.5	0.1350	12.5	30.0
DT:I3	0.1675	6	0.1550	4	10.0
DT:I4	0.1600	12	0.1300	20.5	32.5
DT:I5	0.1725	3	0.1300	20.5	23.5
DT:S1	0.1575	17.5	0.1350	12.5	30.0
DT:S2	0.1575	17.5	0.1350	12.5	30.0
DT:S3	0.1625	9.5	0.1325	18	27.5
DT:S4	0.4375	1	0.4000	2	3.0
DT:C	0.1600	12	0.1275	22.5	34.5
DT:NM	0.1575	17.5	0.1350	12.5	30.0
CS	0.1150	25	0.1400	7.5	32.5
CWS	0.1575	17.5	0.1350	12.5	30.0
SB	0.1450	22	0.1450	5.5	27.5
OR	0.0700	28	0.0800	28	56.0

Table 8.11. Error rates and ranks for the **Normal-mixtures** data

Method	Training		Test		Total
	error	rank	error	rank	rank
Maj	0.116	3	0.096	7.5	10.5
NB	0.120	1.5	0.098	5	6.5
BKS	0.092	26	0.097	6	32.0
Max	0.112	12.5	0.094	22	34.5
Min	0.112	12.5	0.094	22	34.5
Avr	0.112	12.5	0.094	22	34.5
Pr	0.112	12.5	0.095	13.5	26.0
PPr	0.112	12.5	0.095	13.5	26.0
FI	0.108	23	0.108	3	26.0
DS	0.112	12.5	0.095	13.5	26.0
LDC	0.096	25	0.100	4	29.0
QDC	0.120	1.5	0.110	2	3.5
LOG	0.108	23	0.128	1	24.0
DT:I1	0.112	12.5	0.095	13.5	26.0
DT:I2	0.112	12.5	0.095	13.5	26.0
DT:I3	0.112	12.5	0.094	22	34.5
DT:I4	0.112	12.5	0.096	7.5	20.0
DT:I5	0.112	12.5	0.094	22	34.5
DT:S1	0.112	12.5	0.095	13.5	26.0
DT:S2	0.112	12.5	0.095	13.5	26.0
DT:S3	0.112	12.5	0.094	22	34.5
DT:S4	0.112	12.5	0.093	26	38.5
DT:C	0.112	12.5	0.094	22	34.5
DT:NM	0.112	12.5	0.095	13.5	26.0
CS	0.088	27	0.092	27	54.0
CWS	0.112	12.5	0.095	13.5	26.0
SB	0.108	23	0.095	13.5	36.5
OR	0.080	28	0.060	28	56.0

8.4.2 Discussion

As usual, we shall refrain ourselves from definitive judging on the basis of a set of experiments. Bearing this in mind, some observations are detailed below.

Overall classification accuracy. We did not confer special attention on designing the individual first-level classifiers. We were interested in comparing the second-level (fusion and selection) schemes, so the type of first-level classifiers was immaterial. Potentially, the better the individual classifiers, the better the combination scheme.

Improvement over the single-best classifier. The gap between the single best classifier and the oracle presumably shows the "potential" of the pool $\mathcal{D}$. Many of the fusion schemes did not improve very much on the single-best classifier rate. This is probably due to dependencies between the classifiers. If we used a large number of features and built the classifiers on disjoint

Table 8.12. Error rates and ranks for the **Phoneme** data

Method	Training		Test		Total
	error	rank	error	rank	rank
Maj	0.136	21	0.1913	7	28.0
NB	0.160	4	0.2037	3	7.0
BKS	0.082	27	0.2002	4	31.0
Max	0.142	18	0.1921	6	24.0
Min	0.144	15	0.1884	13	28.0
Avr	0.144	15	0.1890	11	26.0
Pr	0.144	15	0.1892	10	25.0
PPr	0.186	1.5	0.1988	5	6.5
FI	0.146	11.5	0.1909	8	19.5
DS	0.148	8	0.1799	18	26.0
LDC	0.114	26	0.1760	25.5	51.5
QDC	0.128	23	0.1756	27	50.0
LOG	0.118	25	0.1809	17	42.0
DT:I1	0.148	8	0.1790	22.5	30.5
DT:I2	0.148	8	0.1790	22.5	30.5
DT:I3	0.186	1.5	0.2290	1	2.5
DT:I4	0.134	22	0.1760	25.5	47.5
DT:I5	0.140	19.5	0.1764	24	43.5
DT:S1	0.148	8	0.1794	20.5	28.5
DT:S2	0.148	8	0.1794	20.5	28.5
DT:S3	0.144	15	0.1905	9	24.0
DT:S4	0.158	5	0.1829	15.5	20.5
DT:C	0.164	3	0.2123	2	5.0
DT:NM	0.146	11.5	0.1796	19	30.5
CS	0.120	24	0.1841	14	38.0
CWS	0.144	15	0.1888	12	27.0
SB	0.140	19.5	0.1829	15.5	35.0
OR	0.046	28	0.0675	28	56.0

data subsets, the chance to obtain good improvement over the single best classifier would have been higher. The best improvement on the test rate for the Cone-torus data (Table 8.10) was 4 % achieved by the linear discriminant classifier over the whole decision profile (LDC, from the "brute force" group). For the Normal mixtures data (Table 8.11), the improvement was virtually non-existing: 0.3 % offered by Clustering and Selection (CS) method. For the Phoneme data (Table 8.12), the improvement was also negligible, 0.73 %, again produced by a method from the "brute-force" group: the quadratic discriminant analysis (QDC).

Overtraining and number of parameters. BKS appeared to be most prone to overtraining because its look-up table needs large data sets in order to be properly calculated. Thus, the BKS method gave good training accuracy but did badly on testing. The BKS overtraining problem is especially severe with a large number of classes and classifiers: the number of parameters (cells in the table) is L^c. Not all combinations will be encountered in practice, but a

Table 8.13. Overall ranks of the combination schemes

Method	Training	Test	Total
Maj	32.0	27.0	59.0
NB	23.0	30.5	53.5
BKS	79.0	28.0	107.0
Max	36.5	52.0	88.5
Min	37.0	40.5	77.5
Avr	33.5	51.0	84.5
Pr	31.5	31.0	62.5
PPr	26.0	21.5	47.5
FI	57.5	36.5	94.0
DS	38.0	44.0	82.0
LDC	75.0	56.5	131.5
QDC	51.5	54.5	106.0
LOG	50.0	19.0	69.0
DT:I1	38.0	48.5	86.5
DT:I2	38.0	48.5	86.5
DT:I3	20.0	27.0	47.0
DT:I4	46.5	53.5	100.0
DT:I5	35.0	66.5	101.5
DT:S1	38.0	46.5	84.5
DT:S2	38.0	46.5	84.5
DT:S3	37.0	49.0	86.0
DT:S4	18.5	43.5	62.0
DT:C	27.5	46.5	74.0
DT:NM	41.5	45.0	86.5
CS	76.0	48.5	124.5
CWS	45.0	38.0	83.0
SB	64.5	34.5	99.0
OR	84.0	84.0	168.0

large number of them might be. To compare the complexity, DT, DS, and NB acquire $L \times c^2$ parameters; FI, $L + 1$; PPR, L; while the simple aggregation techniques need to learn none.

Brute force classifiers. LDC and QDC are not (theoretically) appropriate on the intermediate space because the covariance matrices needed for these designs might be close to singular, as demonstrated in example 8.3.7. However, in our experiments, these designs outperformed most of the other combination methods. The logistic classifier does not share the singularity drawback. On the other hand, LOG might fail when some of the classes are almost separable in the intermediate space (i.e., the classifiers had very distinct decision templates), as it happens with the Cone-Torus data. It would be interesting to look at other conventional classifiers, for example decision trees or nearest neighbor and multiple prototype rules for the intermediate-output space. We may also choose to train a neural network (MLP) as the combiner.

Table 8.14. Fusion schemes sorted by their total ranks (over the three data sets) on test error rates

Method	Rank
LOG	19.0
PPr	21.5
Maj	27.0
DT:I3	27.0
BKS	28.0
NB	30.5
Pr	31.0
SB	34.5
FI	36.5
CWS	38.0
Min	40.5
DT:S4	43.5
DS	44.0
DT:NM	45.0
DT:S1	46.5
DT:S2	46.5
DT:C	46.5
CS	48.5
DT:I1	48.5
DT:I2	48.5
DT:S3	49.0
Avr	51.0
Max	52.0
DT:I4	53.5
QDC	54.5
LDC	56.5
DT:I5	66.5
OR	84.0

Assumption-based schemes. The assumption-based classifier fusion schemes, Naive Bayes and the probabilistic product, did not reach the performance of the other schemes.

Dempster-Shafer method [282]. This method rated moderately high. The calculations that it involves are more complex than any of the DT schemes, but in our experiments DS did not offer a gain in accuracy to justify the increased complexity.

Simple aggregation rules. It is somewhat surprising to see how well the simple aggregation rules compete with the more sophisticated ones. This is probably the reason that simple aggregation continues to be popular [172, 319]. One problem with simple aggregation is, that although they have good overall performance, it is not clear which one is good for a particular data set.

Fuzzy integral. In our experiments the fuzzy integral using a λ-fuzzy measure rates in the middle. Gader et al. [102] report the results from a handwritten word recognition problem, where the fuzzy integral dramatically outperforms various neural networks. The authors attribute this to the efficient way in which the FI fusion model uses the additional information (called here class-consciousness). This shows again that there is no "perfect" classifier or fusion scheme that will surpass the others on all data sets.

Decision templates. In our experiments, DT classifier fusion schemes based on *integral* measures tended to give good results with all data sets. The overall ranking (Table 8.12) puts 5 of them among the best ones. It is proven in [186] that if the output of each individual classifier sums up to 1, the four DT schemes based on S_1, S_2, I_1, and I_2 are indistinguishable (i.e., induce the same order on the set of class labels, and therefore, lead to the same decision about the class label of $\mathbf{x}$). In the experiments here, the NN outputs were not necessarily scaled to sum up to 1, and the four models exhibit different performance. The favourite amongst the DT's appears to be DT:I5 which rates on the top of the list, immediately before the brute-force LDC and QDC. DTs appear to be a robust fusion technique, not relying on questionable assumptions (as do NB,PPR), not suffering from overtraining (as does BKS), or failing for some of the data sets (as happened with LOG, PRO, MIN). The fusion scheme is simple and intuitive, and does not require heavy calculations.

Classifier selection models. From the two designs in this group, better accuracy was shown by the Clustering-and-Selection (CS) one. This model is simple, and also rates amongst the best. We can combine selection and fusion, e.g., by choosing the best group of classifiers and and fusing their outputs only.

Lastly, *why are selection and fusion models considered to be fuzzy classifiers.* At the beginning we stated that any classifier that uses fuzzy sets during its training and/or operating, is a fuzzy classifier. We use fuzzy sets explicitly in many of the combination designs, e.g., 4–7, 9, 14–24. Other designs, such as DS and CWS can be described using fuzzy sets, and therefore enriched by the choice of fuzzy operations. Consensus-based aggregation, discussed in Chapter 4, is another possible direction in classifier combination. The "context-free" application of fuzzy sets here and in many designs in the previous chapters highlights a niche for fuzzy set theory. Since fuzzy sets have been used merely as a mathematical tool and have not been expected to model natural language, vagueness of human reasoning, uncertainty of everyday life, etc., there are no unfulfilled promises and high hopes that do not materialize. From this perspective, fuzzy sets might look less inspiring and romantic but they are certainly useful.

9. Conclusions: What to choose?

Fuzzy or non-fuzzy? This is always the question. Remember the motto on the first page of Chapter 4 (Fuzzy sets)? It continues as follows:

> The people's doctor Praying Mantis started stretching out his dry as grass hands to touch Buratino. 'One of the two,' he whispered, 'the patient is *either* alive *or* dead'.

So, despite the intrinsic vagueness of the opinions (class labels), the problem where the two classes *are* mutually exclusive, like life and death, demands only *one crisp* class label. Such is the case when the label determines the further action. For example, the distinction between benign and malignant tumors is vital as to how to continue the treatment of a patient. A fuzzy label like "rather benign than malignant" is of little use. We may then come to the (wrong) idea that fuzzy classifiers have a very limited scope. However, as we agreed at the beginning, fuzzy classifiers are not necessarily those that produce a fuzzy class label at their output. "Fuzzy" can be hidden in the "black box". Although claiming transparency, many fuzzy if-then designs are actually black boxes using simple concepts and operations from fuzzy set theory. So, let us for a moment forget the claims about interpretability and modeling the human reasoning, and put fuzzy versus non-fuzzy classifiers on the same scale. On the one side are the well-studied statistical models. We know their asymptotic properties and have a good insight about their finite-sample behavior. On the other side are highly ad-hoc and wonderfully rich fuzzy models. I have carried out the experiments devotedly and accurately, as much as I could, with no hidden agenda in mind. (It is much easier when you do not have the thankless task to get better results with *your own* brand new design.) The experimental results do not select a clear winner, suggesting that we cannot afford to dismiss any of the two groups. As in many other disciplines in science, diversity of the models in the toolbox is an asset and not a liability. Apparently, there are models within each of the two groups which are too cumbersome and not adequately accurate. Admittedly, more of such models can be found within the fuzzy group, and this is inevitable given its heuristic nature. On the other hand, greater success can be expected from this group in the finite sample case (the only real-life case), again owing to its heuristic nature. My personal preference goes with the simple models. Edited nearest neighbor and decision templates for classifier fusion are amongst my favorites. However, it is you, my reader, who has to choose from the palette devised in the previous eight chapters. You might be as confused about what to pick as when you opened up the book but at least now you should have a better idea about what is on offer. Do you?

A. Appendix: Numerical results

A.1 Cone-torus data

Data description

Data type:	synthetic
Number of classes c	3
Number of features n:	2
Training set size N_{tr}:	400
Testing set size N_{ts}:	400
Plot:	

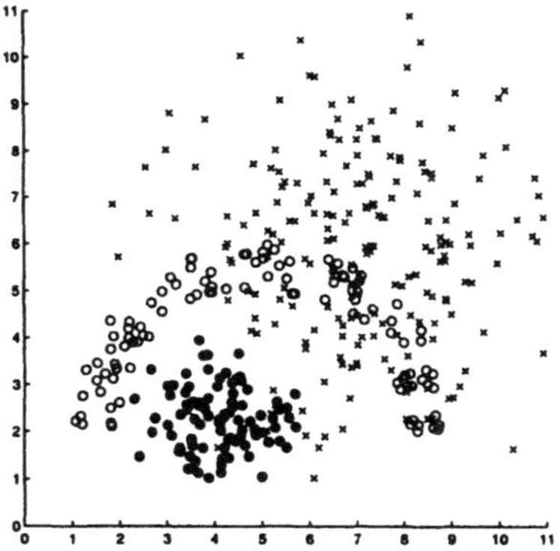

Classical models

Table A.1. Error rates (in %) with five classical pattern recognition models

Classifier	Training error	Testing error
Nearest mean	29.5	26.25
Linear discriminant classifier (LDC)	26.50	23.00
Quadratic discriminant classifier (QDC)	19.50	16.75
Parzen	8.75	12.25
Nearest neighbor (1-nn)	17.25	15.25

Neural network models

Table A.2. MLP classification error (in%). (Fast backpropagation used from the Neural Networks Toolbox of MATLAB 4.2)

MLP configuration	Training error	Testing error
2-5-3	18.00	15.75
2-10-3	15.25	14.25
2-15-3	13.50	12.00
2-20-3	15.75	13.75

Table A.3. RBF network classification error (in%). (Orthogonal Least Squares training used from the Neural Networks Toolbox of MATLAB 4.2)

RBF	$s=1$		$s=1.5$		$s=2$		$s=3$	
configuration	Tr	Ts	Tr	Ts	Tr	Ts	Tr	Ts
2-5-3	17.00	17.00	18.00	17.25	17.00	15.50	19.25	16.50
2-10-3	15.75	13.50	14.00	12.75	14.25	14.25	16.50	13.75
2-15-3	15.25	13.75	12.75	12.00	14.00	13.00	17.00	14.00
2-20-3	11.50	11.00	12.25	10.50	13.75	11.50	14.50	13.75

Table A.4. LVQ1 classification error (in%). (5 independent runs)

Run	10 prototypes		20 prototypes	
number	Tr	Ts	Tr	Ts
1	22.50	22.25	18.50	17.00
2	29.50	28.25	17.00	16.00
3	23.00	22.25	17.00	16.75
4	32.00	32.25	22.75	21.00
5	20.25	17.50	15.50	14.50

Fuzzy TSK models

Table A.5. Training and testing error rates of fuzzy TSK5 model with Average support method for tuning the consequents

Resolution t	Training error [%]	Testing error [%]
2	47.75	53.50
3	31.00	25.75
4	20.00	18.50
5	21.00	18.75
6	17.25	16.50
7	17.50	16.50
8	13.50	12.50
9	15.75	15.00
10	13.75	10.75
11	12.00	15.25
12	13.00	11.00
13	13.50	13.00
14	12.00	12.25
15	12.50	13.25

Table A.6. Training and testing error rates of fuzzy TSK2 model with Maximum support method for tuning the consequents (Wang-Mendel's model)

Resolution t	Training error [%]	Testing error [%]
4	41.50	47.25
9	41.00	47.75
16	24.00	21.75
25	16.50	18.75
36	17.25	15.25
49	18.75	17.75
64	15.75	13.00
81	14.00	16.00
100	13.50	13.00
121	12.75	14.50
144	14.25	11.75
169	12.75	13.75
196	10.50	09.75
225	12.00	11.25
256	12.00	11.75
289	12.00	17.00

Table A.7. Training and testing error rates of fuzzy TSK5 model with the histogram method for tuning the consequents

Resolution t	Training error [%]	Testing error [%]
3	39.00	37.75
8	22.25	24.00
13	15.75	18.25
18	18.25	19.25
23	18.00	19.75
28	16.50	19.25
33	19.00	24.75
38	15.50	18.75
43	14.25	19.00
48	16.25	20.25

Table A.8. Fuzzy TSK classifier with clustering for rule-base design. Shown are the best training and the corresponding test error (displayed underneath) of 10 independent runs in $M = 6$, 9, 12 and 15 clusters (rules). Both values are in %. The **whole data set** is clustered, followed by labeling of the clusters by three labeling procedures: Crisp labeling (Mode=1), Soft singleton labeling (Mode=2) and Fuzzy labeling (Mode=3). Two types of membership functions have been used: triangular($\triangle$) and Gaussian (G) (see the text, Chapter 5, page 182). All ties are counted as errors.

Implication and type of μ	Mode	$M = 6$	$M = 9$	$M = 12$	$M = 15$
minimum $\triangle$	1	25.00 23.25	25.00 21.50	24.25 19.50	26.00 25.75
product $\triangle$	1	25.75 21.00	25.25 23.75	24.25 19.75	27.25 27.00
minimum $\triangle$	2	23.25 22.25	25.25 29.50	25.00 23.75	26.25 23.25
product $\triangle$	2	24.25 22.75	25.50 27.25	24.25 21.50	27.00 23.25
minimum $\triangle$	3	33.75 29.00	28.50 28.25	29.25 28.00	27.25 28.50
product $\triangle$	3	25.50 21.75	24.75 20.25	24.50 19.25	24.50 20.50
minimum G	1	23.00 22.50	21.00 21.25	18.25 20.00	17.25 17.25
product G	1	23.00 24.00	19.50 21.25	18.50 18.25	18.75 18.00
minimum G	2	23.00 22.50	21.00 19.00	18.25 18.00	16.25 16.25
product G	2	21.50 19.50	21.75 23.25	17.75 18.00	17.75 17.00
minimum G	3	30.00 26.25	27.75 27.00	25.25 23.75	21.25 20.25
product G	3	22.75 21.75	21.50 23.75	19.00 18.75	17.00 17.25

Table A.9. Fuzzy TSK classifier with clustering for rule-base design with triangular membership functions (△). Shown are the best training and the corresponding test error (displayed underneath) of 10 independent runs in $M = 6$, 9, 12 and 15 clusters (rules). Both values are in %. The **separate clustering** is used, followed by labeling of the clusters by five labeling procedures: Crisp labeling (Mode=1), Soft singleton labeling (Mode=2), Fuzzy labeling (Mode=3), Crisp own label (Mode=4), Soft singleton own label (Mode=5) (For details see the text, Chapter 5, page 182). All ties are counted as errors.

Implication	Mode	$M = 6$	$M = 9$	$M = 12$	$M = 15$
minimum	1	20.75	23.00	23.50	22.25
		19.25	17.75	16.75	21.00
product	1	21.75	22.50	23.75	24.75
		20.25	17.00	23.50	23.75
minimum	2	21.50	23.00	23.00	24.25
		20.00	17.25	20.50	21.50
product	2	20.75	21.50	23.50	22.50
		18.50	16.25	19.00	26.25
minimum	3	27.75	28.75	27.25	27.50
		24.50	24.50	25.50	25.00
product	3	20.50	20.50	23.25	20.50
		18.00	18.00	21.25	21.50
minimum	4	21.25	22.75	24.75	24.00
		19.25	17.00	23.50	26.75
product	4	21.75	22.75	24.25	24.75
		19.00	17.00	22.50	24.25
minimum	5	22.25	23.75	25.75	26.25
		19.50	16.75	22.00	27.50
product	5	23.00	29.75	27.50	30.50
		21.00	26.50	22.25	28.00

Table A.10. Fuzzy TSK classifier with clustering for rule-base design with Gaussian membership functions (**G**). Shown are the best training and the corresponding test error (displayed underneath) of 10 independent runs in M = 6, 9, 12 and 15 clusters (rules). Both values are in %. The **separate clustering** is used, followed by labeling of the clusters by five labeling procedures: Crisp labeling (Mode=1), Soft singleton labeling (Mode=2), Fuzzy labeling (Mode=3), Crisp own label (Mode=4), Soft singleton own label (Mode=5) (For details see the text, Chapter 5, page 182). No ties have occurred.

Implication	Mode	$M = 6$	$M = 9$	$M = 12$	$M = 15$
minimum	1	19.00	18.50	16.50	16.50
		17.75	16.75	15.50	16.25
product	1	18.75	17.50	17.25	16.75
		18.25	16.25	14.00	19.75
minimum	2	19.50	18.00	18.00	18.25
		18.50	16.25	19.25	19.75
product	2	19.00	17.50	16.50	18.50
		20.00	15.50	15.50	17.25
minimum	3	36.00	30.00	27.75	26.00
		29.25	22.00	25.00	25.75
product	3	18.50	17.50	16.50	17.25
		17.25	16.00	15.50	17.75
minimum	4	19.25	17.75	16.25	17.50
		18.75	16.00	16.50	19.25
product	4	19.25	17.75	16.75	19.25
		18.75	16.00	17.25	18.50
minimum	5	19.25	18.00	17.25	14.50
		19.50	15.25	17.25	19.25
product	5	21.00	21.00	17.00	19.75
		19.00	19.75	16.00	18.25

A.2 Normal mixtures data

Data description

Data type:	synthetic
Number of classes c	2
Number of features n:	2
Training set size N_{tr}:	250
Testing set size N_{ts}:	1000
Plot:	

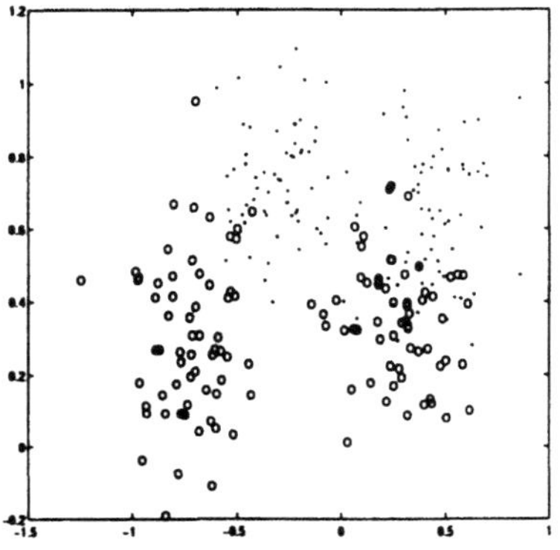

Classical models

Table A.11. Error rates (in %) with five classical pattern recognition models

Classifier	Training error	Testing error
Nearest mean	28.8	28.7
Linear discriminant classifier (LDC)	14.14	10.8
Quadratic discriminant classifier (QDC)	13.6	10.2
Parzen	9.6	10.9
Nearest neighbor (1-nn)	14.8	15.0

Neural networks

Table A.12. MLP classification error (in%). (Fast backpropagation used from the Neural Networks Toolbox of MATLAB 4.2)

MLP configuration	Training error	Testing error
2-5-2	12.00	10.00
2-10-2	12.80	10.30
2-15-2	13.20	10.50
2-20-2	11.60	9.20

Table A.13. RBF network classification error (in%). (Orthogonal Least Squares training used from the Neural Networks Toolbox of MATLAB 4.2)

RBF configuration	$s = 0.5$		$s = 1$		$s = 1.5$	
	Tr	Ts	Tr	Ts	Tr	Ts
2-5-2	12.40	9.10	12.80	9.40	14.00	9.50
2-10-2	11.20	10.10	11.60	9.90	12.40	10.00
2-15-2	11.20	9.50	10.80	10.00	11.60	10.60
2-20-2	9.20	9.20	10.40	9.30	11.60	10.30

Table A.14. LVQ1 classification error (in%). (5 independent runs)

Run number	10 prototypes		20 prototypes	
	Tr	Ts	Tr	Ts
1	12.40	9.70	13.20	13.30
2	10.00	9.70	10.40	14.20
3	14.00	12.80	12.80	15.20
4	11.20	8.80	13.60	12.70
5	13.60	10.00	10.00	8.60

Fuzzy TSK models

Table A.15. Training and testing error rates of fuzzy TSK5 model with Average support method for tuning the consequents

Resolution t	Training error [%]	Testing error [%]
2	24.80	23.10
3	18.40	13.50
4	15.60	11.20
5	13.60	11.10
6	12.00	9.50
7	15.60	10.00
8	11.20	8.60
9	12.00	9.30
10	11.20	9.20
11	12.40	10.30
12	9.20	9.80
13	9.60	12.40
14	10.00	11.80
15	9.20	10.90

Table A.16. Training and testing error rates of fuzzy TSK2 model with Maximum support method for tuning the consequents (Wang-Mendel's model)

Resolution t	Training error [%]	Testing error [%]
4	26.80	27.50
9	17.60	15.20
16	15.20	13.80
25	15.20	10.20
36	12.80	11.00
49	12.40	14.80
64	12.80	8.90
81	13.60	11.00
100	12.40	10.50
121	11.60	8.60
144	13.60	12.30
169	10.40	10.90
196	12.80	14.50
225	8.40	10.10
256	11.20	12.30
289	14.40	12.50

Table A.17. Training and testing error rates of fuzzy TSK5 model with the histogram method for tuning the consequents

Resolution t	Training error [%]	Testing error [%]
3	26.80	24.50
8	13.60	9.40
13	14.00	9.30
18	13.60	9.50
23	11.60	9.30
28	12.80	11.10
33	10.80	10.70
38	10.80	12.20
43	11.60	11.50
48	12.00	13.10

Table A.18. Fuzzy TSK classifier with clustering for rule-base design. Shown are the best training and the corresponding test error (displayed underneath) of 10 independent runs in M = 6, 9, 12 and 15 clusters (rules). Both values are in %. The **whole data set** is clustered, followed by labeling of the clusters by three labeling procedures: Crisp labeling (Mode=1), Soft singleton labeling (Mode=2) and Fuzzy labeling (Mode=3). Two types of membership functions have been used: triangular(Δ) and Gaussian (G) (see the text, Chapter 5, page 182). All ties are counted as errors.

Implication and type of μ	Mode	$M = 4$	$M = 6$	$M = 8$	$M = 10$
minimum Δ	1	18.40 17.70	18.00 15.80	19.60 16.30	22.80 24.80
product Δ	1	18.40 17.70	19.20 16.70	22.00 25.20	22.40 24.00
minimum Δ	2	18.40 17.70	19.60 18.40	22.00 18.10	23.20 26.60
product Δ	2	18.00 17.20	19.20 16.90	19.60 17.20	22.80 20.30
minimum Δ	3	19.60 20.80	20.40 22.80	23.20 29.00	24.00 26.60
product Δ	3	18.10 17.30	19.20 18.90	22.40 23.40	22.00 24.20
minimum G	1	13.20 9.30	12.80 11.80	13.60 13.90	14.00 10.80
product G	1	13.20 9.30	12.80 12.30	12.40 13.20	12.80 13.50
minimum G	2	13.20 9.30	12.00 9.10	12.00 11.20	13.20 11.50
product G	2	13.20 9.20	12.80 9.70	13.20 11.70	11.60 9.60
minimum G	3	18.80 19.40	19.60 21.80	19.60 20.80	18.40 21.20
product G	3	13.20 9.20	11.60 8.90	12.80 12.50	14.40 14.00

Table A.19. Fuzzy TSK classifier with clustering for rule-base design with triangular membership functions (△). Shown are the best training and the corresponding test error (displayed underneath) of 10 independent runs in M = 6, 9, 12 and 15 clusters (rules). Both values are in %. The **separate clustering** is used, followed by labeling of the clusters by five labeling procedures: Crisp labeling (Mode=1), Soft singleton labeling (Mode=2), Fuzzy labeling (Mode=3), Crisp own label (Mode=4), Soft singleton own label (Mode=5) (For details see the text, Chapter 5, page 182). All ties are counted as errors.

Implication	Mode	$M = 4$	$M = 6$	$M = 8$	$M = 10$
minimum	1	16.00 15.70	16.80 16.00	19.20 20.40	21.20 26.50
product	1	16.00 15.70	16.40 15.50	17.60 18.60	21.60 22.90
minimum	2	16.00 15.70	16.80 15.80	18.40 17.30	21.20 23.40
product	2	16.00 15.50	16.10 15.70	18.00 19.40	20.40 28.20
minimum	3	18.40 21.70	20.40 24.80	20.00 25.20	23.60 29.20
product	3	16.00 15.50	16.40 17.40	19.60 21.00	21.20 26.00
minimum	4	16.00 15.70	16.00 15.60	20.40 23.40	21.20 25.10
product	4	16.00 15.70	16.40 15.50	20.40 23.10	19.20 21.80
minimum	5	16.00 15.70	16.40 15.50	19.60 19.10	20.00 25.10
product	5	15.60 15.80	18.40 20.90	20.00 22.00	22.40 22.30

Table A.20. Fuzzy TSK classifier with clustering for rule-base design with Gaussian membership functions (**G**). Shown are the best training and the corresponding test error (displayed underneath) of 10 independent runs in M = 6, 9, 12 and 15 clusters (rules). Both values are in %. The **separate clustering** is used, followed by labeling of the clusters by five labeling procedures: Crisp labeling (Mode=1), Soft singleton labeling (Mode=2), Fuzzy labeling (Mode=3), Crisp own label (Mode=4), Soft singleton own label (Mode=5) (For details see the text, Chapter 5, page 182). No ties have occurred.

Implication	Mode	$M = 4$	$M = 6$	$M = 8$	$M = 10$
minimum	1	10.40	12.00	11.60	10.80
		9.00	10.10	11.90	11.50
product	1	10.40	11.60	9.60	10.40
		9.00	10.00	12.00	13.40
minimum	2	10.40	11.60	10.40	9.60
		9.00	9.80	11.60	11.40
product	2	10.00	10.80	10.40	10.80
		9.10	10.10	10.90	11.20
minimum	3	19.20	16.80	18.80	18.80
		20.90	19.00	22.90	23.00
product	3	10.00	11.60	10.80	11.60
		9.10	9.60	11.80	10.80
minimum	4	10.40	12.40	10.80	10.80
		9.00	11.20	11.70	12.70
product	4	10.40	10.80	10.40	10.80
		9.00	10.10	11.90	13.70
minimum	5	10.40	11.20	10.40	11.20
		9.00	10.60	11.50	10.90
product	5	11.20	12.80	12.40	13.20
		9.50	9.80	13.50	12.50

A.3 Phoneme data

Data type:	real
Number of classes c	2
Number of features n:	5
Training set size N_{tr}:	500
Testing set size N_{ts}:	4904
Plot:	

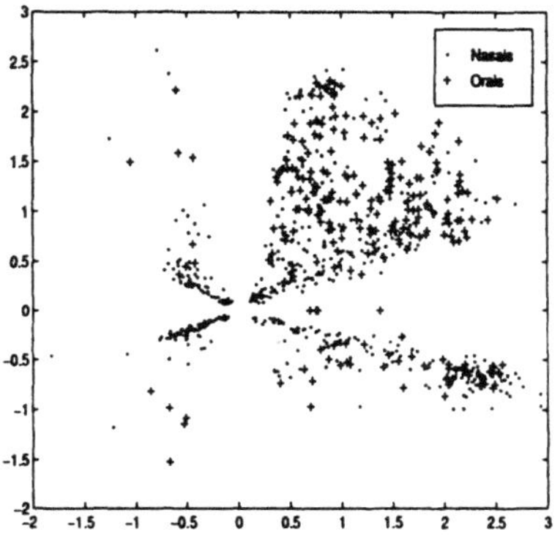

Classical models

Table A.21. Error rates (in %) with five classical pattern recognition models

Classifier	Training error	Testing error
Nearest mean	31.40	28.61
Linear discriminant classifier (LDC)	26.40	25.92
Quadratic discriminant classifier (QDC)	23.40	22.90
Parzen	10.80	18.56
Nearest neighbor (1-nn)	17.60	17.70

Neural networks

Table A.22. MLP classification error (in%). (Fast backpropagation used from the Neural Networks Toolbox of MATLAB 4.2)

MLP configuration	Training error	Testing error
5-5-2	14.00	18.23
5-10-2	16.80	21.04
5-15-2	20.20	23.29
5-20-2	16.60	19.09

Table A.23. RBF network classification error (in%). (Orthogonal Least Squares training used from the Neural Networks Toolbox of MATLAB 4.2)

RBF	$s = 1$		$s = 1.5$		$s = 2$	
configuration	Tr	Ts	Tr	Ts	Tr	Ts
5-5-2	20.00	23.90	21.80	24.37	21.00	23.65
5-10-2	18.20	20.74	17.00	21.55	18.80	21.31
5-15-2	14.80	19.92	15.80	19.72	15.80	18.58
5-20-2	13.60	19.23	13.80	18.37	14.60	18.13

Table A.24. LVQ1 classification error (in%). (5 independent runs)

Run	10 prototypes		20 prototypes	
number	Tr	Ts	Tr	Ts
1	22.40	23.00	16.60	21.80
2	22.00	23.82	15.40	21.53
3	18.60	21.72	15.80	22.43
4	22.00	23.63	16.20	21.94
5	19.00	23.47	17.80	20.68

Table A.25. Training and testing error rates of fuzzy TSK5 model with the histogram method for tuning the consequents

Resolution t	Training error [%]	Testing error [%]
3	32.00	32.40
8	26.20	25.22
13	18.00	20.98
18	20.20	22.74
23	19.00	22.65
28	17.20	21.27
33	16.80	22.15
38	17.60	21.68
43	16.40	22.68
48	16.00	22.55

A.4 Satimage data

Data type:	real
Number of classes c	6
Number of features n:	4
Training set size N_{tr}:	500
Testing set size N_{ts}:	5935
Plot:	

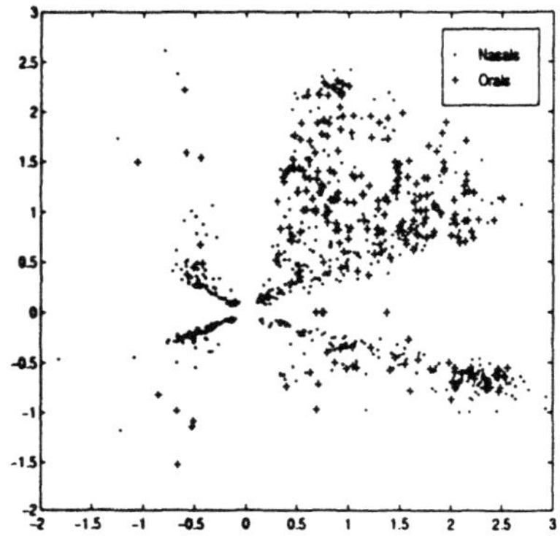

Classical models

Table A.26. Error rates (in %) with five classical pattern recognition models

Classifier	Training error	Testing error
Nearest mean	27.0	24.75
Linear discriminant classifier (LDC)	17.80	17.39
Quadratic discriminant classifier (QDC)	13.0	15.62
Parzen	10.6	15.25
Nearest neighbor (1-nn)	19.8	18.87

Neural networks

Table A.27. MLP classification error (in%). (Fast backpropagation used from the Neural Networks Toolbox of MATLAB 4.2)

MLP configuration	Training error	Testing error
4-5-6	70.00	76.53
4-10-6	49.00	51.26
4-15-6	77.00	76.53
4-20-6	79.20	75.92
4-65-6	24.40	23.08
4-70-6	21.60	21.01
4-75-6	15.00	15.32
4-80-6	64.20	66.76

Table A.28. RBF network classification error (in%). (Orthogonal Least Squares training used from the Neural Networks Toolbox of MATLAB 4.2)

RBF	$s = 10$		$s = 20$		$s = 40$	
configuration	Tr	Ts	Tr	Ts	Tr	Ts
4-5-6	26.40	23.99	19.80	19.81	18.20	17.34
4-10-6	17.20	17.86	14.20	15.89	14.40	15.84
4-15-6	14.80	17.02	13.20	14.83	12.80	15.06
4-20-6	12.80	15.52	12.40	14.52	12.20	14.69

Table A.29. LVQ1 classification error (in%). (5 independent runs)

Run	10 prototypes		20 prototypes	
number	Tr	Ts	Tr	Ts
1	21.80	22.33	16.40	18.15
2	22.20	20.15	15.20	16.68
3	20.60	19.88	18.40	18.64
4	20.60	19.76	16.00	16.60
5	21.80	20.30	17.20	17.52

Table A.30. Training and testing error rates of fuzzy TSK5 model with the histogram method for tuning the consequents

Resolution t	Training error [%]	Testing error [%]
3	33.20	31.58
8	25.80	22.56
13	22.20	21.06
18	22.80	20.86
23	23.00	22.02
28	21.80	22.73
33	21.20	21.77
38	22.80	23.67
43	19.40	21.23
48	20.00	24.21

References

1. S. Abe. Dynamic cluster generation for a fuzzy classifier with ellipsoidal regions. *IEEE Transactions on Systems, Man, and Cybernetics*, 28(6):869–875, 1998.
2. S. Abe and M.-S. Lan. A method for fuzzy rules extraction directly from numerical data and its application to pattern classification. *IEEE Transactions on Fuzzy Systems*, 3(1):18–28, 1995.
3. S. Abe and R. Thawonmas. A fuzzy classifier with ellipsoidal regions. *IEEE Transactions on Fuzzy Systems*, 5:358–368, 1997.
4. K.-P. Adlassnig. Fuzzy set theory in medicine. Part II. *BUSEFAL*, 18:101–105, 1984.
5. K.-P. Adlassnig. Fuzzy set theory in medical diagnosis. *IEEE Transactions on Systems, Man, and Cybernetics*, 16:260–265, 1986.
6. A.A. Afifi and S.P. Azen. *Statistical Analysis. A Computer Oriented Approach.* Academic Press, N.Y., 1979.
7. M. Akay, M. Cohen, and D. Hudson. Fuzzy sets in life sciences. *Fuzzy Sets and Systems*, 90(2):219–224, 1997.
8. I. Aleksander and H. Morton. *Introduction to Neural Computing.* Thomson Computer Press, London Int., 2nd edition, 1995.
9. D. Alfonso, F. Masulli, and A. Sperduti. Competitive learning in a classifier based on an adaptive fuzzy system. In *Proc. Int. Conf. on Soft Computing*, pages C2–8, Reading, UK, 1996.
10. E. Alpaydin. Combining global vs local linear perceptrons for classification. In *International Conference on Soft Computing SOCO'96*, pages B291–B297, Reading, UK, 1996.
11. E. Alpaydin. Voting over multiple condensed nearest neighbors. *Artificial Intelligence Review*, 11:115–132, 1997.
12. E. Alpaydin. Comparison of statistical and neural classifiers and their applications to optical character recognition and speech classification. In C T Leondes, editor, *Image Processing and Pattern Recognition*, volume 5 of *Neural Network Systems*, pages 61–88. Academic Press, 1998.
13. E. Alpaydin. Soft vector quantization and the EM algorithm. *Neural Networks*, 11(3):467–477, 1998.
14. E. Alpaydin and M. I. Jordan. Local linear perceptrons for classification. *IEEE Transactions on Neural Networks*, 7(3):788–792, 1996.
15. C. Von Altrock. *Fuzzy Logic and Neurofuzzy Applications Explained.* Prentice Hall, 1995.
16. J.A. Anderson. Logistic discrimination. In P.R. Krishnaiah and L.N. Kanal, editors, *Classification, Pattern Recognition and Reduction of Dimensionality*, volume 2 of *Handbook of Statistics*, pages 169–191. North Holland, Amsterdam, 1982.
17. J.A. Anderson and E. Rosenfeld. *Neurocomputing. Foundations of Research.* The MIT Press, Cambridge, Massachusetts, 1988.

18. R. Andrews, J. Diederich, and A.B. Tickle. Survey and critiques of techniques for extracting rules from trained artificial neural networks. *Knowledge Based Systems*, 8(6):373–389, 1995.
19. T.M. Apostol. *Calculus*, volume II. Xerox College Publishing, Weltham, Massachusetts, USA, second edition, 1969.
20. P. Arabie, L.J. Hubert, and G. De Soete. *Clustering and Classification*. World Scientific, Singapore, 1996.
21. G.A. Babich and O.I. Camps. Weighted Parzen windows for pattern classification. *IEEE Transactions on Pattern Analysis and Machine Intelligence*, 18(5):567–570, 1996.
22. T. Bailey and A.K. Jain. A note on distance-weighted k-nearest neighbor rules. *IEEE Transactions on Systems, Man, and Cybernetics*, 8(4):311–313, 1978.
23. Y. L. Barabash. *Collective Statistical Decisions in Recognition*. Radio i Sviaz', Moscow, 1983. (In Russian).
24. R. Battiti and A.M. Colla. Democracy in neural nets: Voting schemes for classification. *Neural Networks*, 7:691–707, 1994.
25. B. Beasley, D.R. Bull, and R.R. Martin. An overview of genetic algorithms: Part 2, research topics. *University Computing*, 15:170–181, 1993.
26. R. Bellman, R. Kalaba, and L. Zadeh. Abstraction and pattern classification. *J. Math. Anal. Appl.*, 13:1–7, 1966.
27. M.J. Bender and S.P. Simonovic. Consensus as the measure of sustainability. *Hydrological Sciences Journal*, 42(4):493–500, 1997.
28. J.A. Benediktsson, J.R. Sveinsson, J. I. Ingimundarson, H. Sigurdsson, and O.K. Ersoy. Multistage classifiers optimized by neural networks and genetic algorithms. *Nonlinear Analysis, theory, Methods & Applications*, 30(3):1323–1334, 1997.
29. J.A. Benediktsson and P.H. Swain. Consenus theoretic classification methods. *IEEE Transactions on Systems, Man, and Cybernetics*, 22:688–704, 1992.
30. M. Bereau and B Dubuisson. An adaptive algorithm in a partly supervised environment. In *8th International Conference on Pattern Recognition*, pages 120–122, 1986.
31. M. Bereau and B Dubuisson. A fuzzy extended k-nearest neighbor rule. *Fuzzy Sets and Systems*, 44:17–32, 1991.
32. R.C. Berkan and S.L. Trubatch. *Fuzzy Systems Design Principles. Building Fuzzy IF-THEN Rule Bases*. IEEE Press, NY, 1997.
33. H Bersini and G. Bontempi. Now comes the time to defuzzify neuro-fuzzy models. *Fuzzy Sets and Systems*, 90(2):161–169, 1997.
34. J.C. Bezdek. *Pattern Recognition with Fuzzy Objective Function Algorithms*. Plenum Press, N.Y., 1981.
35. J.C. Bezdek. Fuzzy models and digital signal processing (for pattern recognition): Is this a good marriage? *Digital Signal Processing*, 3:253–270, 1993.
36. J.C. Bezdek. Editorial: Fuzziness vs. probability - again (!?). *IEEE Transactions on Fuzzy Systems*, 2(1):1–3, 1994.
37. J.C. Bezdek, S.K. Chuah, and D. Leep. Generalized k-nearest neighbor rules. *Fuzzy Sets and Systems*, 18:237–256, 1986.
38. J.C. Bezdek, J.M Keller, R. Krishnapuram, and N.R. Pal. *Fuzzy Models and Algorithms for Pattern Recognition and Image Processing*. Kluwer Academic Publishers, 1999. in press.
39. J.C. Bezdek and S.K. Pal. *Fuzzy Models for Pattern Recognition*. IEEE Press, Piscataway, N.J., 1992.
40. J.C. Bezdek, T.R. Reichherzer, G.S. Lim, and Y. Attikiouzel. Multiple-prototype classifier design. *IEEE Transactions on Systems, Man, and Cybernetics*, C-28(1):67–79, 1998.

41. T. Bilgic and I. Türksen. *International Handbook of Fuzzy Systems*, volume 1, chapter Measurement of membership functions: Theoretical and experimental work. Kluwer Academic Publishers, 1999. in press.
42. T. Bilgic and I. B. Türksen. Elicitation of membership functions: How far can they take us? In *Proc FUZZ/IEEE*, pages 1321–1325, Barcelona, Spain, 1997.
43. C.M. Bishop. *Neural Networks for Pattern Recognition*. Clarendon Press, Oxford, 1995.
44. I. Bloch. Information combination operators for data fusion: a comparative review with classification. *IEEE Transactions on Systems, Man, and Cybernetics - Part A: Systems and Humans*, 26:52–67, 1996.
45. J.L. Blue, G.T. Candela, P.J. Grother, R. Chellappa, and C.L. Wilson. Evaluation of pattern classifiers for fingerprint and OCR applications. *Pattern Recognition*, 27(4):485–501, 1994.
46. R.F. Bordley. A multiplicative formula for aggregating probability assessments. *Management Science*, 28:1137–1148, 1982.
47. L.I. Borodkin and O.Ye. Stadnik. Algorithms for construction of decision rules in pattern recognition by using fuzzy sets. *Avtomatika i Telemekhanika*, (11):108–116, 1985.
48. G. Bortolan. An architecture of fuzzy neural networks for linguistic processing. *Fuzzy Sets and Systems*, 100:197–215, 1998.
49. N.K. Bose and A.K. Garga. Neural network design using Voronoi diagrams. *IEEE Transactions on Neural Networks*, 4:778–787, 1993.
50. M. Brown and C.J. Harris. *Neurofuzzy Adaptive Modeling and Control*. Prentice Hall, 1993.
51. A. Browne, editor. *Neural Network Perspectives on Cognition and Adaptive Robotics*. Institute of Physics Publishing, Bristol,UK, 1997.
52. J.J. Buckley and Y. Hayashi. Fuzzy neural networks: A survey. *Fuzzy Sets and Systems*, 66:1–13, 1994.
53. D. Cabello, S. Barro, J.M. Salceda, R. Ruiz, and J. Mira. Fuzzy k-nearest neighbor classifiers for ventricular arrhythmia detection. *International Journal of Biomedical Computing*, 27:77–93, 1991.
54. L.Y. Cai and H.K. Kwan. Fuzzy classification using fuzzy inference networks. *IEEE Transactions on Systems, Man, and Cybernetics*, 28(B)(3):334–347, 1998.
55. B Carse, T.C. Fogarty, and A. Munro. Evolving fuzzy rule-based controllers using genetic algorithms. *Fuzzy Sets and Systems*, 80:273–293, 1996.
56. F. Casalino, F. Masulli, Sperduti, and F. Vannucci. Semantic phase transition in a classifier based on an adaptive fuzzy system. In *IEEE International Conference on Fuzzy Systems IEEE/FUZZ'94*, pages 808–812, Orlando, FL, USA, 1994.
57. F. Casalino, F. Masulli, and A. Sperduti. Rule specialization in networks of fuzzy basis functions. *Intelligent Automation and Soft Computing*, 4(1):73–82, 1998.
58. K. Chen, L. Wang, and H. Chi. Methods of combining multiple classifiers with different features and their applications to text-independent speaker identification. *International Journal on Pattern Recognition and Artificial Intelligence*, 11(3):417–445, 1997.
59. S. Chen, C.F.N. Cowan, and P.M. Grant. Orthogonal least squares learning algorithm for radial basis function networks. *IEEE Transactions on Neural Networks*, 2(2):302–309, 1991.
60. F.-H. Cheng, W.-H. Hsu, and C.-A. Chen. Fuzzy approach to solve the recognition problem of handwritten Chinese characters. *Pattern Recognition*, 22:133–141, 1989.

61. C.-C. Chiang and H.-C. Fu. A divide-and-conquer methodology for modular supervised neural network design. In *IEEE International Conference on Neural Networks*, pages 119–124, Orlando, Florida, 1994.
62. S.-B. Cho and J.H. Kim. Combining multiple neural networks by fuzzy integral and robust classification. *IEEE Transactions on Systems, Man, and Cybernetics*, 25:380–384, 1995.
63. S.B. Cho and J.H. Kim. Multiple network fusion using fuzzy logic. *IEEE Transactions on Neural Networks*, 6:497–501, 1995.
64. K. Cios, W. Pedrycz, and R. Swiniarski. *Data Minimg. Methods for Knowledge Discovery*. Kluwer Academic Publishers, Boston, USA, 1998.
65. O. Cordón, M.J. del Jesus, and F. Herrera. Genetic learning of fuzzy rule-based classification systems cooperating with fuzzy reasoning methods. *International Journal of Intelligent Systems*, 13:1025–1053, 1998.
66. O. Cordón, M.J. del Jesus, and F. Herrera. A proposal on reasoning methods in fuzzy rule-based classification systems. *Int. J. of Approximate Reasoning*, 20(1):21–45, 1999.
67. R.L. Coultrip and R.H. Granger. Sparse random networks with LTP learning rules approximate Bayes classifiers via Parzen's method. *Neural Networks*, 7:463–476, 1994.
68. E. Cox. *The Fuzzy Systems Handbook. A Practitioner's Guide to Building, Using, and Maintaining Fuzzy Systems.* Academic Press, Ltd., London, UK, 1994.
69. B.V. Dasarathy. *Nearest Neighbor (NN) Norms: NN Pattern Classification Techniques.* IEEE Computer Society Press, Los Alamitos, California, 1990.
70. B.V. Dasarathy and B.V. Sheela. A composite classifier system design: concepts and methodology. *Proceedings of IEEE*, 67:708–713, 1978.
71. C. Decaestecker. NNP: A neural net classifier using prototypes. In *Proc. IEEE International Conference on Neural Networks*, pages 822o–824, San Francisco, CA, 1993.
72. C. Decaestecker. Finding prototypes for nearest neighbor classification by means of gradient descent and deterministic annealing. *Pattern Recognition*, 30(2):281–288, 1997.
73. A. DeLuca and S. Termini. A definition of nonprobabilistic entropy in the setting of fuzzy sets theory. *Information and Control*, 20:301–312, 1972.
74. T. Denœ ux. A k-nearest neighbor classification rule based on Dempster-Shafer theory. *IEEE Transactions on Systems, Man, and Cybernetics*, 25(5):804–813, 1995.
75. P.A. Devijver and J. Kittler. *Pattern Recognition: A Statistical Approach.* Prentice-Hall, Inc., Englewood Cliffs, N.J., 1982.
76. A. Di Nola, W. Perdycz, S. Sessa, and E.Sanchez. Designing of classification procedures with the use of equality and difference operators. *Pattern Recognition*, 23:911–917, 1990.
77. C. Diamantini and A. Spalvieri. Certain facts about Kohonen's LVQ1 algorithm. *IEEE Transactions on Circuits and Systems - I: Fundamental Theory and Applications*, 43(5):425–427, 1996.
78. C. Diamantini and A. Spalvieri. Quantizing for minimum average misclassification risk. *IEEE Transactions on Neural Networks*, 9(1):174–182, 1998.
79. H. Drucker, C. Cortes, L.D. Jackel, Y. LeCun, and V. Vapnik. Boosting and other ensemble methods. *Neural Computation*, 6:1289–1301, 1994.
80. D. Dubios and H. Prade. General approach to comparison index determination in fuzzy set theory. In R.R. Yager, editor, *Fuzzy Set and Possibility Theory. Recent Developments*, pages 9–21. Radio i Sviaz', Moscow, 1986.

81. D. Dubois and H. Prade. *Fuzzy Sets and Systems: Theory and Applications*. Academic Press, NY, 1980.
82. D. Dubois and H. Prade. New results about properties and semantics of fuzzy set-theoretic operators. In P.P. Wang and S.K. Chang, editors, *Fuzzy Sets: Theory and Applications to Policy Analysis and Information Systems*, pages 59–75. Plenum Press, N.Y., 1980.
83. D. Dubois and H. Prade. A review of fuzzy set aggregation connectives. *Information Sciences*, 36:85–121, 1985.
84. D. Dubois and H. Prade. Fuzzy relation equations and causal reasoning. *Fuzzy Sets and Systems*, 75:119–134, 1995.
85. D. Dubois and H. Prade. The three semantics of fuzzy sets. *Fuzzy Sets and Systems*, 90:141–150, 1997.
86. T. Duchun. Statistical-fuzzy method and fuzzy-statistical method in pattern recognition. In *7th International Conference on Pattern Recognition*, pages 1329–1331, 1984.
87. R.O. Duda and P.E. Hart. *Pattern Classification and Scene Analysis*. John Wiley & Sons, NY, 1973.
88. S.A. Dudani. The distance-weighted k-nearest-neighbor rule. *IEEE Transactions on Systems, Man, and Cybernetics*, 6(4):325–327, April 1976.
89. R.P.W. Duin. A note on comparing classifiers. *Pattern Recognition Letters*, 17:529–536, 1996.
90. R.P.W. Duin and D.M.J. Tax. Classifier-conditional posterior probabilities. In A. Amin, D. Dori, P. Pudil, and H. Freeman, editors, *Advances in Pattern Recognition*, volume 1451 of *Lecture Notes in Computer Science*, pages 611–619. Springer, Berlin, 1998.
91. T.M. English, M.P. Gomez-Gil, and W.J.B. Oldham. A comparison of neural network and nearest neighbor classifiers of handwritten lowercase letters. In *Proc. IEEE International Conference on Neural Networks*, pages 1618–1621, San Francisco, California, 1993.
92. A.O. Esogbue and R.C. Elder. Measurement and validation of a fuzzy mathematical model for medical diagnosis. *Fuzzy Sets and Systems*, 10:223–242, 1983.
93. L. Fausett. *Fundamentals of Neural Networks*. Prentice Hall Inc., Englewood Cliffs, N.J., 1994.
94. F.J. Ferri, J.V. Albert, and E. Vidal. Considerations about sample size sensitivity of a family of edited nearest neighbor rules. *IEEE Transactions on Systems, Man, and Cybernetics*, B:29(4):667–672, 1999.
95. E. Filippi, M. Costa, and E. Pasero. Multy-layer perceptron ensembles for increased performance and fault-tolerance in pattern recognition tasks. In *IEEE International Conference on Neural Networks*, pages 2901–2906, Orlando, Florida, 1994.
96. D.B. Fogel. *Evolutionary Computation. Towards a New Philosophy of Machine Intelligence*. IEEE Press, NY, 2nd edition, 2000.
97. D.B. Fogel and A. Ghozeli. A note on representation and variation operators. *IEEE Transactions on Evolutionary Computation*, 1(2):159–161, 1997.
98. M.J. Frank. On the simultaneous associativity of $f(x, y)$ and $x + y - f(x, y)$. *Aequationes Math.*, 19:194–226, 1979.
99. K.-S. Fu. *Syntactic Pattern Recognition and Applications*. Englewood Cliffs, NJ, 1982.
100. K. Fukunaga. *Introduction to Statistical Pattern Recognition*. Academic Press, Inc., Orlando, FL, 1972.
101. K. Fukushima. Cognitron: A self-organizing multilayered neural network. *Biological Cybernetics*, 20:121–136, 1975.

102. P.D. Gader, M.A. Mohamed, and J.M. Keller. Fusion of handwritten word classifiers. *Pattern Recognition Letters*, 17:577–584, 1996.
103. S. Gazula and M.R. Kabuka. Design od supervised classifiers using Boolean neural networks. *IEEE Transactions on Pattern Analysis and Machine Intelligence*, 17(12):1239–1246, 1995.
104. T. Gerstenkorn and E. Rakus. A method of calculating the membership degrees fro symptom diagnostic decisions. In *Cybernetics and Systems*, pages D479–486, Vienna, Austria, 1992.
105. T. Gerstenkorn and E. Rakus. On modeling membership function values in diagnostic decisions. *Listy Biomedyczne (Biomedical Letters)*, 30:3–12, 1993.
106. S. Geva and J. Sitte. Adaptive nearest neighbor pattern classification. *IEEE Transactions on Neural Networks*, 2(2):318–322, 1991.
107. N. Glick. Additive estimators for probabilities of correct classification. *Pattern Recognition*, 10:211–222, 1978.
108. D. Goldberg. *Genetic Algorithms in Search, Optimization and Machine Learning*. Addison-Wesley, NY, 1989.
109. A González and R. Pérez. SLAVE: a genetic learning system based on an iterative approach. *IEEE Transactions on Fuzzy Systems*, 7(2):176–191, 1999.
110. A.D. Gordon. *Classification*. Chapman & Hall /CRC, Boca Raton, FL, 1999.
111. M. Grabisch. A new algorithm for identifying fuzzy measures and its application to pattern recognition. In *Proc. FUZZ/IEEE'95*, pages 145–150, Yokohama, Japan, 1995.
112. M. Grabisch. On equivalence classes of fuzzy connectives - the case of fuzzy integrals. *IEEE Transactions on Fuzzy Systems*, 3(1):96–109, 1995.
113. M. Grabisch. Pattern classification and feature extraction by fuzzy integral. In *EUFIT'95*, pages 1465–1469, Aachen, Germany, 1995.
114. M. Grabisch. The representation of importance and interaction of features by fuzzy measures. *Pattern Recognition Letters*, 17:567–575, 1996.
115. M. Grabisch and F. Dispot. A comparison of some for fuzzy classification on real data. In *2nd International Conference on Fuzzy Logic and Neural Networks*, pages 659–662, Iizuka, Japan, 1992.
116. M. Grabisch, H.T. Nguen, and E.A. Walker. *Fundamentals of Uncertainty Calculi, with Applications to Fuzzy Inference*. Kluwer Academic Publishers, Dordrecht, 1995b.
117. M. Grabisch and J.-M. Nicolas. Classification by fuzzy integral. *Fuzzy Sets and Systems*, 65:255–271, 1994.
118. M. Grabisch and M. Sugeno. Multi-attribute classification using fuzzy integral. In *IEEE International Conference on Fuzzy Systems*, pages 47–54, San Diego, California, 1992.
119. T. Gu and B. Dubuisson. A loose pattern process approach to clustering fuzzy data sets. *IEEE Transactions on Pattern Analysis and Machine Intelligence*, 7(3):366–372, 1985.
120. S.K. Halgamuge. Self-evolving neural networks for rule-based data processing. *IEEE Transactions on Signal Processing*, 45(11):2776–2773, 1997.
121. S.K. Halgamuge and M. Glesner. Neural network in designing fuzzy systems for real-world applications. *Fuzzy Sets and Systems*, 65:1–12, 1994.
122. S.K. Halgamuge and M. Glesner. Neural networks in designing fuzzy systems for real world applications. *Fuzzy Sets and Systems*, 65:1–12, 1994.
123. S.K. Halgamuge, W. Pöchmüller, and M. Glesner. An alternative approach for generation of membership functions and fuzzy rules based on radial and cubic basis function networks. *International Journal of Approximate Reasoning*, 12:279–298, 1995.

124. H. Hamacher. über logische Verknupfungen unscharfer Aussagen und deren zugehörige Bewertungs-funktionen. In R. Trappl, G.J. Klir, and L. Riccardi, editors, *Progress in Cybernetics and Systems Research*, pages 276–287. Hemisphere, N.Y., 1978.
125. Y. Hamamoto, S. Uchimura, and S. Tomita. On the behavior of artificial neural network classifiers in high-dimensional spaces. *IEEE Transactions on Pattern Analysis and Machine Intelligence*, 18(5):571–574, 1996.
126. Y. Hamamoto, S. Uchimura, and S. Tomita. A bootstrap technique for nearest neighbor classifier design. *IEEE Transactions on Pattern Analysis and Machine Intelligence*, 19(1):73–79, 1997.
127. P.E. Hart. The condensed nearest neighbor rule. *IEEE Transactions on Information Theory*, 16:515–516, 1968.
128. E.J. Hartman, J.D. Keeler, and J.M. Kowalski. Layered neural networks wuth Gaussian hidden units as universal approximations. *Neural Computation*, 2:210–215, 1990.
129. M.H. Hassoun. *Fundamentals of Artificial Neural Networks*. MIT Press, Cambridge, Massachusetts, 1995.
130. K. Hattori and M. Takahashi. A new nearest neighbor rule in pattern classification problem. *Pattern Recognition*, 32:425–432, 1999.
131. S. Haykin. *Neural Networks. A Comprehensive Foundation*. Macmillan College Publishing Company, N.Y., 1994.
132. F. Herrera and L. Magdalena. Introduction: genetic fuzzy systems. *International Journal of Intelligent Systems*, 13:887–890, 1998.
133. K. Hirota and W. Pedrycz. Geometrical-logical pattern classification. In *2nd International Conference on Fuzzy Logic and Neural Networks*, pages 675–678, Iizuka, Japan, 1992.
134. K. Hirota and W. Pedrycz. OR/AND neuron in modeling fuzzy set connectives. *IEEE Transactions on Fuzzy Systems*, 2(2):151–161, 1994.
135. E. Hisdal. Are grades of membership probabilities? *Fuzzy Sets and Systems*, 25:325–348, 1988.
136. L. Holmstrom, P. Koistinen, J. Laaksonen, and E. Oja. Neural and statistical classifiers - taxonomy and two case studies. *IEEE Transactions on Neural Networks*, 8(1):5–17, 1997.
137. B.-G. Hu, R.G. Gosine, L.X. Cao, and C.W. de Silva. Application of fuzzy clustering technique in computer grading of fish products. *IEEE Transactions on Fuzzy Systems*, 6(1):144–152, 1998.
138. W.Y. Huang and R.P. Lippmann. Comparisons between neural nets and conventional classifiers. In *Proc. IEEE First International Conference on Neural Networks*, pages IV–485–IV–493, San Diego, California, 1987.
139. Y.S. Huang and C.Y. Suen. A method of combining multiple experts for the recognition of unconstrained handwritten numerals. *IEEE Transactions on Pattern Analysis and Machine Intelligence*, 17:90–93, 1995.
140. D.L. Hudson and M.E. Cohen. *Neural Networks and Artificial Intelligence for Biomedical Engineering*. Biomedical Engineering. IEEE Press, N.Y., 2000.
141. B. Hussein, R. McLaren, and S. Bleha. An application of fuzzy algorithms in a computer access security system. *Pattern Recognition Letters*, 9:39–43, 1989.
142. H. Ishibuch, T. Murata, and sen I.B. Türk˙Single-objective and two-objective genetic algorithms for selectin linguistic rules for pattern classification problems. *Fuzzy Sets and Systems*, 89:135–150, 1997.
143. H. Ishibuch, T. Nakashima, and T. Murata. Comparison of the Michigan and Pittsburgh approaches to the design of fuzzy classification systems. *Electronics and Communications in Japan, Part 3*, 80(12):379–387, 1997.

144. H. Ishibuchi, R. Fujioka, and H. Tanaka. Neural networks that learn from fuzzy if-then rules. *IEEE Transactions on Fuzzy Systems*, 1(2):85–97, 1993.
145. H. Ishibuchi and T. Nakashima. A study on generating fuzzy classification rules using histograms. In L.C. Jain and R.K. Jain, editors, *Proc. 2nd Int. Conf. on Knowledge-Based Intelligent Electronic Systems*, pages 132–140, Adelaide, Australia, 1998.
146. H. Ishibuchi, K. Nozaki, and H. Tanaka. Distributed representation of fuzzy rules and its application to pattern classification. *Fuzzy Sets and Systems*, 52:21–32, 1992.
147. H. Ishibuchi, K. Nozaki, and H. Tanaka. Efficient fuzzy partition on pattern space for classification problems. In *2nd International Conference on Fuzzy Logic and Neural Networks*, pages 671–674, Iizuka, Japan, 1992.
148. H. Ishibuchi, K. Nozaki, N. Yamamoto, and H. Tanaka. Construction of fuzzy classification systems with rectangular fuzzy rules using genetic algorithms. *Fuzzy Sets and Systems*, 65:237–253, 1994.
149. H. Ishibuchi, K. Nozaki, N. Yamamoto, and H. Tanaka. Selecting fuzzy if-then rules for classification problems using genetic algorithms. *IEEE Transactions on Fuzzy Systems*, 3(3):260–270, 1995.
150. R.A. Jacobs. Methods for combining experts' probability assessments. *Neural Computation*, 7:867–888, 1995.
151. R.A. Jacobs, M.I. Jordan, S.J. Nowlan, and G.E. Hinton. Adaptive mixtures of local experts. *Neural Computation*, 3:79–87, 1991.
152. A.K. Jain, R.C. Dubes, and C-C. Chen. Bootstrap techniques for error estimation. *IEEE Transactions on Pattern Analysis and Machine Intelligence*, 9(5):628–633, 1987.
153. A.K. Jain and J. Mao. Guest editorial: Special issue on artificial neural networks and statistical pattern recognition. *IEEE Transactions on Neural Networks*, 8(1):1–3, 1997.
154. J.-S. R. Jang. ANFIS: adaptive-network-based fuzzy inference system. *IEEE Transactions on Systems, Man, and Cybernetics*, 23(3):665–685, 1993.
155. J.-S. R. Jang and C.T. Sun. Functional equivalence between radial basis function networks and fuzzy inference systems. *IEEE Transactions on Neural Networks*, 4(1):156–159, 1993.
156. M.I. Jordan and L. Xu. Convergence results for the EM approach to mixtures of experts architectures. *Neural Networks*, 8:1409–1431, 1995.
157. A. Joshi, N. Ramakrishman, E.N. Houstis, and J.R. Rice. On neurobiological, neuro-fuzzy, machine learning and statistical pattern recognition techniques. *IEEE Transactions on Neural Networks*, 8(1):18–31, 1997.
158. A. Joussellin and B. Dubuisson. A link between *k*-nearest neighbor rules and knowledge based systems by sequence analysis. *Pattern Recognition Letters*, 6:287–295, 1987.
159. A. Jóźwik. A learning scheme for a fuzzy *k*-nn rule. *Pattern Recognition Letters*, 1:287–289, 1983.
160. Z. Kabala and I. Wrocinski. A short notice on Bayes rule and related topics. *BUSEFAL*, 18:92–99, 1984.
161. A. Kaufmann. *Introduction to Fuzzy Set Theory*. Radio i Sviaz', Moscow, 1982. (In Russian. Translation from French).
162. J.M. Keller and Z. Chen. Learning in fuzzy neural network utilizing additive hybrid operators. In *2nd International Conference on Fuzzy Logic and Neural Networks*, pages 85–87, Iizuka, Japan, 1992.
163. J.M. Keller and Z. Chen. Learning in fuzzy neural networks utilizing additive hybrid operators. In *Proc. 2nd Int. Conf. on Fuzzy Logic and Neural Networks*, pages 85–87, Iizuka, Japan, 1992.

164. J.M. Keller, P. Gader, H. Tahani, J.-H. Chiang, and M. Mohamed. Advances in fuzzy integration for pattern recognition. *Fuzzy Sets and Systems*, 65:273–283, 1994.
165. J.M. Keller and J.A. Givens. Membership function issues in fuzzy pattern recognition. In *IEEE International Conference on Cybernetics and Society*, pages 210–214, 1985.
166. J.M. Keller, M.R. Gray, and J.A. Givens. A fuzzy k-nearest neighbor algorithm. *IEEE Transactions on Systems, Man, and Cybernetics*, 15:580–585, 1985.
167. J.M. Keller and D.J. Hunt. Incorporating fuzzy membership function into the perceptron algorithm. *IEEE Transactions on Pattern Analysis and Machine Intelligence*, 7:693–699, 1985.
168. J.M. Keller and H. Tahani. Implementation of conjunctive and disjunctive fuzzy rules with neural networks. *Int. Journal on Approximate Reasoning*, 6:221–240, 1992.
169. H.M. Kim and J.M. Mendel. Fuzzy basis functions: Comparison with other basis functions. *IEEE Transactions on Fuzzy Systems*, 3(2):158–168, 1995.
170. Y.K. Kim and J.H. Han. Fuzzy K-NN algorithm using modified k-selection. In *Proc. FUZZY/IEEE'95*, pages 1673–1680, Yokohama, Japan, 1995.
171. V.T. Kissiov and S.T. Hadjitodorov. A fuzzy version of the k-nn method. *Fuzzy Sets and Systems*, 49:323–329, 1992.
172. J. Kittler, M. Hatef, R.P.W. Duin, and J. Matas. On combining classifiers. *IEEE Transactions on Pattern Analysis and Machine Intelligence*, 20(3):226–239, 1998.
173. F. Klawon and P.E. Klement. Mathematical analysis of fuzzy classifiers. *Lecture Notes in Computer Science*, 1280:359–370, 1997.
174. G.J. Klir and T.A. Folger. *Fuzzy Sets, Uncertainty and Information*. Prentice-Hall International, Inc., Englewood Cliffs, N.J., 1988.
175. T. Kohonen. Improved versions of learning vector quantization. In *Proc. International Joint Conference on Neural Networks*, pages I–545–550, San Diego, CA, June 1990.
176. T. Kohonen. *Self-Organizing Maps*. Springer, Germany, 1995.
177. L.J. Kohout and M. Kallala. Design criteria of a handwriting monitor with a short survey. *BUSEFAL*, 25:133–140, 1986.
178. A. König, R.J. Raschhofer, and M. Glesner. A novel method for the design of radial-basis-function networks and its application for knowledge extraction. In *Proc. IEEE Int. Conf. on Neural Networks*, pages 1804–1809, Orlando, FL, 1994.
179. K. Koutroumbas and N. Kalouptsidis. Nearest neighbor pattern classification neural networks. In *Proc. IEEE International Conference on Neural Networks*, pages 2911–2915, Orlando, Florida, 1994.
180. V. L. Kreinovich, C.G. Mouzouris, and H.T. Nguyen. Fuzzy rule based modeling as a universal approximation tool. In H. T. Nguyen and M. Sugeno, editors, *Fuzzy Systems : Modeling and Control*, pages 135–196. Kluwer, Boston, MA, 1998.
181. F. Kretzschmar. A fuzzy logical model of nutrition diagnosis. In *Medical Informatics in Europe*, volume 45 of *Lecture Notes in Medical Informatics*, pages 939–943, Vienna, Austria, 1991.
182. E. Kruśińska. Discriminant functions: classical and fuzzy. A comparison of two approaches in medical diagnosis. In *Polish Symposium on Interval and Fuzzy Mathematics*, pages 111–114, Poznań, Poland, 1983.

183. E. Kruśińska and J. Liebhart. A note on the usefulness of linguistic variables for differentiating between some respiratory diseases. *Fuzzy Sets and Systems*, 18:131-142, 1986.
184. A. Krzyżak, T. Linder, and G. Lugosi. Nonparametric classification using radial basis function nets and empirical risk minimzation. In *Proc. 12th Int. Conference on Pattern Recognition*, pages 72–76, Jerusalem, Israel, 1994.
185. A. Krzyżak, T. Linder, and G. Lugosi. Nonparametric estimation and classification using radial basis function nets and empirical risk minimization. *IEEE Transactions on Neural Networks*, 7(2):475–487, 1996.
186. L.I. Kuncheva. Using measures of similarity and inclusion for multiple classifier fusion by decision templates. *Fuzzy Sets and Systems*, 2000, (to appear).
187. L.I. Kuncheva. Additional criteria in fuzzy two-level pattern recognition. *Fuzzy Sets and Systems*, 47:281–289, 1992.
188. L.I. Kuncheva. An aggregation of pro and con evidence for medical decision support systems. *Computers in Biology and Medicine*, 23:417–424, 1993.
189. L.I. Kuncheva. Change-glasses approach in pattern recognition. *Pattern Recognition Letters*, 14:619–623, 1993.
190. L.I. Kuncheva. Pattern recognition with a model of fuzzy neuron using degree of consensus. *Fuzzy Sets and Systems*, 66:241–250, 1994.
191. L.I. Kuncheva. Editing for the k-nearest neighbors rule by a genetic algorithm. *Pattern Recognition Letters*, 16:809–814, 1995.
192. L.I. Kuncheva. A fuzzy aggregation of multiple classification decisions. *Control and Cybernetics*, 25:337–352, 1996.
193. L.I. Kuncheva. On the equivalence between fuzzy and statistical classifiers. *International Journal of Uncertainty, Fuzziness and Knowledge-Based Systems*, 4.3:245–253, 1996.
194. L.I. Kuncheva. An application of OWA operators to the aggregation of multiple classification decisions. In R.R. Yager and J. Kacprzyk, editors, *The Ordered Weighted Averaging operators. Theory and Applications*, pages 330–343. Kluwer Academic Publishers, USA, 1997.
195. L.I. Kuncheva. Fitness functions in editing k-nn reference set by genetic algorithms. *Pattern Recognition*, 30:1041–1049, 1997.
196. L.I. Kuncheva. Initializing of an rbf network by a genetic algorithm. *Neurocomputing*, 14:273–288, 1997.
197. L.I. Kuncheva. On combining multiple classifiers. In *Proc. 7th International Conference on Information Processing and Management of Uncertainty (IPMU'98)*, pages 1890–1891, Paris, France, 1998.
198. L.I. Kuncheva and J.C. Bezdek. A fuzzy generalized nearest prototype classifier. In *7th IFSA World Congress*, pages 217–222, Prague, Czech Republic, 1997.
199. L.I. Kuncheva and J.C. Bezdek. An integrated framework for generalized nearest prototype classifier design. *International Journal of Uncertainty, Fuzziness and Knowledge-Based Systems*, 6(5):437–457, 1998.
200. L.I. Kuncheva and J.C. Bezdek. On prototype selection: Genetic algorithms or random search? *IEEE Transactions on Systems, Man, and Cybernetics*, C28(1):160–164, 1998.
201. L.I. Kuncheva, J.C. Bezdek, and R.P.W. Duin. Decision templates for multiple classifier fusion: an experimental comparison. *Pattern Recognition*, 1999. (accepted).
202. L.I. Kuncheva and L.C. Jain. Nearest nieghbor classifier: simultaneous editing and feature selection. *Pattern Recognition Letters*, 1999, 20:1149–1156.

203. L.I. Kuncheva, R.K. Kounchev, and R.Z. Zlatev. Aggregation of multiple classification decisions by fuzzy templates. In *Third European Congress on Intelligent Technologies and Soft Computing EUFIT'95*, pages 1470–1474, Aachen, Germany, August 1995.
204. L.I. Kuncheva and R. Krishnapuram. A fuzzy consensus aggregation operator. *Fuzzy Sets and Systems*, 79:347–356, 1996.
205. L.I. Kuncheva and F. Steimann. Fuzzy diagnosis (editorial). *Atrificial Intelligence in Medicine*, 16:121–128, 1999.
206. L.I. Kuncheva, C.J. Whitaker, C.A. Shipp, and R.P.W. Duin. Limits on the majority vote accuracy in classifier fusion. *IEEE Transactions on Pattern Analysis and Machine Intelligence*, 2000. (submitted).
207. M.S. Kupriyanov and O.N. Yarygin. Design of relation and measure of similarity of fuzzy objects. *Bulletin of the USSR Academy of Sciences: Technicheskaya Kibernetika*, (3):147–154, 1988.
208. H.K. Kwan and Y. Cai. A fuzzy neural network and its application to pattern recognition. *IEEE Transactions on Fuzzy Systems*, 2(3):185–193, 1994.
209. L. Lam and C.Y. Suen. Optimal combination of pattern classifiers. *Pattern Recognition Letters*, 16:945–954, 1995.
210. L. Lam and C.Y. Suen. Application of majority voting to pattern recognition: An analysis of its behavior and performance. *IEEE Transactions on Systems, Man, and Cybernetics*, 27(5):553–568, 1997.
211. M. Lamotte, L. Bour, and G. Hirsch. Fuzzy phoneme recognition. *Fuzzy Sets and Systems*, 28:363–374, 1988.
212. S. C. Lee and E. T. Lee. Fuzzy neural networks. *Int. Journal on Math. Biosciences*, 23:151–177, 1975.
213. Y. Lee. Handwritten digit recognition using k-nearest neighbor, radial-basis function, and backpropagation neural networks. *Neural Computation*, 3:440–449, 1991.
214. W.V. Leekwijck and E.E. Kerre. Defuzzification: criteria and classification. *Fuzzy Sets and Systems*, 108(2):159–178, 1999.
215. K. Leszczynski, S. Cosby, R. Bissett, D. Provost, S. Boyko, S. Loose, and E. Mvilongo. Application of a fuzzy pattern classifier to decision making in portal verification radiotherapy. *Phys. Med. Biol.*, 44:253–269, 1999.
216. R.-P. Li and M. Mukaidono. A new approach to rule learning based on fusion of fuzzy logic and neural networks. *IEICE Transactions on Information and Systems*, E78-D(11):1509–1514, 1995.
217. T.W. Liao and D. Li. Two manufacturing applications of the fuzzy K-NN algorithm. *Fuzzy Sets and Systems*, 92:289–303, 1997.
218. R.P. Lippmann. An introduction to computing with neural nets. *IEEE ASSP Magazine*, pages 4–22, April 1987.
219. R.P. Lippmann. Pattern classification using neural networks. *IEEE Communication Magazine*, pages 47–64, November 1989.
220. R.P. Lippmann. A critical overview of neural network pattern classifiers. In *Proc. IEEE Workshop on Neural Networks for Signal Processing*, pages 266–275, 1991.
221. I.D. Longstaff and J.F. Gross. A pattern recognition approach to understanding the multi-layer perceptron. *Pattern Recognition Letters*, 5:315–319, 1987.
222. C.G. Looney. *Pattern Recognition Using Neural Networks. Theory and Algorithms for Engineers and Scientists.* Oxford University Press, Oxford, 1997.
223. S.W. Looney. A statistical technique for comparing the accuracies of several classifiers. *Pattern Recognition Letters*, 8:5–9, 1988.
224. Y. Lu. Knowledge integration in a multiple classifier system. *Applied Intelligence*, 6:75–86, 1996.

225. Y. Lu and X.-L. Fan. Fuzzy weighted distance and its application in pattern recognition and classification. In *9th International Conference on Pattern Recognition*, pages 1065–1067, Rome, Italy, 1988.
226. W.S. MacCulloch and W. Pitts. A logical calculus of the ideas immanent in nervous activity. *Bulletin of Mathematical Biophysics*, 5:115–133, 1943.
227. J.E.S. Macleod, A. Luk, and D.M. Titterington. A re-examination of the distance-weighted k-nearest neighbor classification rule. *IEEE Transactions on Systems, Man, and Cybernetics*, 17(4):689–696, 1987.
228. D.P. Mandal, C.A. Murthy, and S.K. Pal. Formulation of a multivalued recognition system. *IEEE Transactions on Systems, Man, and Cybernetics*, 22:607–620, 1992.
229. D.P. Mandal and H. Tanaka. Use of fuzzy if-then rules for pattern classification. In *FUZZ/IEEE'95*, pages 1649–1656, Yokohama, Japan, 1995.
230. J. Mantas. The maximum σ-count as a decision operator. *Fuzzy Sets and Systems*, 27:149–158, 1988.
231. Z.-H. Mao, Y.-D. Li, and X.-F. Zhang. Approximation capability of fuzzy systems using translations and dilations of one fixed function as membership functions. *IEEE Transactions on Fuzzy Systems*, 5(3):468–473, 1997.
232. R. Marin and J. Mira. On knowledge-based fuzzy classifiers: A medical case study. *Fuzzy Sets and Systems*, 44:421–430, 1991.
233. E. Masson and Y.-J. Wang. Introduction to computation and learning in artificial neural networks. *European Journal on Operational Research*, 47:1–28, 1990.
234. F. Masulli, F. Casalino, and F. Vannucci. Bayesian properties and performances of adaptive fuzzy systems in pattern recognition problems. In *ICANN'94*, pages 189–192, Sorrento, Italy, 1994.
235. S. Medasani, J. Kim, and R. Krishnapuram. An overview of membership function generation techniques for pattern recognition. *International Journal of Approximate Reasoning*, 19:391–417, 1998.
236. M.L. Menendez and L. Prado. A class of bounds of probability of error based on information energy for the fuzzy discrimination problem. *BUSEFAL*, 39:41–52, 1989.
237. A. Mitiche and J.K. Aggarwal. Pattern category assignment by neural networks and nearest neighbor rule: A synopsis and a characterization. *International Journal of Pattern Recognition and Artificial Intelligence*, 10(5):393–408, 1996.
238. S. Mitra. Fuzzy MLP based expert system for medical diagnosis. *Fuzzy Sets and Systems*, 64, 1994.
239. S. Mitra, R.K. De, and S.K. Pal. Knowledge based fuzzy MLP for classification and rule generation. *IEEE Transactions on Neural Networks*, 8(6):1338–1350, 1997.
240. S. Mitra and S.K. Pal. Fuzzy multi–layer perceptron, inferencing and rule generation. *IEEE Transactions on Neural Networks*, 6:51–63, 1995.
241. J. Moody and C.J. Darken. Fast learning in neural networks of locally-tuned processing units. *Neural Computation*, 1:281–294, 1989.
242. R.L. Morin and D.E. Raeside. A reappraisal of distance-weighted k-nearest neighbor classification for pattern recognition with missing data. *IEEE Transactions on Systems, Man, and Cybernetics*, 11(3):241–243, 1981.
243. O.J. Murphy. Nearest neighbor pattern classification perceptrons. *Proceedings of IEEE*, 78(10):1595–1598, 1990.
244. G. Nagy. Candide's practical principles of experimental pattern recognition. *IEEE Transactions on Pattern Analysis and Machine Intelligence*, 5(2):199–200, 1993.

245. H. Narazaki and A.L. Ralescu. Implementation of fuzzy systems using multilayered neural network. In *Proc. IEEE Int. Conf. on Neural Networks*, pages 317–322, San Diego, CA, 1993.
246. D. Nauck. Neuro-fuzzy systems: review and prospects. In *Proc. 5th European Congress on Intelligent Techniques and Soft Computing, EUFIT'97*, pages 1044–1053, Aachen, Germany, 1997.
247. D. Nauck, F. Klawonn, and R. Kruse. *Foundations of Neuro-Fuzzy Systems*. Wiley, Chichester, 1997.
248. D. Nauck and R. Kruse. NEFCLASS – a neuro–fuzzy approach for the classification of data. In K.M. George, J. H. Carrol, E. Deaton, D. Oppenheim, and J. Hightower, editors, *Applied Computing 1995. Proc. of the 1995 ACM Symposium on Applied Computing, Nashville, Feb. 26–28*, pages 461–465. ACM Press, New York, 1995.
249. D. Nauck and R. Kruse. A neuro-fuzzy method to learn fuzzy classification rules from data. *Fuzzy Sets and Systems*, 89:277–288, 1997.
250. D. Nauck and R. Kruse. New learning strategies for NEFCLASS. In *Proc. 7th World Congress of IFSA, IFSA97*, volume IV, pages 50–55, Prague, Czech Republic, 1997.
251. D. Nauck and R. Kruse. What are neuro-fuzzy classifiers? In *Proc. 7th World Congress of IFSA, IFSA97*, volume III, pages 228–233, Prague, Czech Republic, 1997.
252. K.-C. Ng and B. Abramson. Consensus diagnosis: A simulation study. *IEEE Transactions on Systems, Man, and Cybernetics*, 22:916–928, 1992.
253. H.T. Nguyen. Fuzzy sets and probability. *Fuzzy Sets and Systems*, 90:129–132, 1997.
254. A. Nigrin. *Neural Networks for Pattern Recognition*. MIT Press, USA, 1993.
255. D. Norris, B.W. Pilsworth, and J.F. Baldwin. Medical diagnosis from patient records – a method using fuzzy discrimination and connectivity analysis. *Fuzzy Sets and Systems*, 23:73–87, 1987.
256. S.J. Nowlan and G.E. Hinton. Evaluation of adaptive mixtures of competing experts. In R.P. Lippmann, J.E. Moody, and D.S. Touretzky, editors, *Advances in Neural Infprmation Processing Systems 3*, pages 774–780, 1991.
257. R. Odorico. Learning vector quantization with training count (lvqtc). *Neural Networks*, 10(6):1083–1088, 1997.
258. V. Cerverón and F.J. Ferri. Another move towards the minimum consistent subset: A tabu search approach to the condensed nearest neighbor rule. *IEEE Transactions on Systems, Man, and Cybernetics*.
259. N.R. Pal and J.C. Bezdek. Measuring fuzzy uncertainty. *IEEE Transactions on Fuzzy Systems*, 2(2):107–118, 1994.
260. S.K. Pal and B. Chakraborty. Intraclass and interclass ambiguities (fuzziness) in feature evaluation. *Pattern Recognition Letters*, 2:275–279, 1984.
261. S.K. Pal and S. Mitra. Multi–layer perceptron, fuzzy sets and classification. *IEEE Transactions on Neural Networks*, 3:683–697, 1992.
262. Y.-H. Pao. *Adaptive Pattern Recognition and Neural Networks*. Addison-Wesley, Reading, Massachusetts, 1989.
263. H.-S. Park and S.-W. Lee. Off-line recognition of large-set handwritten characters with multiple hidden Markov models. *Pattern Recognition*, 29(2):231–244, 1996.
264. J. Park and I.W. Sandberg. Universal approximation using radial-basis-function networks. *Neural Computation*, 3:246–257, 1991.
265. J. Park and I.W. Sandberg. Approximation and radial basis function networks. *Neural Computation*, 5:305–316, 1993.

266. G. Parthasarathy and B.N. Chatterji. A class of new KNN methods for low sample problems. *IEEE Transactions on Systems, Man, and Cybernetics*, 20(3):715–718, 1990.
267. S.K. Parui and D.D. Majumder. Estimation of fuzzy membership functions. In *Proc. 6th Int. Conf. on Pattern Recognition*, pages 423–426, Munich, Germany, 1982.
268. E.A. Patrick. *Fundamentals of Pattern Recognition*. Prentice-Hall, Inc., Englewood Cliffs, N.J., 1972.
269. D.W. Patterson. *Artificial Neural Networks. Theory and Applications*. Prentice Hall, Simon & Schuster, Singapore, 1996.
270. M. Pawlak and M.F. Yat Fung Ng. On kernel and radial basis function techniques for classification and function recovering. In *Proc. 12th Int. Conference on Pattern Recognition*, pages 454–456, Jerusalem, Israel, 1994.
271. W. Pedrycz. Classification in a fuzzy environment. *Pattern Recognition Letters*, 3:303–308, 1985.
272. W. Pedrycz. A method of fuzzy classifier for fuzzy data. *BUSEFAL*, 21:113–116, 1985.
273. W. Pedrycz. Fuzzy sets in pattern recognition: Methodology and methods. *Pattern Recognition*, 23:121–146, 1990.
274. W. Pedrycz and A. Rocha. Fuzzy-set based models of neurons and knowledge-based networks. *IEEE Transactions on Fuzzy Systems*, 1(4):254–266, 1993.
275. K. Piasecki. New concept of separated fuzzy subsets. In *Polish Symposium on Interval and Fuzzy Mathematics*, pages 193–195, Poznań, Poland, 1983.
276. N. Pizzi. Fuzzy preprocessing of gold standard as applied to biomedical spectra classification. *Artificial Intelligence in Medicine*, 16(2):171–182, 1999.
277. L. Prechelt. PROBEN1 - A set of neural network benchmark problems and benchmarking rules. Technical Report 21/94, University of Karlsruhe, Karlsruhe, Germany, 1994.
278. L.A. Rastrigin and R.H. Erenstein. *Method of Collective Recognition*. Energoizdat, Moscow, 1981. (In Russian).
279. M.D. Richard and R.P. Lippmann. Neural network classifiers estimate Bayesian a posteriori probabilities. *Neural Computation*, 3:461–483, 1991.
280. B.D. Ripley. *Pattern Recognition and Neural Networks*. University Press, Cambridge, 1996.
281. A. Rizzi, F.M.F. Mascioli, and G. Martinelli. Adaptive resolution min-max classifier. In *Proc. FUZZ-IEEE'98*, pages 1435–1440, Anchorage, Alaska, 1998.
282. G. Rogova. Combining the results of several neural network classifiers. *Neural Networks*, 7:777–781, 1994.
283. R. Rojas. *Neural Networks. A Systematic Introduction*. Springer, Berlin, 1995.
284. F. Rosenblatt. *Principles of Neurodynamics*. Spartan Books, Washington, 1962.
285. A. Rosenfeld. Distances between fuzzy sets. *Pattern Recognition Letters*, 3:229–233, 1985.
286. R. Rovatti. Fuzzy piecewise multilinear and liecewise linear systems as universal approximators. *IEEE Transactions on Fuzzy Systems*, 6(2):235–249, 1998.
287. R. Rovatti, R. Ragazzoni, Zs. M. Kovàcs, and R. Guerrieri. Adaptive voting rules for *k*-nearest neighbors classifiers. *Neural Computation*, 7:594–605, 1995.
288. M.K. Roy and R. Biswas. I-v fuzzy relations and Sanchez' approach to medical diagnosis. *Fuzzy Sets and Systems*, 47:35–38, 1992.
289. D.W. Ruck, S.K. Rojers, M. Kabrisky, M.E. Oxley, and B.W. Suter. The multilayer perceptron as an approximation to a bayes optimal discriminant function. *IEEE Transactions on Neural Networks*, 1(4):296–298, 1990.

290. M. Russo. FuGeNeSys - a fuzzy genetic neural system for fuzzy modeling. *IEEE Transactions on Fuzzy Systems*, 6(3):373–388, 1998.
291. R. Thawonmas S. Abe and M. Kayama. A fuzzy classifier with ellipsoidal regions for diagnosis problems. *IEEE Transactions on Systems, Man, and Cybernetics*, 29(C)(1):140–149, 1999.
292. L. Saitta and P. Torasso. Fuzzy characterization of coronary disease. *Fuzzy Sets and Systems*, 5:245–258, 1981.
293. R. Salomon. Implicit independence assumtions; a notoroius problem for genetic algorithms. In *Proc. Int. Conf. on Soft Computing*, pages B93–99, Reading, UK, 1996.
294. E. Sanchez. Inverses of fuzzy relations: Application to possibility distributions and medical diagnosis. *Fuzzy Sets and Systems*, 2:75–86, 1979.
295. A. Sato and J. Tsukumo. A criterion for training reference vectors and improved vector quantization. In *Proc. IEEE Int. Conference on Neural Networks*, pages 161–166, Orlando,FL, 1994.
296. F. Scarselli and A.C. Tsoi. Universal approximation using feedforward neural networks: A survey of some existing methods, and some new results. *Neural Networks*, 11(1):15–37, 1998.
297. B. Schölkopf, K.-K. Sung, C.J.C. Burges, F. Girosi, P. Niyogi, T. Poggio, and V. Vapnik. Comparing support vector machines with Gaussian kernels to radial basis function classifiers. *IEEE Transactions on Signal Processing*, 45(11):2758–2765, 1997.
298. J. Schürmann. *Pattern Classification. A Unified View of Statistical and Neural Approaches.* John Wiley & Sons, Inc., N.Y., 1996.
299. B. Schweizer and A. Sklar. Associative functions and statistical triangle inequalities. *Publ. Math. Debrecen*, 8:169–186, 1961.
300. M. Setnes and R. Babuška. Fuzzy relational classifier trained by fuzzy clustering. *IEEE Transactions on Fuzzy Systems*, 29(5):619–625, 1996.
301. M. Setnes, R. Babuška, and H.B. Verbruggen. Rule-based modeling: precision and transparency. *IEEE Transactions on Systems, Man, and Cybernetics*, 28(C)(1):165–169, 1998.
302. S.-Y. Shao and W.-M. Wu. A method of graph and fuzzy techniques for Chinese character recognition. *Fuzzy Sets and Systems*, 36:97–102, 1990.
303. A. Sherstinsky and R.W. Picard. On the efficiency of the orthogonal least squares training method for radial basis function networks. *IEEE Transactions on Neural Networks*, 7(1):195–200, 1996.
304. Y. Shi, R. Eberhart, and Y. Chen. Implementation of evolutionary fuzzy systems. *IEEE Transactions on Fuzzy Systems*, 7(2):109–119, 1999.
305. K. Shimojima, T. Fukuda, and Y. Hasegawa. RBF-fuzzy system with GA based unsupervised/supervised learning method. In *Proc. FUZZ/IEEE'95*, pages 253–258, Yokohama, Japan, 1995.
306. G.F. Simmons. *Introduction to Topology and Modern Analysis.* McGraw-Hill Book Company, Inc., Tokyo, Japan, 1963.
307. P.K. Simpson. Fuzzy min-max neural networks - part 1: classification. *IEEE Transactions on Neural Networks*, 3(5):776–786, 1992.
308. D.B. Skalak. Prototype and feature selection by sampling and random mutation hill climbing algorithms. In *Proc. 11th Int. Conf. on Machine Learning*, pages 293–301, New Brunswick, NJ, 1994. Morgan Kaufmann,Los Altos, CA.
309. F. Smieja. The pandemonium system of reflective agents. *IEEE Transactions on Neural Networks*, 7:97–106, 1996.
310. D.F. Specht. Probabilistic neural networks. *Neural Networks*, 3:109–118, 1990.
311. O.Ye. Stadnik. Fuzzy recognition rules in medical diagnosis and prediction. *Avtomatika i Telemekhanika*, (3):129–135, 1986.

312. M. Sugeno. Fuzzy measures and fuzzy integrals: a survey. In M.M. Gupta, G.N. Saridis, and B.R. Gaines, editors, *Fuzzy Automata and Decision Processes*, pages 89–102. North-Holland, Amsterdam, 1977.
313. C.-T. Sun and J.-S. Jang. A neuro-fuzzy classifier and its applications. In *Second IEEE International Conference on Fuzzy Systems*, pages 94–98, San Francisco, CA, USA, 1993.
314. L.M. Sztandera and Y. Hayashi. Automated extraction of fuzzy if-then rules and its application to medical diagnosis. In *Proc. IPMU'98*, pages 1907–1908, Paris, France, 1998.
315. T. Takagi and M. Sugeno. Fuzzy identification of systems and its application to modeling and control. *IEEE Transactions on Systems, Man, and Cybernetics*, 15(1):116–132, 1985.
316. G. Tao and B. Dubuisson. A loose-pattern process approach to clustering fuzzy data sets. *IEEE Transactions on Pattern Analysis and Machine Intelligence*, 7:366–372, 1985.
317. L. Tarassenko and S. Roberts. Supervised and unsupervised learning in radial basis function classifiers. *IEEE Proc.-Vis. Image Signal Process.*, 141(4):210–216, 1994.
318. G.D. Tattersall and K. Yi. Packed hyper-ellipsoid classifiers. *Electronic Letters*, 30:427–428, 1994.
319. D.M.J. Tax, R.P.W. Duin, and M. ban Breukelen. Comparison between product and mean classifier combination rules. In *Proc. Workshop on Statistical Pattern Recognition*, Prague, Czech, 1997.
320. T. Terano, K. Asai, and M. Sugeno. *Fuzzy Systems Theory and its Application*. Academic Press, Inc., San Diego, USA, 1991.
321. R. Thawonmas and S. Abe. Extraction of fuzzy rules for classification based on partitioned hyperboxes. *Journal of Intelligent and Fuzzy Systems*, 4:215–226, 1996.
322. A.B. Tickle, R. Andrews, M. Golea, and J. Diederich. Rule extraction from trained artificial neural networks. In A. Browne, editor, *Neural Network Analysis, Architectures and Applications*, pages 61–99. Institute of Physics Publishing, Bristol, UK, 1997.
323. F.T.B. toe Water and R.P.W. Duin. Dealing with *a priori* knowledge by fuzzy labels. *Pattern Recognition*, 14:111–115, 1981.
324. J.T. Tou and R.C. Gonzalez. *Pattern Recognition Principles*. Reading, MA: Addison-Wesley, 1974.
325. G.T. Toussaint. Bibliography on estimation of misclassification. *IEEE Transactions on Information Theory*, 20:472–479, 1974.
326. H.G.C. Traven. A neural network approach to statistical pattern classification by "semiparametric" estimation of probability density functions. *IEEE Transactions on Neural Networks*, 2(3):366–377, 1991.
327. I.B. Türk(s)en, V. Kreinovich, and R.R. Yager. A new class of fuzzy implications. axioms of fuzzy implications revisited. *Fuzzy Sets and Systems*, 100:267–272, 1998.
328. S. Umejama. The complementary process of fuzzy medical diagnosis and its properties. *Information Sciences*, 38:229–242, 1986.
329. M. Valenzuela-Rendón. The fuzzy classifier system. motivations and first results. *Lecture Notes in Computer Science*, 496:338–342, 1991.
330. M. van Breukelen, R.P.W Duin, D.M.J. Tax, and J.E. den Hartog. Combining classifiers for the recognition of handwritten digits. In *I-st IAPR TC1 Workshop on Statistical Techniques in Pattern Recognition*, pages 13–18, Prague, Czech Republic, 1997.

331. A. Verikas, A. Lipnickas, K. Malmqvist, M. Bacauskiene, and A. Gelzinis. Soft combination of neural classifiers: A comparative study. *Pattern Recognition Letters*, 20:429–444, 1999.
332. M.A. Vila and M. Delgado. On medical diagnosis using possibility measures. *Fuzzy Sets and Systems*, 10:211–222, 1983.
333. B. von Schmidt and F. Klawonn. Fuzzy max-min classifiers decide locally on the basis of two attributes. *Mathware & Soft Computing*, 1999.
334. E.A. Wan. Neural network classification: A bayesian interpretation. *IEEE Transactions on Neural Networks*, 1(4):303–305, 1990.
335. D. Wang, J. M. Keller, C.A. Carson, K.K. McAdoo-Edwards, and C.W. Bailey. Use of fuzzy-logic-inspired features to improve bacterial recognition through classifier fusion. *IEEE Transactions on Systems, Man, and Cybernetics*, 28B(4):583–591, 1998.
336. H.-F. Wang, C.-W. Wu, C.-H. Ho, and M.-J. Hsieh. Diagnosis of gastric cancer by fuzzy pattern recognition. *Journal of Systems Engineering*, 2:151–163, 1992.
337. L.-X. Wang. Fuzzy systems are universal approximators. In *Proc. IEEE Int. Conf. on Fuzzy Systems*, pages 1163–1170, San Diego, CA, 1992.
338. L.-X. Wang and J.M. Mendel. Generating fuzzy rules by learning from examples. *IEEE Transactions on Systems, Man, and Cybernetics*, 22(6):1414–1427, 1992.
339. L.X. Wang and J.M. Mendel. Fuzzy basis functions, universal approximation and orthogonal least squares learning. *IEEE Transactions on Neural Networks*, 3(5):807–814, 1992.
340. P.P. Wang and C.Y. Wang. Experiment on character recognition using fuzzy filters. In P.P Wang and S.K. Chang, editors, *Fuzzy Sets: Theory and Application to Policy Analysis and Information Systems*, pages 195–221. Plenum Press, NY, 1980.
341. P.D. Wasserman. *Neural Computing*. Van Nostrand Reinhold, N.Y., 1989.
342. P.D. Wasserman. *Advanced Methods in Neural Computing*. Van Nostrand Reinhold, USA, 1993.
343. J. Watada, K. Montonami, H. Tanaka, and K. Asay. Discriminant analysis based on fuzzy distance. In *IFAC: Fuzzy Information, Knowledge Representation, Decision Analysis*, pages 177–182, Marseille, France, 1983.
344. J. Watada, H. Tanaka, and K. Asay. Fuzzy discriminant analysis in fuzzy groups. *Fuzzy Sets and Systems*, 19:261–271, 1986.
345. W.E. Weideman, M.T. Manry, H.-C. Yan, and W. Gong. Comparison of neural networks and a nearest neighbor classifier via the numeric handprint recognition problem. *IEEE Transactions on Neural Networks*, 6(6):1524–1530, 1995.
346. B. Whitehead. Genetic evolution of radial basis function coverage using orthogonal niches. *IEEE Transactions on Neural Networks*, 7(6):1525–1528, 1996.
347. B.A. Whitehead and T.D. Choate. Cooperative-competitive genetic evolution of radial basis function centers and widths for time series prediction. *IEEE Transactions on Neural Networks*, 7(4):869–880, 1996.
348. D.L. Wilson. Asymptotic properties of nearest neighbor rules using edited data. *IEEE Transactions on Systems, Man, and Cybernetics*, SMC-2:408–421, 1972.
349. D.H. Wolpert and W.G. Macready. No free lunch theorems for optimization. *IEEE Transactions on Evolutionary Computation*, 1(1):67–82, 1997.
350. K. Woods, W.P. Kegelmeyer, and K. Bowyer. Combination of multiple classifiers using local accuracy estimates. *IEEE Transactions on Pattern Analysis and Machine Intelligence*, 19:405–410, 1997.

351. Q. Xie, C.A. Laszlo, and R.K. Ward. Vector quantization technique for nonparametric classifier design. *IEEE Transactions on Pattern Analysis and Machine Intelligence*, 15(12):1326–1330, 1993.
352. L. Xu, A. Krzyzak, and C.Y. Suen. Methods of combining multiple classifiers and their application to handwriting recognition. *IEEE Transactions on Systems, Man, and Cybernetics*, 22:418–435, 1992.
353. R.R. Yager. On a general class of fuzzy connectives. *Fuzzy Sets and Systems*, 4:235–242, 1980.
354. R.R. Yager. On the aggregation of processing units in neural networks. In *IEEE 1-st International Conference on Neural Networks*, pages II–327–II–333, San Diego, California, 1987.
355. R.R. Yager. On Ordered Weighted Averaging operators in multicriteria decision making. *IEEE Transactions on Systems, Man, and Cybernetics*, 18:183–193, 1988.
356. R.R Yager and D.P. Filev. *Essentials of Fuzzy Modeling and Control*. John Wiley & Sons, N.Y., 1994.
357. R.R. Yager and J. Kacprzyk, editors. *The Ordered Weighted Averaging Operators. Theory and Applications*. Kluwer Academic Publishers, Boston, 1997.
358. T. Yamakawa and M. Furukawa. A design algorithm of membership functions for a fuzzy neuron using example-based learning. In *Proc. IEEE Int. Conf. on Fuzzy Systems*, pages 75–82, San Diego, CA, 1992.
359. H. Yan. Handwritten digit recognition using an optimized nearest neighbor classifier. *Pattern Recognition Letters*, 15:207–211, 1994.
360. M.-S. Yang and C.-H. Chen. On the edited fuzzy k-nearest neighbor rule. *IEEE Transactions on Systems, Man, and Cybernetics*, B 28(3):461–466, 1998.
361. M.-S. Yang and C.-T. Chen. On strong consistency of the fuzzy generalized nearest neighbor rule. *Fuzzy Sets and Systems*, 60:273–281, 1993.
362. H.-C. Yau and M.T. Manry. Iterative improvement of a nearest neighbor classifier. *Neural Networks*, 4:517–524, 1991.
363. P. Yee and S. Haykin. Pattern classification as an ill-posed inverse problem: A regularization approach. In *Proc. IEEE Int. Conference on Acoustics, Speech and Signal Processing*, pages I-597–600, Minneapolis, USA, 1993.
364. H. Ying. General SISO Takagi-Sugeno fuzzy systems with linear rue consequent are universal approximators. *IEEE Transactions on Fuzzy Systems*, 6(4):582–587, 1998.
365. H. Ying. Sufficient conditions on uniform approximation of multivariate functions by general Takagi-Sugeno fuzzy systems with linear rule consequent. *IEEE Transactions on Systems, Man, and Cybernetics*, 28 (A)(4):515–520, 1998.
366. L.A. Zadeh. Fuzzy sets. *Information and Control*, 8:338–353, 1965.
367. L.A. Zadeh. Outline of a new approach to the analysis of complex systems and decision processes. *IEEE Transactions on Systems, Man, and Cybernetics*, 3(1):28–44, 1973.
368. L.A. Zadeh. Is possibility different from probability? *Human Systems Management*, 3:253–254, 1983.
369. X.-J. Zeng and M.G. Singh. Approximation properties of fuzzy systems generated by the min inference. *IEEE Transactions on Systems, Man, and Cybernetics*, 26 (B)(1):187–193, 1996.
370. Y. Zhang, K. Shigeta, and E. Shimizu. A pattern classifier – adaptive fuzzy classification neural network. In *Proc. IEEE Int. Conf. on Neural Networks*, pages 1706–1709, Orlando, FL, 1994.
371. Q. Zhao and T. Higuchi. Evolutionary learning of nearest neighbor MLP. *IEEE Transactions on Neural Networks*, 7(3):762–767, 1996.

372. H.-J. Zimmermann. Fuzzy sets in pattern recognition. In P.A. Devijver and J. Kittler, editors, *Pattern Recognition Theory and Applications*, volume F30 of *NATO ASI*, pages 383–391. Springer-Verlag, Berlin, 1987.
373. L. Zuoyong and W. Peishuang. Making classification forecasting of weather by multivariable membership function. *BUSEFAL*, 25:141–148, 1986.

Index

Studies in Fuzziness and Soft Computing

Vol. 25. J. Buckley and Th. Feuring
Fuzzy and Neural: Interactions and Applications, 1999
ISBN 3-7908-1170-X

Vol. 26. A. Yazici and R. George
Fuzzy Database Modeling, 1999
ISBN 3-7908-1171-8

Vol. 27. M. Zaus
Crisp and Soft Computing with Hypercubical Calculus, 1999
ISBN 3-7908-1172-6

Vol. 28. R.A. Ribeiro, H.-J. Zimmermann, R.R. Yager and J. Kacprzyk (Eds.)
Soft Computing in Financial Engineering, 1999
ISBN 3-7908-1173-4

Vol. 29. H. Tanaka and P. Guo
Possibilistic Data Analysis for Operations Research, 1999
ISBN 3-7908-1183-1

Vol. 30. N. Kasabov and R. Kozma (Eds.)
Neuro-Fuzzy Techniques for Intelligent Information Systems, 1999
ISBN 3-7908-1187-4

Vol. 31. B. Kostek
Soft Computing in Acoustics, 1999
ISBN 3-7908-1190-4

Vol. 32. K. Hirota and T. Fukuda
Soft Computing in Mechatronics, 1999
ISBN 3-7908-1212-9

Vol. 33. L.A. Zadeh and J. Kacprzyk (Eds.)
Computing with Words in Information/ Intelligent Systems 1, 1999
ISBN 3-7908-1217-X

Vol. 34. L.A. Zadeh and J. Kacprzyk (Eds.)
Computing with Words in Information/ Intelligent Systems 2, 1999
ISBN 3-7908-1218-8

Vol. 35. K.T. Atanassov
Intuitionistic Fuzzy Sets, 1999
ISBN 3-7908-1228-5

Vol. 36. L.C. Jain (Ed.)
Innovative Teaching and Learning, 2000
ISBN 3-7908-1246-3

Vol. 37. R. Słowiński and M. Hapke (Eds.)
Scheduling Under Fuzziness, 2000
ISBN 3-7908-1249-8

Vol. 38. D. Ruan (Ed.)
Fuzzy Systems and Soft Computing in Nuclear Engineering, 2000
ISBN 3-7908-1251-X

Vol. 39. O. Pons, M.A. Vila and J. Kacprzyk (Eds.)
Knowledge Management in Fuzzy Databases, 2000
ISBN 3-7908-1255-2

Vol. 40. M. Grabisch, T. Murofushi and M. Sugeno (Eds.)
Fuzzy Measures and Integrals, 2000
ISBN 3-7908-1255-2

Vol. 41. P. Szczepaniak, P. Lisboa and J. Kacprzyk (Eds.)
Fuzzy Systems in Medicine, 2000
ISBN 3-7908-1263-4

Vol. 42. S. Pal, G. Ashish and M. Kundu (Eds.)
Soft Computing for Image Processing, 2000
ISBN 3-7908-1217-X

Vol 43. L.C. Jain, B. Lazzerini and U. Halici (Eds.)
Innovations in ART Neural Networks, 2000
ISBN 3-7908-1270-6

Vol. 44. J. Aracil, F. Gordillo (Eds.)
Stability Issues in Fuzzy Control
ISBN 3-7908-1277-3

Vol. 45. N. Kasabov (Ed.)
Future Directions for Intelligent Information Systems on Information Sciences
ISBN 3-7908-1276-5

Vol. 46. J.N. Mordeson, P.S. Nair
Fuzzy Graphs and Fuzzy Hypergraphs
ISBN 3-7908-1286-2

Vol. 47. Ernest Czogała†, Jacek Łęski
Fuzzy and Neuro-Fuzzy Intelligent Systems
ISBN 3-7908-1289-7

Vol. 48, Masatoshi Sakawa
Large Scale Interactive Fuzzy Multiobjective Programming
ISBN 3-7908-1293-5

Printing: Weihert-Druck GmbH, Darmstadt
Binding: Buchbinderei Schäffer, Grünstadt

GPSR Compliance

The European Union's (EU) General Product Safety Regulation (GPSR) is a set of rules that requires consumer products to be safe and our obligations to ensure this.

If you have any concerns about our products, you can contact us on ProductSafety@springernature.com

In case Publisher is established outside the EU, the EU authorized representative is:

Springer Nature Customer Service Center GmbH
Europaplatz 3
69115 Heidelberg, Germany

Batch number: 10725800

Printed by Printforce, the Netherlands